ELECTRONIC AND PHOTONIC CIRCUITS AND DEVICES

IEEE Press
445 Hoes Lane, P.O. Box 1331
Piscataway, NJ 08855-1331

ELECTRONIC AND PHOTONIC CIRCUITS AND DEVICES

Edited by

Ronald W. Waynant
Food and Drug Administration, U. S. Government

John K. Lowell
Jekyll Consulting, Dallas, Texas

IEEE Circuits and Systems Society, *Sponsor*
IEEE Components, Packaging, and Manufacturing Technology Society, *Sponsor*

A Selected Reprint Volume

IEEE Press Series on Microelectronic Systems
Stuart K. Tewksbury, *Series Editor*

The Institute of Electrical and Electronics Engineers, Inc., New York

This book may be purchased at a discount from the publisher
when ordered in bulk quantities. Contact:

IEEE Press Marketing
Attn: Special Sales
445 Hoes Lane, P.O. Box 1331
Piscataway, NJ 08855-1331
Fax: (732) 981-9334

For more information on the IEEE Press,
visit the IEEE home page: http://www.ieee.org/

Printed in the United States of America

10 9 8 7 6 5 4 3 2 1

ISBN 0-7803-3496-5

IEEE Order Number: PP5748

Library of Congress Cataloging-in-Publication Data

Electronic and photonic circuits and devices / edited by Ronald W.
Waynant, John K. Lowell.
p. cm.—(IEEE Press series on microelectronic systems)
"IEEE Circuits and Systems Society, sponsor; IEEE Components,
Packaging, and Manufacturing Technology Society, sponsor."
"A selected reprint volume."
ISBN 0-7803-3496-5
1. Electronic circuits. 2. Electronic apparatus and appliances.
3. Optoelectronic devices. I. Waynant, Ronald W. II. Lowell,
John. III. IEEE Circuits and Systems Society. IV. IEEE Components,
Packaging & Manufacturing Technology Society. V. Series.
TK7867.E36 1998
621.3815—dc21 98-41717
CIP

CONTENTS

PREFACE

For nearly 14 years the IEEE has published *Circuits and Devices Magazine,* making it available first to the Societies of Division I but later to all IEEE members. We are proud of the accomplishments attributed to the magazine by its readers and its designation among the magazines with the greatest impact based on citations. During that time, we have solicited and published numerous tutorial papers describing new technology and new ideas. But the theme of what we believe is the future of our field of engineering was so well described by an article written by Anthony DeMaria that we wanted to preserve it and a number of supporting papers by others in a special collection. This collection expresses the state of the art and the thinking of the 1990s, and it projects the likely path that developers and engineers will follow for the next 25 or more years. It details the gradual, but inevitable, shift from electronic devices now being devised and improved to optoelectronics, in which a combination of optical and electronic devices predominate and on to the future systems in which photonics will be dominant. This kind of perspective, given in well-written, understandable form by experts in their field, can help direct students in choosing their future work in science.

Ronald W. Waynant
John K. Lowell

ACKNOWLEDGMENTS

We thank the authors who have kindly given permission for the reprinting of their original or revised papers and the IEEE for its permission and encouragement. We also thank Eve Protic and Mark Protic for their assistance at the beginning of this project and Marcia Patchan for pushing it through to completion.

Ronald W. Waynant
John K. Lowell

THEME PAPER

Maxwell's Children Light the Way

Photonics, as a complement to electronics, will initially capture markets where connection to the electronic interface is easy

Twentieth-century electronics is the child of 19th-century electromechanics and the parent of 21st-century photonics. The electronics industry, and its growth, is certainly indebted in large part to E. H. Armstrong, and his first electronic oscillator in 1912. The first optical oscillator, i.e., Ted Maiman's ruby laser, didn't glow until 48 years later (Fig. 1). It took electronics approximately 80 years to develop its markets and technology to where they are today. In contrast, last year marked the 30th anniversary of Maiman's first laser. If we assume that photonics will experience the same rate of technology, market, and manufacturing development that electronics did, we can predict that the photonics market will equal that of its electronics counterpart as we approach midway into the 21st century. Unfortunately, the prediction may be too optimistic, because electronics possesses the lower-cost advantage associated with a mature technology. Consequently, photonics will initially be most successful when we use it (1) to perform functions that cannot be performed as well, or at all, by electronics; and (2) in applications where photonics can easily interface with electronics.

Three major technological generations have their genealogical roots in the equations James Clerk Maxwell formulated in 1864. Technologies based on electromagnetics and electromechanics such as motors, generators, magnetics, and power transmission, experienced their major growth in the 19th century and began to mature in the first part of the 20th century. (The membership of the American Institute of Electrical Engineers—AIEE—reached a peak of 18,344 in 1927. In 1955 or 1956, the membership of the Institute of Radio Engineers (IRE) surpassed the AIEE's membership. In 1963, the two organizations merged to form the IEEE.)

The last three quarters of the 20th century have been a period of dramatic growth for electronics-based technologies. The rate of development does not yet show evidence of slowing down, but these technologies can be expected to mature in the first half of the 21st century.

In the family tree of major technologies that are based on Maxwell's equations, one can think of electrically and magnetically based technologies as the grandparents; electronics hardware and software as the parents; and the subgroups of photonics technology as the children (Fig. 2). The growth of photonics is just beginning. Its maturity will probably begin in the latter half of the 21st century.

Reprinted from *IEEE Circuits and Devices Magazine,* Vol. 7, No. 2, pp. 36-43, March 1991.

by Anthony J. DeMaria

Electronics Overview

Ferdinand Braun, working in Marburg, Germany, in 1874, discovered the first metal semiconductor junction by establishing the rectifying properties of galena (i.e., lead sulfide). These "cat whisker" diode detectors played an important role in researching radio waves and their propagation before the invention of vacuum tubes, but their physics was not understood until the development of semiconductor solid-state physics in this century.

While working on his carbon filament lamp in 1883, Thomas Edison discovered that electrical current could be made to flow in a vacuum when an electrically positive charged plate was positioned near a heated filament within a vacuum envelope. This discovery remained essentially unused until John Ambrose Fleming invented the vacuum diode rectifier for converting alternating current to direct current in 1904 [1]. This invention was followed two years later by Lee DeForest's vacuum triode— the first electronic amplifier. Armonstrong's oscillator followed, the first generation of temporal coherent electromagnetic radiation.

Following Armstrong's invention, the electronics industry developed very rapidly. Electronic vacuum-tube devices became the heart of the industry and were responsible for the rapid development of radio, radar, television, electronic controls, telecommunications, and electronic information processing, among many other technologies. Twenty-nine years elapsed from Edison's discovery to the operation of the first electronic oscillator. McGraw-Hill published the first trade magazine with the title *Electronics* beginning on April 30, 1930. (The word "electronic" did not appear in dictionaries until many years later.) The term is now used to denote the broad range of technologies dependent on controlling the flow of electrical charges in a vacuum, solid, liquid, or plasma. But until the invention of the transistor in 1948, the field of electronics was primarily dependent on vacuum tubes, which exerted control over the flow of an electron stream in a vacuum.

In 1948, John Bardeen, Walter H. Brattain and William Shockley of Bell Labs announced the invention of the transistor. Transistor-type devices control the flow of either positive charged particles (holes) or negative charged particles (electrons) in a crystalline solid. At first germanium was the crystal of choice, but silicon soon came into almost exclusive use because of its superior mechanical and electrical properties. In addition, it was soon recognized that silicon's easily grown native protective oxide has exceptional electrical properties. For a short while, transistor-type devices were resisted by the electronics industry. But their small

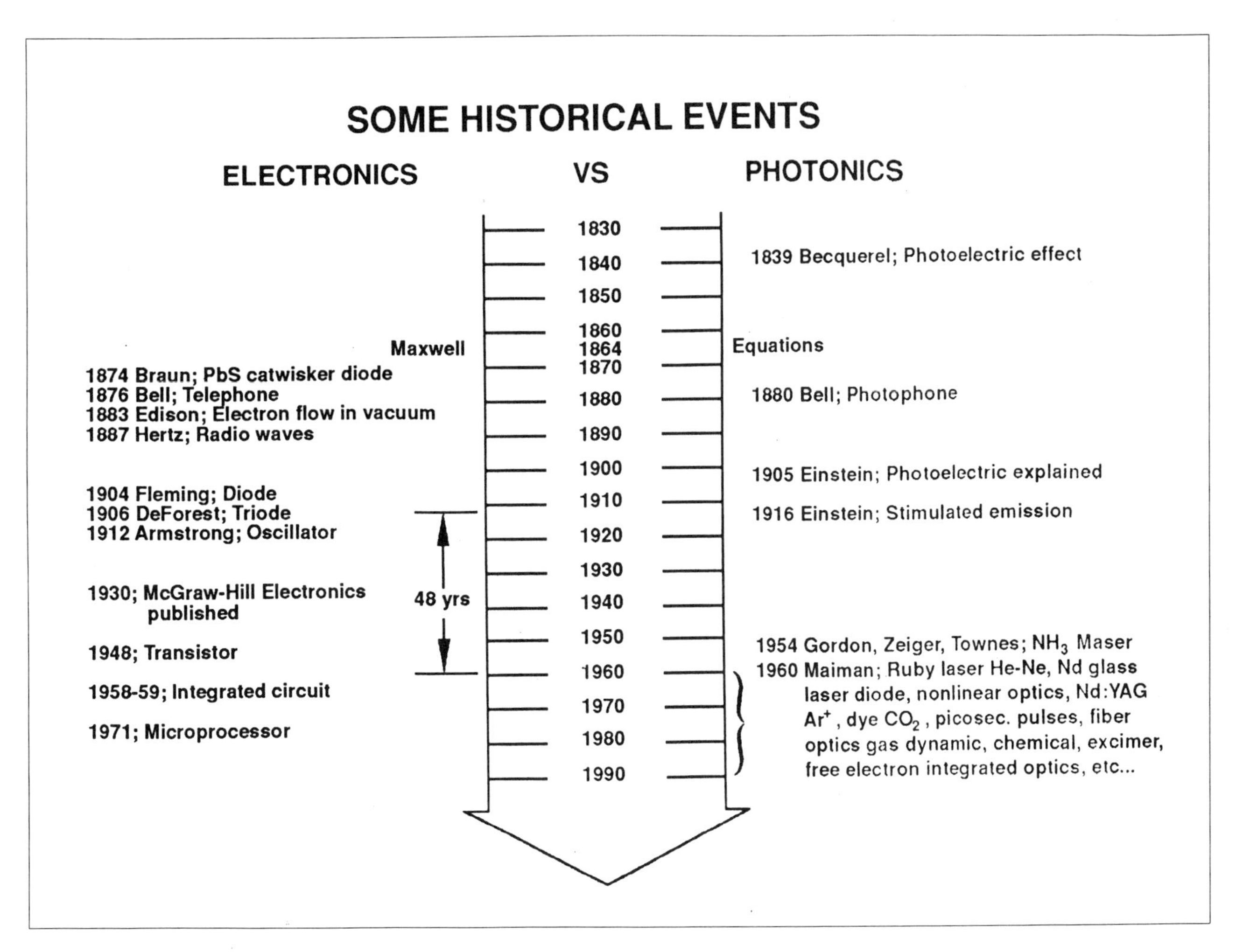

1. Major historical events in the development of the electronic and photonic fields.

size, high efficiency, mass producibility, lower unit cost, lower voltage requirements, and higher reliability in comparison with vacuum tubes led to their eventual dominance in the electronics field, a dominance that occurred even though the manufacturability of transistors was considerably more capital intensive and difficult to master.

In 1958-59, the semiconductor integrated circuit was invented by Jack Kilby of Texas Instruments and Robert Noyce of Fairchild Semiconductor [2]. The integrated circuit chip enabled an entire electronic system to be miniaturized on a small piece of silicon and launched the modern information processing revolution. The development of the microprocessor by Intel in 1971 gave birth to the pocket calculator and, eventually, to personal computers and workstations.

If the invention of the triode in 1906 marked the start of the electronics revolution, then the field of electronics is today a little over eighty years old. Clearly, its tempo is still increasing. In 1970, $1,000 could purchase approximately 400 barrels of oil or 40 kilobytes of semiconductor random access memory (RAM). In 1985, $1,000 could purchase approximately 30 barrels of oil or 25 megabytes of RAM. The amount of oil one could buy per dollar decreased by over an order of magnitude over a 15 year period while the amount of RAM bytes per dollar increased by over two orders of magnitude. The long-term cost of semiconductor memory circuits is still dropping rapidly, which indicates that the information revolution is still continuing at a rapid pace.

The semiconductor industry has been on an exponential growth path for the last 30 years. Worldwide revenues now total about $30 billion. The industry produces the equivalent of 250,000 transistors per year for every person on earth. Over the last two decades, the numbers of transistors produced per year has doubled every year. An integrated circuit with a million transis-

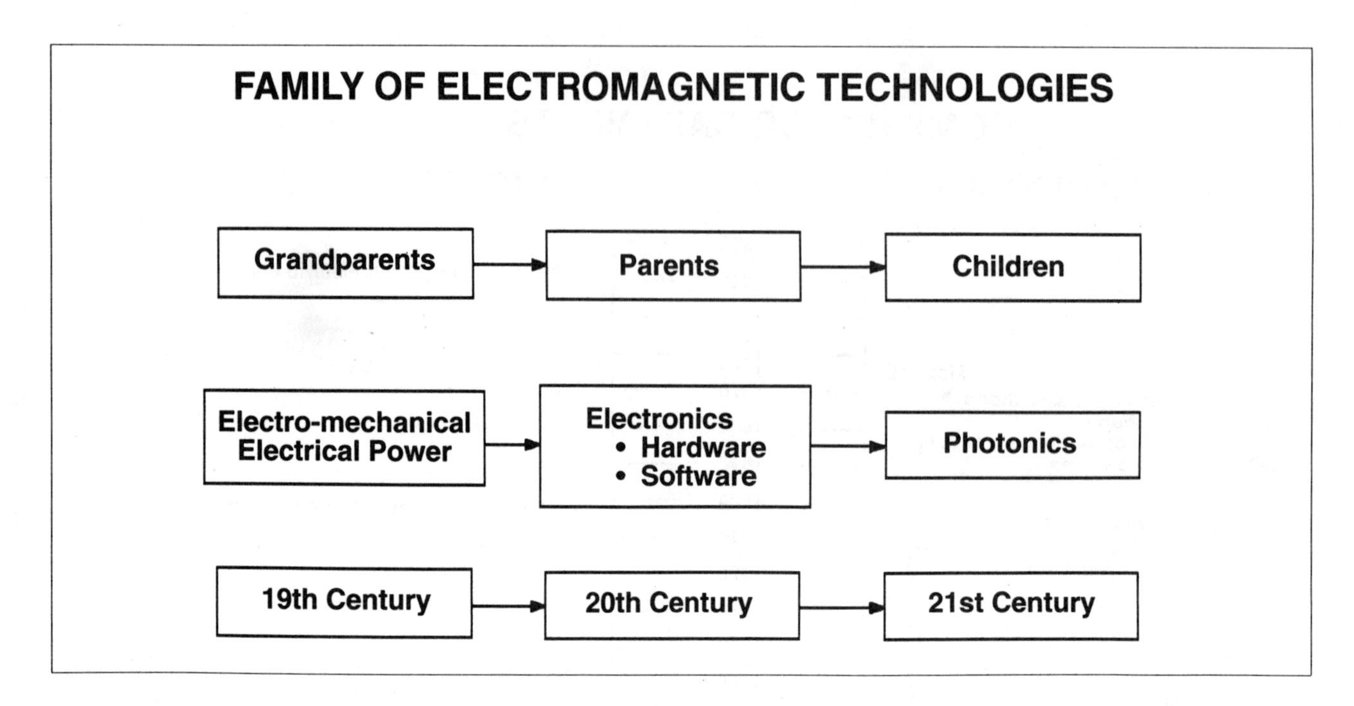

2. *In the family tree of major technologies that are based on Maxwell's equations, electrically and magnetically based technologies may be viewed as the grandparents; electronics hardware and software as the parents; and the subgroups of photonics technology as the children.*

tors today sells for the same price as did a single transistor in the early days of the industry. The layout print of a 4-megabit dynamic random access memory (DRAM) chip is equivalent to the task of drawing a road map of the entire United States showing every side street. A deficiency, during the manufacturing process of such a chip, is equivalent to a large pothole showing up on any street on such a map, and such a defect would necessitate rejecting the whole chip. Such chips will enter production in the near future. Work has started on 16- and 64-megabit DRAM chips. The layout for these chips is like drawing a map of the entire United States in such detail that even sidewalks are shown in the drawing. Such designs could not be effectively accomplished without sophisticated computers.

Advances in integrated circuits produce advances in computers, which in turn produce more advances in integrated circuits. This process continues to repeat itself, and electronics provides effective technological feedback that contributes to it's own explosive growth.

Parallel Advances in Photonics

In 1839, Alexandre Edmond Becquerel of France discovered the photodetector—the first opto-electronic device— when he generated a voltage by irradiating the junction of an electrolyte. The effect was not understood until Albert Einstein explained it in 1905.

Even though photoelectric devices were not understood, they were used by researchers. Alexander Graham Bell used a photoelectric device in his photophone of 1880, which was the first electro-optical system [1]. Einstein presented his formulation of stimulated and spontaneous emission processes and their relationship to the radiation absorption process in atomic and molecular systems in 1916. After 1916, many scientists investigating the spectroscopic properties of gases realized that population inversion in atomic and molecular systems would result in amplification of radiation through the stimulated emission process, but they ignored its development.

Why didn't scientists in the 1916-to-1954 time period ask themselves how they could exploit stimulated emission to obtain an optical oscillator? The most likely answer is that the researchers investigating the spectroscopic properties of gases did not adequately appreciate the principle of positive feedback, revealed by Armstrong in 1912 with electronic oscillators.

The cultural divide between physicists and electrical engineers was bridged by World War II. The importance of microwave radar during the war caused many scientists to work on microwave oscillators and amplifiers. One of these scientists was C. H. Townes, whose research exposed him to the principles of positive feedback in

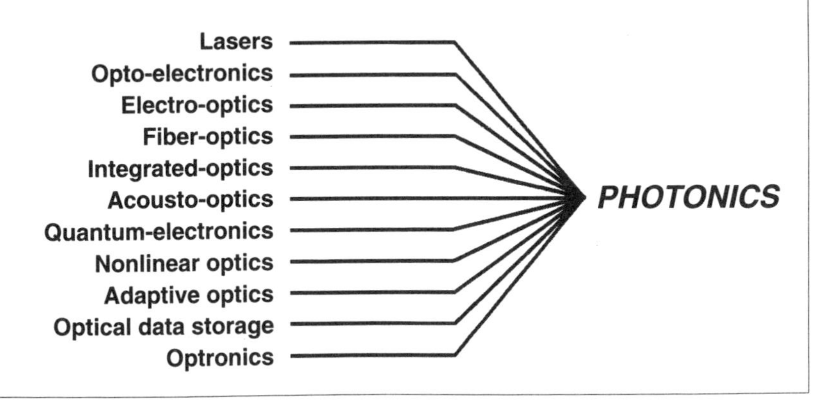

3. The term "photonics" now denotes a large number of optically related technologies.

electronic microwave oscillators. This experience, coupled with his spectroscopic training, provided him with the technical foundation required for the invention of the microwave NH_3 maser in 1954 [3]. This device was the first quantum electronic oscillator. In retrospect, providing positive feedback in the microwave region in a population-inverted molecular system was a relatively straightforward undertaking. But doing the same thing in the optical region required another new invention: the use of a Fabry-Perot interferometer as a multiple-axial-mode, positive-feedback optical cavity for population-inverted atomic or molecular amplifiers [4].

T. H. Maiman's ruby laser of 1960 was the first operation of a coherent electromagnetic oscillator in the optical region [5]. Contrary to the relatively mild attention given the announcement of the transistor in 1948, the laser was greeted with great excitement and expectation by the scientific and engineering communities, as well as the general public. The technical community realized that the laser would make possible the transportation of techniques and technologies from the audio, video, radio-frequency, and microwave regions into the optical portion of the electromagnetic spectrum.

Researchers quickly made the transition from maser to laser research, and there was an explosion of scientific activity and discovery. In less than 20 years, the following lasers were rapidly developed: helium-neon, glass, argon, argon ion, dye, CO_2, chemical, excimer, free-electron, gas dynamic, Nd^{3+}: glass, Nd^{3+}:YAG, and semiconductor laser diodes. Broad advances were secured in the areas of optical parametric amplifiers; picosecond laser-pulse generation; fiber optics; Q-switching, holography, nonlinear optics, optical mixing, integrated optics; optical phase conjugation; and optically squeezed states. Laser devices provided the major stimulation for the broad field of photonics after 1960 in the same way that the vacuum tube stimulate electronics after 1906.

After the invention of the laser, new terms such as quantum electronics, electro-optics, lidar, and optical information processing, were coined to name the numerous new specialties that depended on laser devices. It soon became apparent that an umbrella term was needed to refer to all of these specialties, much as the term "electronics" covered the large number of fields dependent on electronic devices. Since these optical specialties are dependent on the control of a light beam, i.e., a stream of photons, the term "photonic" was formed by adding "ic" to the word "photon." The term is now used to refer to the technologies for generating, amplifying, detecting, guiding, and modulating coherent optical radiation and applying these technologies to fields such as energy generation, communications, and information processing (Fig. 3).

Going Down the Same Road

The parallelism is apparent in the development of electronics and photonics. Optically pumped solid-state lasers, gas lasers, and dye lasers, for example, perform the same role in photonics as vacuum-tube devices perform in electronics. Both devices were responsible for the initial rapid development of their respective fields, and they all enjoy much the same advantages and disadvantages. Lasers, like vacuum tubes, are capable of the highest power and operating frequency within their electromagnetic regions. And, like vacuum tubes, they suffer from heavier weight, larger size, higher power consumption, and lower life expectancy than their solid-state counterparts.

Semiconducting laser and transistor devices also play parallel roles in their respective fields. As in electronics, it is the semiconductor lasers and their associated applications that offer the largest market opportunities because of their small size, lower weight and power consumption, mass producibility, lower cost, higher reliability, and ability to be integrated on monolithic substrates utilizing microelectronic fabrication techniques. Semiconductor lasers also interface easily with transistors. Consequently, semiconductor lasers can take advantage of the rapid market growth and the large existing technology infrastructure that transistor devices presently enjoy.

In electronics, waveguides such as coaxial cables, strip lines, and metal guides were used before active semiconductor devices became dominant. In photonics, semiconductor laser devices were invented a few years after the announcement of the ruby laser. Until K. C. Kao proposed it in 1966, researchers did not appreciate that losses could be greatly reduced in glass fibers [6]. Then, research programs were launched in earnest by several organizations to develop glass fibers for telecommunication optical-waveguide applications. In 1970, Corning Glass Works achieved losses below 20 dB/km, a level considered critical to the extensive application of fiber optics to telecommunication. Today, glass-fiber losses have been reduced to near the theoretical limit, and glass fibers are the transmission line of choice for telecommunication applications. Telecommunication is presently the largest market for photonics.

Researchers today are busy developing integrated optical technology to achieve the benefits for the photonics field that semiconductor integrated circuits have achieved for the electronics field. It is important not to confuse the huge role that integrated digital electronic circuits have played in digital data processing with the role that integrated optics chips are expected to play in photonics. At present, it appears that the appropriate parallelism is between the role that microwave integrated circuits, microwave/millimeter-wave strip lines, and microwave/millimeter- wave hybrid circuits play in the lower frequency region, and the role that optical integrated circuits are expected to play in the optical region. With present technology, microwave monolithic integrated circuits (MIMICs) cannot match the density of digital integrated circuits. And optical integrated circuits will not match the density of MIMIC chips with presently foreseen optical technology.

Compound semiconductor materials such as GaAs and GaAlAs are unique when compared to silicon and germanium, because they can be used to generate optical radiation. They are also electro-optically and piezoelectrically active, and have good semiconductor properties for the fabrication of transistors. Consequently, current research efforts are being aimed at merging integrated electronic circuits with integrated-optic, opto-electronic, electro-optic, and even acousto-optic devices on monolithic substrates of GaAs and similar compound materials (Fig. 4). Monolithic optical chips do exist today, typically combining a few optical components on a $LiNbO_3$ substrate (Fig. 5).

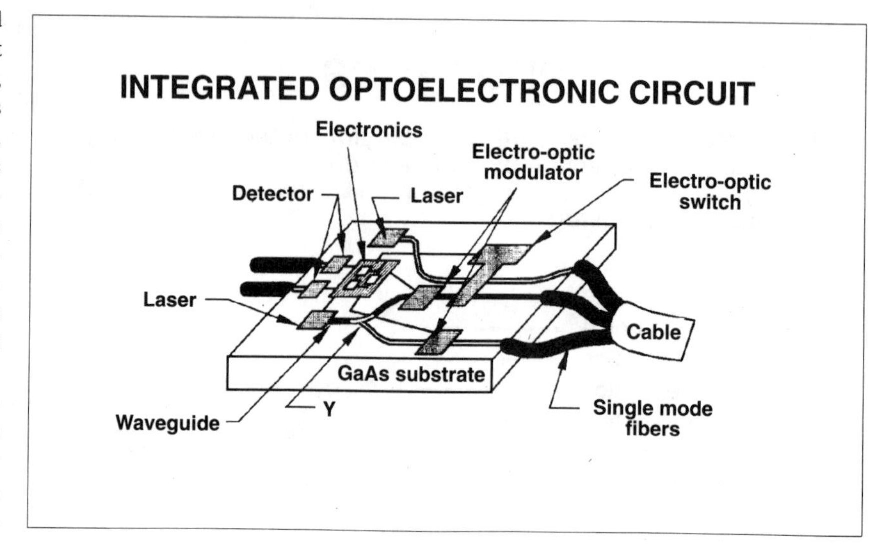

4. An example of a potential integrated optoelectronic circuit that researchers may some day realize [7]. Optical signals can be brought to such a chip by optical fibers. Optical radiation from these fibers can be coupled into planar waveguides embedded within the chip. Photodetectors fabricated on the chip can convert the optical signals to electrical signals, which can be digitally processed by electronic devices also fabricated on the chip. These electrical signals can then be used to modulate semiconductor lasers directly or to energize electro-optic modulators that are also fabricated directly on the chip. The modified optical signals are then transported through the chip by embedded planar waveguides and eventually coupled to optical fibers for transmission to remote locations. Only small portions of such a chip have been reported in the technical literature thus far, but the potential is clear.

Microelectronics and Photonics

Light emitting diodes fabricated from GaAs for numerical display applications appeared in the late 1950s. In 1962, GaAs lasers operating in the near infrared were first reported by General Electric, IBM, and Lincoln Lab researchers. These reports ensured that semiconductor devices would play a major role in photonics. Today, lead-salt semiconductor lasers have extended the operating-wavelength range of semiconductor lasers into the far infrared region, and GaAs-based semiconductor lasers have operated in the red portion of the visible spectrum. In the future, higher-bandgap materials may permit semiconductor lasers in the blue portion of the spectrum.

The compatibility of semiconductor lasers with microelectronic devices has

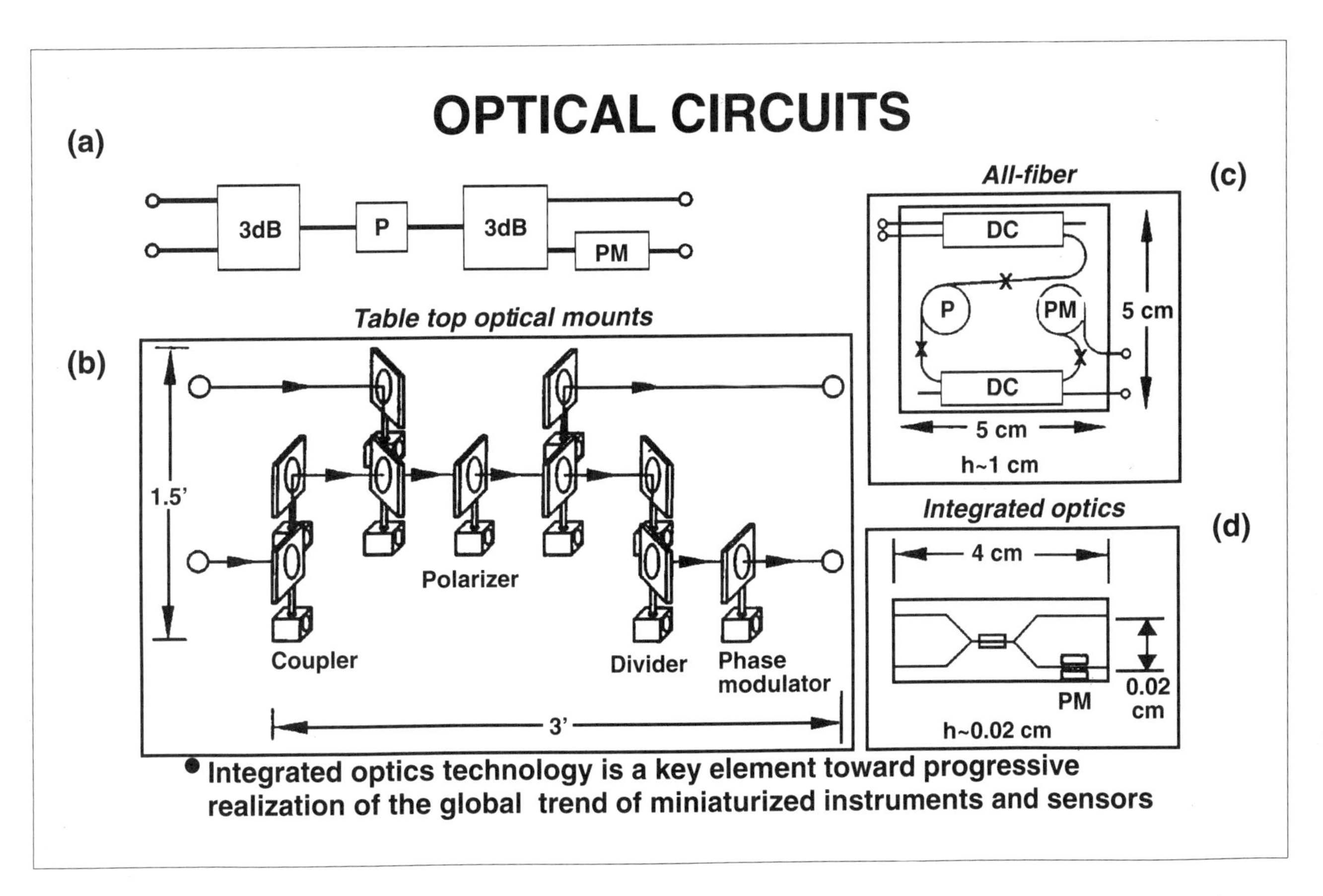

5. This typical optical circuit (a) contains four optical components: a 3 dB power combiner, a polarizer (P), a 3 dB power splitter, and a phase modulator (PM). It is useful in realizing miniature, low cost fiber-optic gyros for navigational applications. It is possible to fabricate a discrete-optical-component version consisting of mirrors, a polarizer and an electro-optics phase modulator (b). The all-fiber configuration (c) is much smaller, with typical dimensions of 5 x 5 x 1 cm. An integrated-optics-chip configuration (d) can be as small as 4 x 0.02 x 0.02 centimeters.

launched major segments of existing markets. Among these are fiber-optic telecommunication, optical data recording, audio disks, video disks, laser printers, and bar-code readers. These markets are served by the semiconductor laser. (The He-Ne laser is now a far-distant second in annual production.) Manufacturers are producing several hundred thousand semiconductor lasers per month.

Laser weapons, controlled fusion with lasers, and semiconductor laser development are the three areas of laser technology that have received continuous and extensive R & D support over the last 25 years. The first two are still a long way from making a national impact, but they have received extensive government support in many countries because of their perceived importance to national security and economical well-being. The semiconductor laser, on the other hand, has been developed primarily with industrial R&D funds, and it has been the most successful in spawning new sectors of existing industries. It is therefore important to have a manufacturing competitiveness in semiconductor lasers if a nation is to maintain competitiveness in photonics.

Microelectronics processing technology has developed a large assortment of processing techniques that yield feature dimensions that fall within the range of optical wavelengths. Consequently, the application of semiconductor processing technology such as epitaxial film deposition, etching, milling, photolithography, and ion implantation, advances photonics by advancing optical technology and originating new subfields. These subfields include binary/diffractive optics, microlens fabrication, new optical coatings, integrated optics, and super-lattice devices.

Such contributions are enabling photonics to interface easily with microelectronics and are producing many instances in which photonics are being integrated directly with microelectronics devices and subsystems. This "partnership" between photonics and electronics contributes to the explosive growth of photonics in telecommunications, signal and data processing, information storage, entertainment products, computing, data buses, interconnections of electronic subsystems, sensors, medical applications, and scientific instrumentation. The growth is pushing photonics into large-volume, low-cost, mass-production manufacturing that requires large capitalization rivaling that required for manufacturing of electronic components and subsystems.

The strong interdependence of photonics and electronics has educational effects, too; more and more university-based programs offer a wide selection of graduate and undergraduate optics-related courses. As a result, the center of mass for training in optics has shifted from physics departments to electrical engineering departments.

Things to come

Photonics is still in a state of infancy. It is young and robust with a highly promising and exciting future. The field is now spawning new products and opening major new segments of basic industries. Accelerating growth in fiber-optic telecommunications, optical data buses, optical data storage, optical interconnects between electronic subsystems, and many other applications, will ensure the growth of photonics well into the next century.

Photonics and electronics are complementary and not competitive; indeed, photonics is heavily dependent on electronics. Consequently, the most success in the near term will be in applications where photonics interfaces easily with electronics and advantages can be taken from the large infrastructure and momentum for growth that the electronic revolution still enjoys. And by its own growth, photonics will reciprocate and further expand the growth of electronics.

The most serious challenge facing photonics is the shortage of the photonics engineers required to develop the numerous new and rapidly evolving products, and to work at the interface between electronics and photonics. Offering a formal undergraduate engineering curriculum in photonics engineering would be a big boost to this important, though still emerging, field. A formal curriculum would ensure that the photonics revolution will succeed the electronics revolution just as the electronics revolution succeeded the old industrial revolution.

Just how large can photonics get? Can it exceed the field of electronics in market size? The answer will probably be "no" for a very long time, because of the relative maturity of electronics. Electronics has a half-century head start on photonics. It has a very large infrastructure, and large capital and knowledge bases that will difficult to abandon or overcome with a new technology. Indeed, comparing the memberships of electronics- and photonics-oriented professional societies illustrates the entrenchment of electronics. The Institute of Electrical and Electronics Engineers has approximately 300,000 members. The total membership of the Optical Society of America, the Society of Photo-optical and Instrumentation Engineers, the IEEE Lasers and Electro-optics Society, the Laser Association of America (LAA), and the Laser Institute of America (LIA), which together represent the field of photonics, have only about 30,000 members (and many of these have joint memberships.) It will be some time before the number of photonics engineers exceeds the the world's present number of electronics engineers. The arrival of photonics, however, will be no less important to society than was electronics.

Acknowledgment

This article was first presented as an invited paper at the *First International Workshop on Photonic Networks, Components and Applications*, held in Montebello, Quebec, Canada on October 10-13, 1990, under the title *Electronics and Photonics: Partners or Competitors*.

Biography

Dr. Anthony J. DeMaria [F] is Assistant Director of Research for Electronics and Photonics Technologies at the United Technologies Research Center, East Hartford, Connecticut. **CD**

References

1. Special Commemorative Issue, *50 Years of Electronics* (McGraw-Hill Publications, 1980).
2. The 30th Anniversary of the Integrated Circuit, *VLSI System Design*, Vol. **IX**, No. 9A, (1988).
3. J. P. Gordon, H. J. Zeiger, and C. H. Townes, *Physical Review* 95, (1954) p. 282.
4. A. L. Schawlow and C. H. Townes, *Physical Review* **112**, (1958) p. 1940.
5. T. H. Maiman, *Physical Review* **4**, (1960) p. 564.
6. K C. Kao and G. A. Hockham, Proceedings of the IEEE, 133, (1966) p. 1151.
7. Photonics, Maintaining Competitiveness in The Information Age, National Academy Press, Washington, DC, (1988).

CHAPTER 1

INTEGRATED CIRCUIT TECHNOLOGY

Modeling GaAs/AlGaAs Devices: A Critical Review

Herbert S. Bennett

Abstract

Device models for GaAs devices and GaAs/AlGaAs heterostructures are much less advanced than those for silicon devices. This paper critically reviews recent advances in the modeling of GaAs/AlGaAs devices. It is based on the examination of five selected device models that contain features common to the majority of device models for heterostructure bipolar and field effect transistors. Areas requiring improved measurement techniques on processed GaAs and improved physical concepts for GaAs/AlGaAs device models are identified.

Introduction

This review summarizes recent advances in modeling GaAs/AlGaAs heterostructure bipolar transistors (HBT) and GaAs and GaAs/AlGaAs field effect transistors (FET). It identifies those physical concepts that are not adequately included in present device models for HBTs, for conventional metal semiconductor FETs (MESFETs), and for heterostructure FETs (HFETs) such as two-dimensional electron gas FETs (TEGFETs), modulation doped FETs (MODFETs), high electron mobility FETs (HEMTs), and selectively doped FETs (SDFETs). This review also identifies areas requiring increased efforts for measurement techniques on processed GaAs. The term *processed GaAs* refers to material that is representative of the active regions of devices and not to bulk material. This review is based on the examination of five device models that have been selected from among several available GaAs device models. The five device models discussed contain features and assumptions found in the majority of device models reported in the archival literature.

Developing computationally efficient models that simulate the operation of solid-state devices is one goal of workers in this area. Achieving this goal requires compromises between the sophistication of solid-state physics and the pragmatic demands of electrical engineering. There are two classes of computer models:

(a) Compact models, which use the methods of Gummel-Poon or Ebers-Moll, based on closed form solutions to approximate device equations.

(b) Detailed models based on doping profiles and numerical solutions to the coupled nonlinear semiconductor device equations with appropriate boundary conditions [1]. These equations are usually solved self-consistently by either finite element or finite difference procedures and include Poisson's equation, current-density equations for holes and electrons, and continuity equations for holes and electrons.

The discussion that follows pertains to detailed device models.

Typical Device Structures

Figure 1 gives the structure for a device similar to the one investigated by Asbeck et al. [2]. This HBT avoids many of the tradeoffs in the design of homojunction bipolar transistors. The HBT has a wider bandgap emitter (layer 3). The greater bandgap of the emitter compared to the base reduces substantially the hole injection from the base into the emitter. Reducing the basewidth decreases the electron transit time in the device and thereby increases f_T. The conduction band spike is reduced by the compositional grading [3]. This increases the injected electron current and, therefore, the gain of the HBT. Such HBT devices with thin bases and high f_Ts are fabricated by molecular beam epitaxy (MBE) or by organometallic vapor phase epitaxy (OMVPE).

Circuits containing HBTs, which are similar to the HBT in Fig. 1, and operating at 300 K or 77 K may compete with or exceed the ultrahigh performance of Josephson junction circuits at 4K [4]. However, before HBTs can be used in high-speed integrated circuits, improved fabrication technology needs to be devised. In particular, the number of defects in the substrate and the number of interface states at junctions need to be reduced.

Figure 2 shows a typical heterostructure HFET, which is similar to the one reported in Ref. [5]. The main feature of these devices is to have the donors in a wider bandgap $GaAl_xAs_{1-x}$ layer (layer 2) and to have the electrons move in a nearby undoped GaAs channel (layer 4). These devices frequently have a spacer layer of undoped GaAs (layer 3) to shield the two-dimensional electron gas from the fields of the donors in layer 2. When the donor densities in layer 2 exceed about 10^{17} cm^{-3}, the electrons do not have a bound state associated with the donor ions. The electron-ion scattering becomes significant in layer 2 and lowers the electron mobility. The undoped layers 3 and 4 provide the required high mobility, active region of the device.

The nonlinearity of the HFET is good for L_gs $\sim 1\mu m$ and very good for L_gs $\sim 0.5\mu m$ [4]. This nonlinearity is the increase of g_m with small $(V_{gs} - V_T)$ and increases as μ_n/L_g increases, where μ_n is the mobility of the electron in the channel. The current gain-band-

The views stated in this paper are those of the author and do not represent necessarily the views of the National Bureau of Standards.

Reprinted from *IEEE Circuits and Devices Magazine*, Herbert S. Bennett, "Modeling GaAs/AlGaAs Devices: A critical review, Vol. 1, No. 1, pp. 35-42, January 1985.

width product increases as L_g decreases. Hence, to increase g_ms and f_Ts, shorter gate lengths and higher mobilities are needed. Gate lengths less than 0.5μm place unacceptable demands on lithographic precision for commercial production. Heterostructure HFETs offer ways to achieve mobilities that are higher than those for GaAs alone. Such FETs are called TEGFETs [5], MODFETs, HEMTs [6], and SDFETs. One reason that similar devices are referred to by many names is that the understanding of how the devices function is incomplete. Most workers concentrate on how the carriers move in the channel and place less emphasis on how the carriers move from the source to the channel and from the channel to the drain.

Representative Device Models and Associated Assumptions

Many HBT, MESFET, and HFET models exist in the literature. Most authors use the predictions from such models to perform numerical experiments and thereby to suggest ways to improve device performance. Few compare the predictions with measurements from devices. In this section, five representative device models are presented to illustrate the recent advances in modeling GaAs devices. These five device models have been selected from among the many models reported in the archival literature, since as a set, they contain most of the essential features and assumptions found in HBT, MESFET, and HFET models. The many other models that are not discussed or referenced here are, with few exceptions, variations on those outlined below. Including discussions of additional models would detract from the main purposes of this review; namely, identifying areas for which improved physical concepts and measurement techniques are needed.

Formulation of Lundstrom and Schuelke

Lundstrom and Schuelke have developed a numerical method for analyzing heterostructure semiconductor devices (HBTs and HFETs)[7]. Their analysis is based on a macroscopic description of semiconductors with nonuniform composition. The Lundstrom-Schuelke (LS) model contains conventional device equations. It discretizes the basic equations by the finite difference technique and uses the Scharfetter-Gummel [8] formulation for the current densities.

These authors have modified conventional device analysis programs to evaluate the two band parameters V_n and V_p. They describe the nonuniform composition by position dependent $\kappa_s(x)$, $V_n(x)$, and $V_p(x)$. These modifications are valid only for heterostructures in equilibrium. Also, the LS model is strictly valid only when the material composition changes slowly and the concept of a position dependent effective mass is reasonable [9].

However, the above restrictions may not be appropriate for processed GaAs.

The main assumptions of the LS class of models are summarized below. The equilibrium *pn* product is assumed to have the form

$$n_0 p_0 = n_i^2 \exp(-\Delta E_g/\kappa T) \quad (1)$$

where $\Delta E_g = -q(V_n + V_p)$. Recent work [10] has shown that Eq. (1) gives incorrect descriptions for heavily doped silicon devices with emitter widths less than 3 μm. Preliminary results indicate that similar difficulties occur with Eq. (1) for GaAs above $10^{17} cm^{-3}$. Also, this and most other models assume that the temperature is uniform throughout the device and that no strains are present. However, Moglestue has shown that local heating occurs between the gate and drain of *n*-type GaAs FETs [11].

Lundstrom and Schuelke have applied their model to both the HBT in Ref. [2] and to the TEGFET in Refs. [5,12]. They have not compared the predictions of the LS model for HBTs with measured I-V data. Only numerical experiments to understand better the behavior of HBTs are reported. They have compared the LS model predictions with measured quasi-static capacitance versus reverse gate voltage for the TEGFET [12]. The agreement is good except at high voltages (> 8V), for which breakdown may occur.

Formulation of Asbeck et al.

Asbeck et al. [13] have modified the one-dimensional, finite difference code SEDAN [14] to be applicable for HBTs. The basic semiconductor equations are similar to those for the LS model in the "Formulation of Lundstrom and Schuelke," section except the constitutive relation for $\mathbf{J}_n$

$$\mathbf{J}_n = -nq\mu_n \nabla(V + V_n) + D_n \nabla_n \quad (2)$$

is replaced with the equation

$$\mathbf{J}_n = -qn\mathbf{v}(\mathbf{x}) \quad (3)$$

The electron velocities are obtained from equilibrium and ballistic transport relations for **v(E)**. They have considered the dependence of f_T on current density and on various **v(E)** relations [15] by performing numerical experiments on HBTs [2]. They have used these predictions to suggest new designs for devices.

Because adequate measurements and theories for the dependence of n_{ie}^2/n_i^2 on the high carrier and dopant densities present in their HBTs do not exist, HBT models contain the assumptions that $n_{ie}^2/n_i^2 = 1$ and that $\mu_p(\text{maj}) = \mu_p(\text{min})$ and $\mu_n(\text{maj}) = \mu_n(\text{min})$ at the same doping densities. These physically questionable assumptions exist also in the recent Monte Carlo simulations reported by Tomizawa et al. [16].

Using the current crowding under the emitter as a variational parameter, Asbeck et al. have compared the dc common emitter gain versus collector current with measured values. The agreement is marginally acceptable. Continuing additional numerical experiments without better input data for mobilities, band edge changes, and effective intrinsic carrier concentrations has limited value to assist in improving HBTs.

Formulation of Riemenschneider and Wang

Riemenschneider and Wang have developed a two-dimensional, finite-element program for analyzing transient and steady-state characteristics of GaAs MESFETs [17]. The devices that they have analyzed are dominated by electrons, and the contribution of holes to the total current flow is negligible. The holes would be important if the model were to include parasitic effects, such as backgating or other mechanisms for minority carrier injection. But, few, if any, two-dimensional, finite-element models consider parasitics. The generation and recombination terms in the relation for the conservation of electrons are set to zero. The electron current in the model is derived from classical transport theory

$$\mathbf{J}_n = -qn\mathbf{v}_n + D_n \nabla n \quad (4)$$

Equation (4) contains transport by diffusion, whereas Eq. (3) does not.

The values of the electron velocity are based on steady-state data from the Monte Carlo calculations of Ref. [18]. The diffusivity D_n values are interpolated between the low field values computed by Einstein's relationship and the high field values [19]. They do not include the effect of dynamic velocity overshoot. Since most velocity overshoot formulations such as Cook and Frey [20] do not consider the multi-valley nature of electron transport in GaAs in detail, Riemenschneider and Wang consider the accuracy of the models that use them to be doubtful.

Riemenschneider and Wang have reported only numerical experiments on comparing the predicted performance of planar MESFETs with recessed gate MESFETs. They have no experimental verification of their calculations. This tends to be the rule and not the exception for modeling GaAs devices. Verification of models to the extent accomplished for silicon devices is rare for GaAs device models.

Formulation of Cook and Frey

Cook and Frey [20] have presented computer simulations of GaAs MESFETs that include transport effects (velocity overshoot). They offer an engineering-level description of hot electron effects in GaAs MESFETs. Their transport model contains many assumptions. Some of the more significant ones are as follows:

(1) Including the upper and lower valleys in the transport equations makes the procedure too complicated, so they have used an equivalent single valley model [21].

List of Symbols

$\mathbf{D}$	Dielectric displacement	R	Recombination rate
D	Defect or trap density	t	Time
D_n	Electron diffusivity	T	Temperature
E	Carrier energy	V_{gs}	Gate-source voltage
E_F	Fermi energy	V_{BE}	Base-emitter voltage
E_G	Bandgap	V_T	Threshold voltage
$\mathbf{E}$	Electric field	V_n and V_p	Band parameters
f_r	Current-gain bandwidth products	W	Carrier energy
g_m	Transconductance	W_B	Base width
G	Generation rate	W_E	Emitter width
$\hbar$	Planck's Constant/2π	$\mathbf{x}$	Position vector
I–V	Current–voltage characteristic	β	Gain
J	Current density	Γ, L, X	Symmetry points in Brillouin Zone
k	Boltzmann Constant	$\Delta_{\Gamma L}$	Subband energy difference between Γ and *L* points
$\mathbf{k}$	Carrier wave vector	$\Delta_{\Gamma X}$	Subband energy difference between Γ and *X* points
L_g	Gate length	ϵ	Dielectric constant
m^*	Carrier effective mass	κ_s	Position-dependent dielectric constant
n	Electron density	μ_n(maj)	Electron mobility n-type
n_0	Electron equilibrium density	μ_n(min)	Electron mobility p-type
n_i	Intrinsic carrier concentration	μ_p(maj)	Hole mobility p-type
n_{ie}	Effective intrinsic carrier concentration	μ_p(min)	Hole mobility n-type
N_D	Donor density	τ_p(W)	Momentum relaxation time
N_A	Acceptor density	$\tau_W(W)$	Energy relaxation time
p	Hole density		
p_0	Hole equilibrium density		
q	Electronic charge		

(2) The electron-phonon and electron-ionized impurity scattering processes can be described by energy-dependent relaxation times for electron momentum and energy, $\tau_p(W)$ and $\tau_w(W)$, respectively, i.e.,

$$(dv/dt)\text{coll} = v/\tau_p(W) \quad (5)$$

$$(dW/dt)\text{coll} = (W - W_o)/\tau_w(W) \quad (6)$$

(3) Equations (5) and (6) require the assumption that $\tau_p(W(t_1)) = \tau_p(W(t_2))$ and $wt_w(W(t_1)) = \tau_w(W(t_2))$.

Most discussions of energy transport in GaAs use expressions such as those given by Eqs. (5) and (6). However, these expressions may not be valid for GaAs. Nougier et al. have shown in Ref. [22] that the transient velocity in a semiconductor is the solution of a relaxation time equation, such as Eq.(5), when impurity and polar (longitudinal long range) optical phonon scatterings are negligible. However, impurity and optical phonon scatterings [23] are large for the structures shown in Figs. 1 and 2.

Hence, a very significant question remains to be answered in the modeling of state-of-the-art GaAs devices. Namely, do the two assumptions (5) and (6) compensate each other and lead to good quantitative simulations or do they add and lead to additional errors? Before this question may be answered, however, the quantitative correctness of the boundary conditions used in GaAs device models must be determined. The convoluted nature of boundary conditions representing the physical device and of assumptions (5) and (6) makes verifying GaAs device models a challenge. Others [24] state that, at present, the boundary conditions used in most GaAs models may be quantitatively in much greater error than the errors associated with any of the above assumptions. If this is true, then comparing the predictions of energy transport and conventional models with measured I-V data is not of value until knowledge of the boundary conditions improves.

Cook and Frey have compared the predictions of conventional models with their energy transport model for the case of a uniformly doped planar MESFET. They have concluded that whenever W (E) increases rapidly with E, then the transport model should be quantitatively superior to conventional models. However, they have performed only numerical experiments.

Formulation of Yoshii, Tomizawa, and Yokoyama

Using a Monte Carlo scattering description, which is similar to that given by W. Hockney et al. [25], Yoshii, Tomizawa, and Yokoyama [26] follow in space and time the trajectories for all particles under the inhomogeneous, local electric field. Their method is a valid way to solve the Boltzmann transport equation. As is the case for the formulation of Cook and Frey, the formulation of Yoshii et al. requires that the scattering mechanisms and the band structure be quantitatively given as functions of carrier concentrations and/or doping densities. However, since it also requires substantial computer resources compared to the other methods, it is used rarely to optimize devices. Instead, full Monte Carlo particle simulations may provide a sound physical basis for parameters in the relaxation time approximation. The latter then may be used with improved confidence to optimize device performance.

Using their two-dimensional full Monte Carlo particle simulations, the authors of reference [26] have concluded that nonstationary carrier transport influences considerably the device characteristics of submicrometer-gate GaAs MESFETs and that the relaxation time approproximation may overestimate the nonstationary effects.

Physical Concepts for GaAs Devices

Reducing the number of unknown parameters in device models increases the effectiveness of computer models for product development. The extent to which manufacturers of GaAs ICs will be successful in "forward engineering" with computer models depends, in part, on the correctness of the physical concepts used. An example of "forward engineering" based on detailed models is the recently reported improvement in the performance of short-channel NMOS devices [27].

It is essential that any changes in the device physics be anticipated at the time that computer codes for solving, numerically, the device model are developed. The numerical stability of the solutions depends greatly on the algorithms employed [24]. The latter may be sensitive to the dependences of the device parameters on densities, electric fields, temperatures, and position.

The remaining parts of this section contain examples of how improved device physics might contribute to better performance of GaAs/AlGaAs HBTs and FETs.

GaAs/AlGaAs Bipolar Heterostructures

The potential for high-frequency performance has increased the effort devoted to GaAs/AlGaAs heterojunction bipolar transistors (HBT). Proposed devices [28] have p^+ bases with Be acceptor densities of $10^{19} cm^{-3}$ and n^+ emitters with Si donor densities of $5 \times 17^{17} cm^{-3}$. Estimates for GaAs indicate that its band structure is perturbed whenever dopant ion or carrier concentrations exceed $6 \times 10^{18} cm^{-3}$ in p-type GaAs and exceed $10^{17} cm^{-3}$ in n-type GaAs [29]. Above these concentrations, theory suggests that quantities such as E_G, $\Delta_{\tau L}$, $\Delta_{\tau X}$, n_{ie}, mobilities, and lifetimes should be calculated in terms of the perturbed band structures and not in terms of the band structure for the intrinsic material.

Any bipolar-detailed device model planned for the immediate application to GaAs/AlGaAs HBTs, such

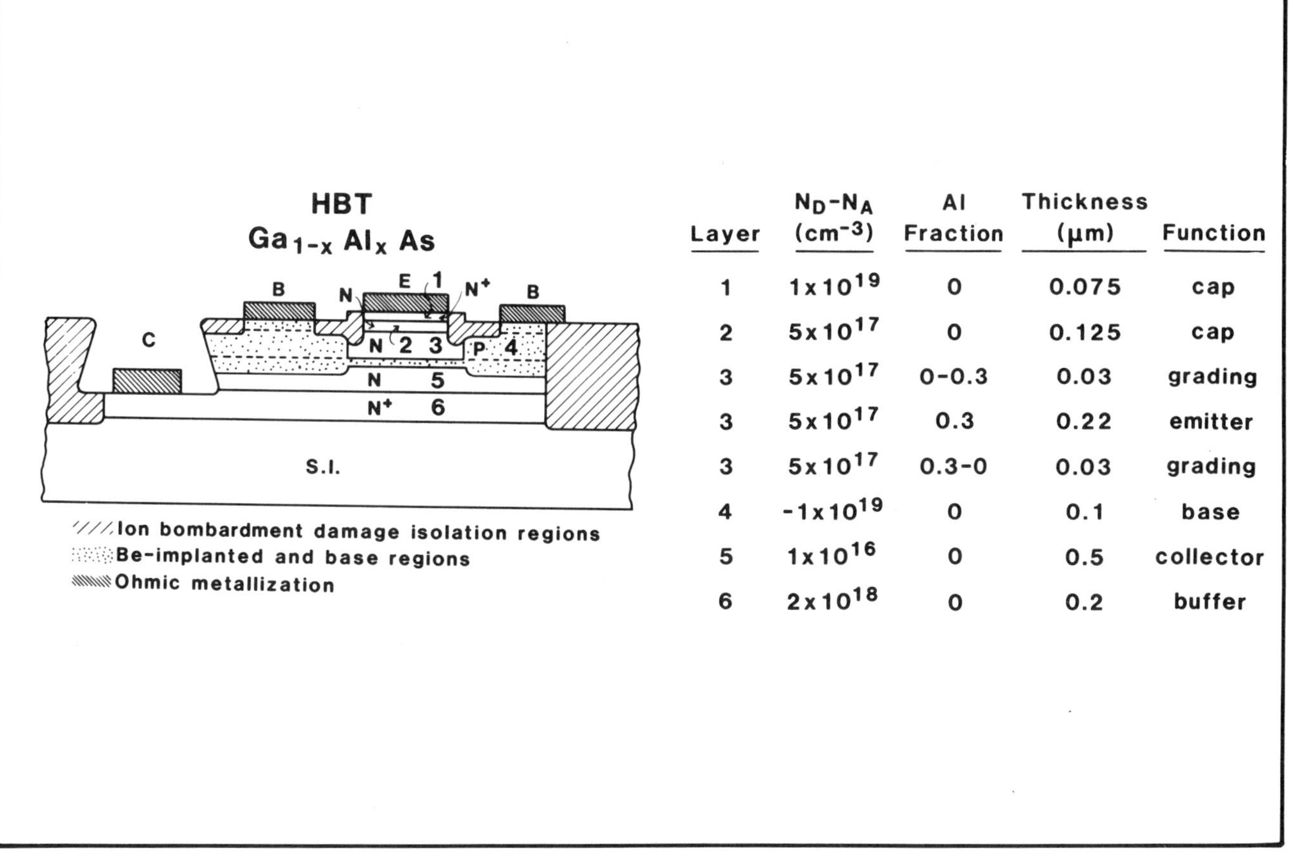

Layer	N_D-N_A (cm^{-3})	Al Fraction	Thickness (μm)	Function
1	1×10^{19}	0	0.075	cap
2	5×10^{17}	0	0.125	cap
3	5×10^{17}	0-0.3	0.03	grading
3	5×10^{17}	0.3	0.22	emitter
3	5×10^{17}	0.3-0	0.03	grading
4	-1×10^{19}	0	0.1	base
5	1×10^{16}	0	0.5	collector
6	2×10^{18}	0	0.2	buffer

Fig. 1 Cross section of the HBT described in Ref. 2.

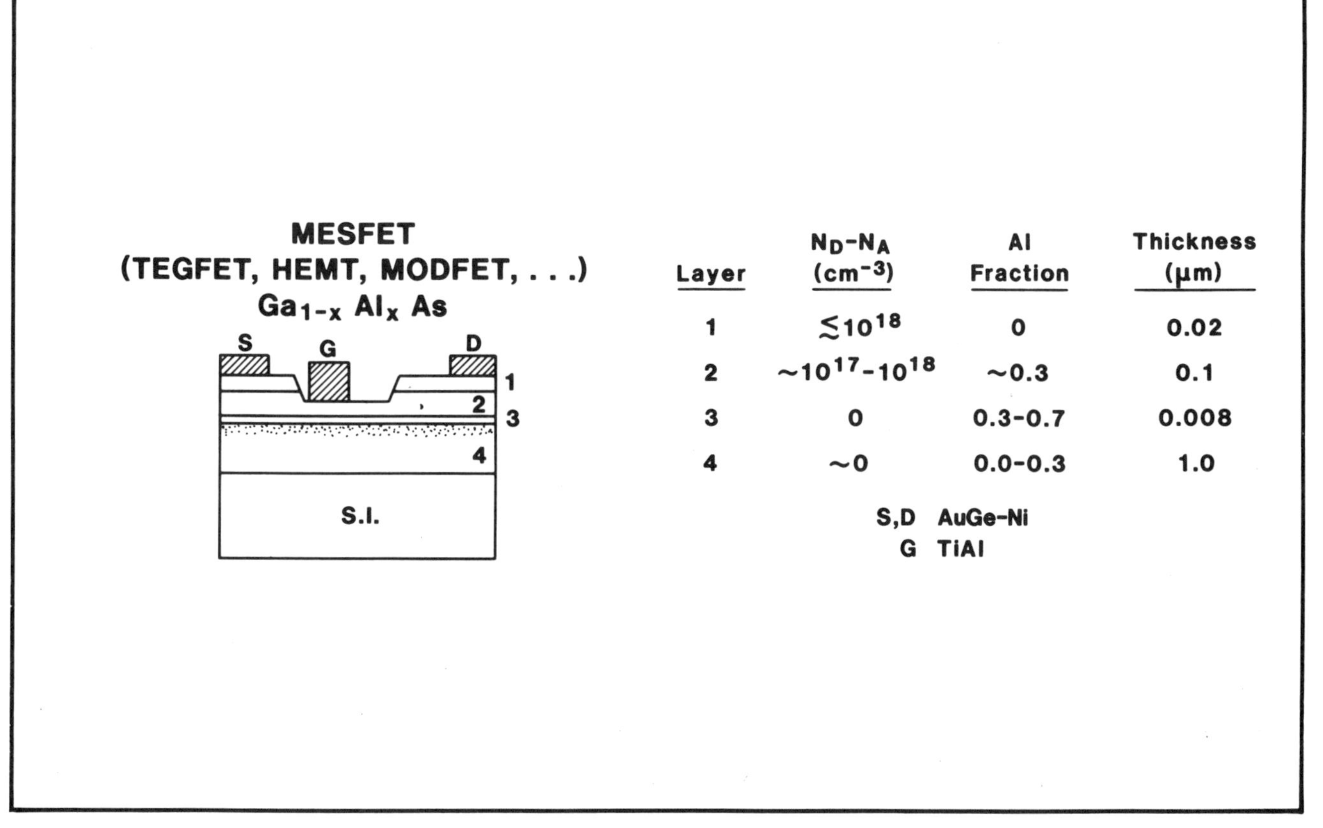

Layer	N_D-N_A (cm^{-3})	Al Fraction	Thickness (μm)
1	$\lesssim10^{18}$	0	0.02
2	$\sim10^{17}-10^{18}$	~0.3	0.1
3	0	0.3-0.7	0.008
4	~0	0.0-0.3	1.0

Fig. 2 Cross section of the enhancement mode TEGFET described in Ref. 12.

as those described in Ref. [13], would not incorporate these effects because of high dopant and carrier concentrations. This would occur because expressions for such quantities as n_{ie} as a function of doping or carrier density are not available in a form suitable for use in detailed models. The usual expressions for the majority carrier mobilities and drift velocities would also be used, because acceptable expressions for minority carriers do not exist. A third area of difficulty is that knowledge of recombination centers and minority carrier lifetimes is very meager in GaAs. For example, if the number of recombination centers is too great, carrier transport becomes impeded. As a result of this lack of verified input data for GaAs bipolar device models, no discernible correlation between measurements and model predictions should be expected.

FETs

Numerical simulations for the carrier transport in the channel of MESFETs reveal three distinct effective field regions: contact, channel, and a rapidly forming space charge dipole [30]. For a $1\mu m$-gate with an n-type dopant density of $2 \times 10^{17} cm^{-3}$ in the channel of a conventional MESFET, the carrier density in the accumulation region of the dipole region is about $3.5 \times 10^{17} cm^{-3}$ and the carrier density in the depletion region is about $10^{17} cm^{-3}$. This dipole forms in about 10ps when ideal voltage sources are assumed [31].

Device performance characteristics, such as transconductance, are sensitive to changes in the carrier mobility. The changes in mobilities for the channel and dipole regions depend in part on the band structure parameters, particularly on $\Delta_{\Gamma L}$ [29,24]. These band parameters are expected to differ from the intrinsic band parameters due to carrier-dopant ion and carrier-carrier interactions.

Backgating and light sensitivity, which degrade circuit performance and may limit packing densities [32,33] in MESFETs, have been shown by C. P. Lee and coworkers [34,35] to be related to carrier injection from the substrate and to carrier transport from the substrate to the active region. Many proposals for reducing the effects of backgating [34,36,37] involve the transport of minority carriers.

The source and drain contacts and the Schottky-barrier gate provide other areas where physical concepts are not adequate. For example, the transition from the n-type channel, through the n^+ drain, to the highly degenerate alloy contact has abrupt concentration changes and large mobility variations. Also, the presence of trapping states at the Schottky-barrier interface greatly influences the Fermi energy and the work function and may contribute a new mechanism of charge storage. The latter may limit device speed. Design engineers need detailed models for ohmic contacts and Schottky barriers to predict their high-frequency behavior. They are particularly interested in knowing before fabrication whether any upper limits for the frequency responses of contacts or gates will affect their design considerations.

Breakdown in the region between the drain side of the gate and the drain has been observed. Device engineers need to know more precisely where the avalanche process begins and how the carrier injection mechanism takes place in their search for strategies to increase the gate to drain breakdown voltage.

New Measurements for Processed GaAs/AlGaAs

As devices become faster and smaller, understanding the ultrafast behavior and nonequilibrium transport must improve. Designers of high-speed devices lack accurate techniques to measure device performance. An underlying principle is that the measurement technique must be faster than the device under test. Electrical measurements are usually used to test devices. This approach is successful when the electrical techniques for device characterization are faster than the devices under development. However, when the goal is to build the fastest device, the usual electrical techniques use the device to measure itself. An example of the latter approach is the ring oscillator technique applied to MESFETs. Since the time resolution of the electrical measurements based on a ring oscillator is determined by the individual devices themselves, only estimates of device speed are possible and little understanding of ultrafast processes and nonequilibrium transport result.

Pulsed optical or electron beam techniques offer alternative methods by which to measure many of the parameters listed in the Appendix and to increase understanding ultrafast devices. Pulsed optical techniques are faster than most high-speed technologies such as semiconductor and superconducting electronics. Pulsed optical techniques may, therefore, be the preferred way to measure key parameters for device models and to measure the electrical behavior of fast devices. One new technique [38] introduces discrete charge packets by a pulsed laser. This time of flight method has been applied to measure high-field transport at $Si\text{-}SiO_2$ interfaces. Whether such techniques are applicable to interfaces involving GaAs needs to be determined. A basic problem is that a technology for building a resistive gate in GaAs does not exist. Other challenges include: (1) pulsed optical or electron beam techniques may not be fast enough for GaAs; (2) sample quality may not be adequate over the path length needed [24]; and (3) applying a uniform, lateral electric field to the sample over the entire path length has not been demonstrated for GaAs.

Conclusions

The major conclusions from the discussions above are

(1) The measured and theoretical data for many of the electrical and material properties of GaAs devices are not adequate for reliable engineering without detailed verification by measurements. The input quantities for detailed device models contain many unknown parameters.

(2) The interdependence of numerical stability and device physics dictates that any changes in the device physics must be anticipated at the time the algorithms and computer codes are developed.

(3) Many physical concepts are not adequate for GaAs devices. These include carrier scattering rates due to ionized impurities and polar optical phonons and carrier transport when carrier or doping concentrations exceed $10^{17} cm^{-3}$ in n-type and $10^{19} cm^{-3}$ in p-type GaAs/AlGaAs. Incorporating adequate physical concepts in GaAs device models requires new measurement techniques, perhaps based on ultrafast spectroscopy, for mobilities and lifetimes as functions of electric fields, dopant density, carrier energy, and carrier density in processed GaAs. Specially designed test structures will be needed to resolve many of the uncertainties concerning the physical concepts to be incorporated into GaAs device models.

Appendix: Input Parameters for GaAs Device Modeling

Several of the input parameters that are needed for modeling GaAs devices are listed below.

Band Structure

Bandgap $E_G(N_D)$ and valley separations $\Delta_{\Gamma L}(N_D)$ and $\Delta_{\Gamma x}(N_D)$

Equilibrium Carrier Densities

$n_{ie}^2(N_D) = np$ and $n_{ie}^2(N_A) = np$

Transport Parameters for Majority and Minority Carriers

$\mu_n(\mathbf{E},N_D)$ and $\mu_n(\mathbf{E},N_A)$
$v_n(\mathbf{E},N_D)$ and $v_n(\mathbf{E},N_A)$
$\mu_p(\mathbf{E},N_D)$ and $\mu_p(\mathbf{E},N_A)$
$v_p(\mathbf{E},N_D)$ and $v_p(\mathbf{E},N_A)$
where E is the electric field

Recombination Parameters

Auger recombination lifetimes for holes and electrons

Shockley-Read-Hall (defect) lifetimes for holes and electrons

Generation Parameters

Impact ionization rates and tunnel rates

Dopant Density Profiles

$N_D(x,y,z)$ and $N_A(x,y,z)$

Defect Density Profiles

$D(x,y,z)$ and the extent to which $D(x,y,z)$ depends on N_D and N_A

Contact and Interface Parameters

Interface trap density, Fermi energy, and work function at Schottky-barrier interface

Interface trap densities at ohmic contacts and other interfaces

Acknowledgments

The author thanks the staff members of the Semiconductor Devices and Circuits Division who have provided helpful comments during the preparation of this article.

References

[1] J. L. Blue and C.L. Wilson, "Two-Dimensional Analysis of Semiconductor Devices Using General Purpose PDE Software," *IEEE Trans. Electron Devices*, vol. 30, p. 1056, 1983.

[2] P. Asbeck, D. L. Miller, R. A. Milano, J. S. Harris, G. R. Kaelin, and R. Zucca, "(Ga,A1)As/GaAs Bipolar Transistors for Digital Integrated Circuits," *Technical Digest*, 1981 International Electronic Devices Meeting (IEEE, New York), pp. 629–632.

[3] H. Kroemer, "Heterostructure Bipolar Transistor and Integrated Circuits," *Proc. IEEE*, vol. 70, p. 13, 1982.

[4] R. C. Eden, "Comparison of GaAs Device Approaches for Ultrahigh-Speed VLSI," *Proc. IEEE*, vol. 70, p. 5, 1982.

[5] D. Delagebeaudeuf, P. Delescluse, P. Etienne, M. Laviron, J. Chaplant, and N. T. Linh, "Two-Dimensional Electron Gas MESFET Structure," *Electron. Lett.*, vol. 16, p. 667, 1980.

[6] T. Mimura, S. Hiyamizu, and K. Nanbu, "A New Field-Effect Transistor with Selectively Doped GaAs/n-$Al_xGa_{1-x}As$ Heterojunctions," *Jap. J. Appl. Phys. Lett.*, vol. 19, p. 225, 1980.

[7] M. S. Lundstrom and R. J. Schuelke, "Numerical Analysis of Heterostructure Semiconductor Devices," *IEEE Trans. Electron Devices*, ED-30, p. 1151, 1983.

[8] D. L. Scharfetter and H. K. Gummel, "Large-Signal Analysis of a Silicon Read Diode Oscillator," *IEEE Trans. Electron Devices*, ED-16, p. 64, 1969.

[9] K. M. van Vliet and A. H. Marshak, "Wannier-Slater Theorem for Solids With Position Dependent Band Structure," *Phys. Rev.*, B26, p. 6734, 1982.

[10] H. S. Bennett, "Improved Concepts for Predicting the Electrical Behavior of Bipolar Structures in Silicon," *IEEE Trans. Electron Devices*, ED-30, p. 920, 1983.

[11] C. Moglestue, *IEE Proc.*, vol. 128, p. 131, 1981.

[12] M. Le Brun, P. R. Jay, C. Rumelhard, G. Rey and P. Delescluse, "Low Noise Performance of Two-Dimensional Electron Gas FETs," *Proc. IEEE/Cornell Conference on High-Speed Semiconductor Devices and Circuits*, IEEE Cat. No. 83CH1959-6, p. 187, Aug. 1983.

[13] P. M. Asbeck, D. L. Miller, R. Asatourian, and C. G. Kirkpatrick, "Numerical Simulation of GaAs/AlGaAs Heterojunction Bipolar Transistors," *IEEE Electron Device Lett.*, EDL-3, p., 402, 1982.

[14] Semiconductor Device Analysis (Stanford University), Jan. 1980 version. Certain materials and computer codes are identified in this paper in order to specify the procedures adequately. Such identification does not imply recommendation or endorsement by the National Bureau of Standards, nor does it imply that the materials or computer codes identified are necessarily the best available for the purpose.

[15] S. Katzer and J. Frey, "Transient Velocity Characteristics of Electrons in GaAs With Γ-L-X Conduction Band Ordering," *J. Appl. Phys.*, vol. 49, p. 4064, 1978.

[16] K. Tomizawa, Y. Awano, and N. Hashizume, "Monte Carlo Simulation of AlGaAs/GaAs Heterojunction Bipolar Transistors," *IEEEE Electron Device Lett.*, EDL-5, p. 362, 1984.

[17] P. R. H. Riemenschneider and K. L. Wang, "A Finite-Element Program for Modeling Transient Phenomena in GaAs MESFETs," *IEEE Trans. Electron Devices*, ED-30, p. 1142, 1983.

[18] M. A. Littlejohn, J. R. Hauser, and T. H. Glisson, "Velocity-Field Characteristics of GaAs with Γ-L-X Conduction Band Ordering," *J. App. Phys.*, vol. 48, p. 4587, 1977.

[19] J. Pozela and A. Reklaitis, "Electron Transport Properties in GaAs at High Electric Fields," *Solid-State Electronics*, vol. 23, p. 927, 1980.

[20] R. K. Cook and J. Frey, "Two-Dimensional Numerical Simulation of Energy Transport Effects in Si and GaAs MESFETs," *IEEE Trans. Electron Devices*, ED-29, p. 970, 1982.

[21] D. E. McCumber and A. G. Chynoweth, "Theory of Negative Conductance Amplication and of Gunn Instabilities in 'Two Valley' Semiconductors," *IEEE Trans. Electron Devices*, ED-13, p. 4, 1966.

[22] J. P. Nougier, J. C. Vaissiere, and D. Gasquet, "Determination of Transient Regime of Hot Carriers in Semiconductors, Using the Relaxation Time Approximations," *J. Appl. Phys.*, vol. 52, p. 825, 1981.

[23] J. W. Harrison and J. R. Hauser, "Theoretical Calculations of Electron Mobility in Ternary III-V Compounds," *J. Appl. Phys.*, vol. 47, p. 292, 1976.

[24] C. L. Wilson, "High Accuracy Physical Modeling of Submicron MOSFET's," *IEEE Trans. Electron Devices*, ED-30, p. 1579, 1983.

[25] R. W. Hockney, R. A. Warriner, and M. Reiser, "Two-Dimensional Particle Models in Semiconductor-Device Analysis," *Electron. Lett.*, vol. 10, p. 484, 1974.

[26] A. Yoshii, M. Tomizawa, and K. Yokoyama, "Accurate Modeling for Submicrometer-Gate Si and GaAs MESFET's Using Two-Dimensional Particle Simulation," *IEEE Trans. Electron Devices*, ED-30, p. 1376, 1983.

[27] W. Fichtner, "Physics and Simulation of Small MOS Devices," *Technical Digest*, 1982 International Electron Devices Meeting (IEEE, New York), p. 638.

[28] P. M. Asbeck, D. L. Miller, W. C. Petersen and C. G. Kirkpatrick, "GaAs/AIGaAs Heterojunction Bipolar Transistors with Cutoff Frequencies Above 10 GHz," *IEEE Electron Device Lett.*, EDL-3, p. 366, 1982.

[29] H. S. Bennett, to be published.

[30] S. Swierkowski and L. F. Jelsma, "Two-Dimensional Computer Modeling of GaAs Devices," *IEEE Trans. Electron Devices*, ED-28, p. 1219, 1981.

[31] S. Swierkowski, private communication.

[32] M. S. Birrittella, W. C. Seelbach, and H. Goronkin, "The Effect of Backgating on the Design and Performance of GaAs Digital Integrated Circuits," *IEEE Trans. Electron Devices*, ED-29, p. 1135, 1982.

[33] H. Goronkin, M. S. Birrittella, W. C. Seelbach, and R. Vaitkus, "Backgating and Light Sensitivity in Ion-Implanted GaAs Integrated Circuits," *IEEE Trans. Electron Devices*, ED-29, p. 845, 1982.

[34] C. P. Lee and B. M. Welch, "GaAs MESFETs with Partial p-Type Drain Regions," *IEEE Trans. Electron Devices*, ED-29, p. 1687, 1982 and "GaAs MESFETs with Partial p-Type Drain Regions," *IEEE Electron Device Lett.*, EDL-3, p. 200, 1982.

[35] C. P. Lee, S. J. Lee, and B. M. Welch, "Carrier Injection and Backgating Effects in GaAs MESFETs," *IEEE Electron Device Lett.*, EDL-3, p. 97, 1982.

[36] C. Kocot and C. A. Stolte, "Backgating in GaAs MESFETs," *IEEE Trans. Electron Devices*, ED-29, p. 1059, 1982.

[37] D. C. D'Avanzo, "Proton Isolation for GaAs Integrated Circuits," *IEEE Trans. Elecron Devices*, ED-29, p. 1051, 1982.

[38] J. A. Cooper, Jr. and D. F. Nelson, "High-Field Drift Velocity of Electrons at the Si-SiO_2 Interface as Determined by a Time-of-Flight Technique," *J. Appl. Phys.*, vol. 54, p. 1445, 1983.

Update. . .

Since the publication of the Critical Review in 1985, much progress has occurred in the sophistication and integration of computer programs to simulate the processing and operation of devices made from compound semiconductors such as GaAs and AlGaAs. This progress includes not only simulations for transistors but also simulations for optoelectronic devices such as lasers, sensors, and components for flat panel displays and for microelectromechanical structures. The abstracts and proceedings of the three international conferences on process and devices modeling [1-3] and the modeling and simulations session of the International Electron Devices Meetings (IEDM) are excellent sources and guides for additional archival information.

Some of the major challenges for process and devices simulations are:

1. Verifying independently the chemical and physical models that these simulations contain.
2. Comparing or benchmarking the predictions of different approaches to simulating the same process, device or structure.
3. Calibrating or validating the simulations by comparing their prediction with measurements.

The latter is crucial if such simulations are to be used effectively in competitive design and manufacturing. At the 1993 IEDM Evening Panel on Funding Technology Development, one panel member stated that process and device simulations are among the top three highest priority items from a list of 12 priority items for reducing the high costs of transferring technologies from research and development to the marketplace [4].

Some progress also has occurred in first principles calculations of the input parameters listed in the Appendix of the foregoing Critical Review. For example, researchers [5 and 6] have completed calculations and experimental verifications for minority and majority mobilities in heavily doped GaAs. This was an international effort. The next tasks include such calculations and verifications for AlGaAs. Supporting this effort will require more complete experimental data on the optical and electronic properties of AlGaAs as functions of the Al concentration.

Until more resources become available for advancing the knowledge and understanding that underlies the infrastructure of computer assisted design and manufacturing, those who develop computer simulations of manufacturing process and devices using these materials will in the interim have to rely more on empirical models than on first principles calculations. Also, parametric fits to the results of theoretical calculations will continue to be useful for interpolating between calculated points on curves and for reducing the costs and times requires for process and devices simulations.

References:

1. International Workshop on Numerical Modeling of Processes and Devices for Integrated Circuits. This meeting is known as NUPAD and is held during May or June in even numbered years.
2. International Workshop on VLSI Process and Device Modeling. This meeting is known as VPAD and is held during May or June of odd numbered years.
3. International Conference on Simulation of Semiconductor Devices and Processes. This meeting is known as SISDEP and is held during September in odd numbered years.
4. B. Cambou, from presentation given at Session 25 of the 1993 International Electron Devices Meeting (IEDM): Evening IEDM Panel on Funding Technology Development, December 7, 1993, Washington Hilton and Towers, International Ballroom Center, Washington, DC.
5. H.S. Bennett, J.R. Lowney, M. Tomizawa, and T. Ishibashi, "Experimentally Verified Majority and Minority Mobilites in Heavily Doped GaAs for Device Simulations," IEICE Transactions Electronics, Volume E75-C, Number 2, February 1992, pp. 161-171, and references 1, 2, 4, 11, and 24-27.
6. M. Tomizawa, T. Ishibashi, H.S. Bennett, and J.R. Lowney, "Effects of Heavy Doping on Numerical Simulations on Gallium Arsenide Bipolar Transistors," Solid State Electronics, Volume 35, Number 6, 1992, pp. 865-874, and references 1-8, 12-18, 20, 25, 26 and 28-34.

Overview

Harry T. Weaver, Guest Editor

Background

Silicon-on-insulator (SOI) materials provide an emerging technology for high-density, high-performance, special-purpose integrated circuits (ICs). These useful SOI properties are fundamentally derived from the capability of total electrical isolation of silicon areas and from the qualitative reduction of junction areas. We illustrate this in Fig. 1, where a bulk complementary MOS cross section is compared to an equivalent SOI structure. Bulk CMOS is used for comparison since the most likely initial impact of SOI is special-purpose integrated circuits, such as three-dimensional circuits, high-voltage transistors, or radiation-tolerant devices. The extraordinary success for dynamic RAM technology presents a formidable barrier for any new competitor in packing density, whereas, in other areas, SOI has special characteristics that will provide it with some advantage.

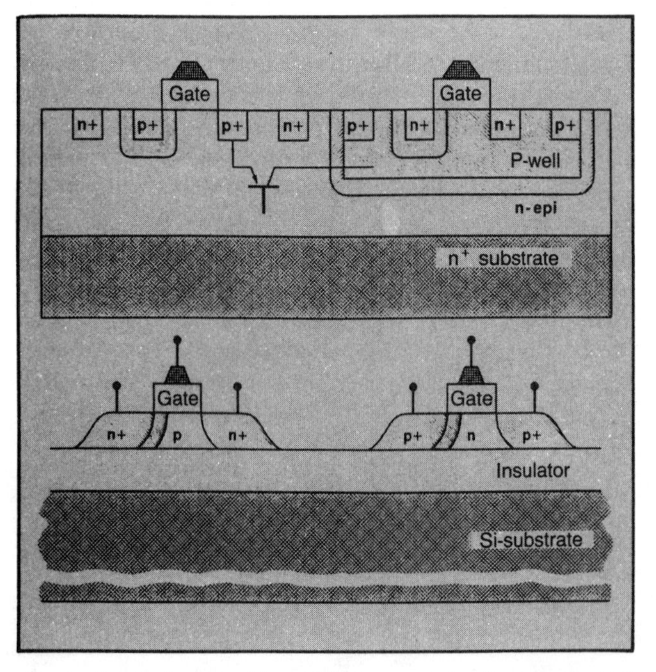

Fig. 1 Sketch of a bulk silicon CMOS cross section (top) and an SOI structure (bottom). The reversed-biased depletion regions are indicated in both sketches. Also illustrated is a parasitic bipolar transistor in the bulk silicon structure. Activation of combinations of such parasitics lead to "latch-up." Note that this p-n-p configuration does not exist in the SOI.

Figure 1 illustrates a p-well CMOS technology in which a reversed-biased well-substrate junction isolates n- from p-channel transistors. This isolation is not perfect, in that transistor operating voltages are limited by the breakdown properties of the doped silicon and, more importantly, various parasitic elements exist. We show, in Fig. 1, a commonly referenced parasitic bipolar transistor formed by p-n-p regions within this structure. Under proper operation, such parasitics are biased off, but random electrical perturbations can trigger bipolar action. A major consideration for CMOS technology is to minimize the effects of parasitic elements such as these. Junction areas, present in the CMOS structure, are also illustrated. These contribute capacitance and reduce circuit speed.

The SOI structure inherently overcomes these CMOS deficiencies. First, the parasitic elements do not exist. Second, the junction area is drastically reduced. And, third, breakdown is limited only by the oxide thickness. This allows close placement of the transistors, high-speed operation, and the potential of high-voltage power and low-voltage logic on the same chip.

Inherent difficulties with SOI derive from the sidewalls and the interface between the buried insulator and silicon (backgate). Both these regions provide leakage paths that can degrade device performance. Sidewall and backgate leakage are, therefore, new processing technology issues that must be satisfactorily addressed. However, the appearance of working ICs on SOI suggests that these problems have been largely overcome. There remains, primarily, material quality as the barrier to common SOI application and, of course, cost.

The first SOI work, begun two decades ago, used sapphire for both substrate and insulator, hence the designation silicon-on-sapphire (SOS). Sapphire has the advantage of being fairly well lattice-matched to silicon. Although there have been notable successes with SOS over the years, technology and material issues have prevented major commercial use. Recent advances in SOS technology, however, continue to preserve the viability of this material, and it remains today as the only commercial SOI technology.

The term SOI is generally associated with structures such as illustrated in Fig. 1, where the insulator is roughly a 1-μm layer of silicon dioxide. Some advantages over SOS are the possibility of a much cleaner interface (fewer interface states), the ability to bias the substrate (backgate), and more variety in the design of novel devices. The backgate bias can be used to control leakage along the substrate-silicon interface, particularly when it is introduced by ionizing radiation.

Construction of quality SOI material is required to achieve the potential for this technique in IC fabrication. There are several known methods for making SOI layers, but, at present, three techniques are generally believed to hold the most promise. Figure 2 crudely illustrates these methods: (1) ion implantation of oxygen below the silicon surface, (2) recrystallization of polysilicon deposited on a thermally grown oxide, and (3) oxidation of porous silicon layers within a substrate.

Ion implantation of oxygen, normally designated SIMOX (Separation by IMplanted OXygen), has the advantage that the material formation is simply another step in an IC fabrication sequence, a step well understood by process engineers. The problems involve setting of implant parameters and anneal schedules for optimum IC performance. There is the further problem that high-volt-

Reprinted from *IEEE Circuits and Devices Magazine,* Vol. 3, No. 4, pp. 3-5, July 1987.

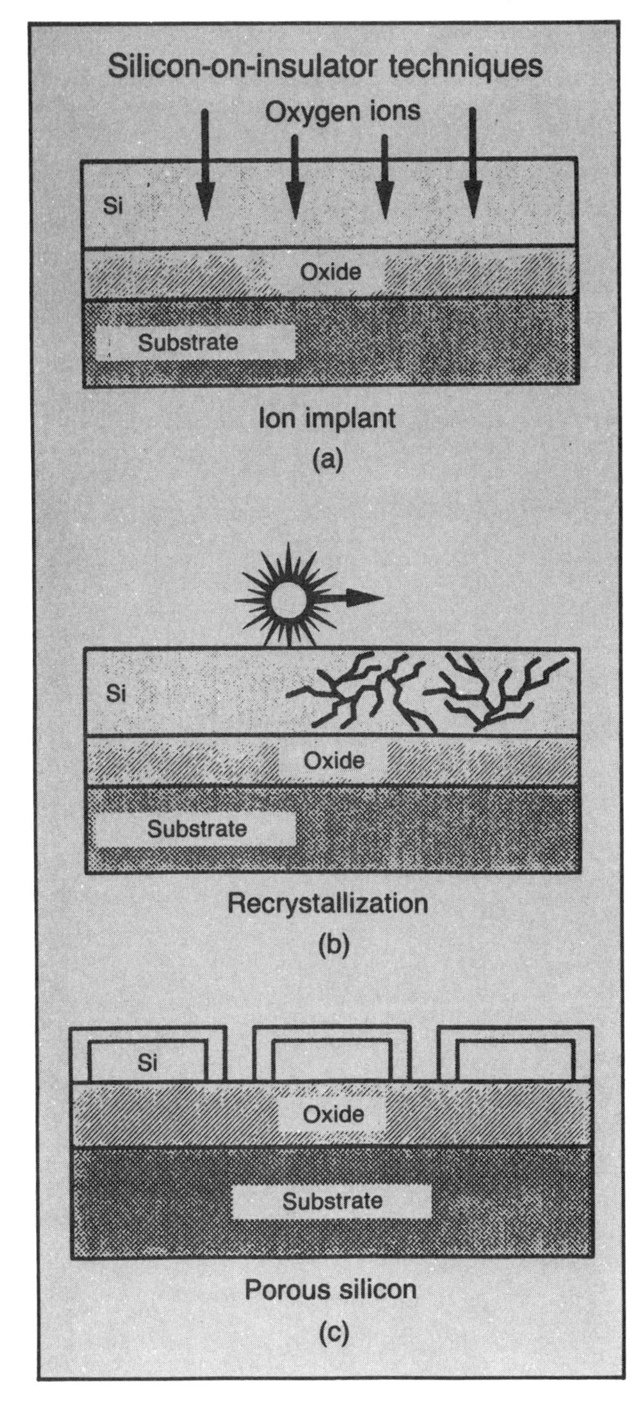

Fig. 2 Sketch illustrating the three principal SOI techniques: (a) ion-implanted oxygen (SIMOX), (b) zone-melt recrystallization (ZMR), and (c) oxidized porous silicon (FIPOS).

age, high-current accelerators are needed to form the oxide layers. This is clear from noting that of order 1×10^{18} cm^{-2} oxygen atoms are needed for a 1-μm silicon dioxide layer. An ion accelerator with 10-mA capability would require one day to implant a 50-wafer lot. Large (100-mA) machines are becoming available to solve this part of the problem, but layer thickness is limited by this consideration.

A second SOI technique—zone-melt recrystallization (ZMR)—forms the active silicon region by recrystallizing polysilicon layers that are deposited onto thermally grown oxides. The recrystallization can be accomplished with or without a "seed." This method has advantages of simplicity and the ability to form any thickness for the buried layer. Development of ZMR focuses on reducing the number of electrically active defects. There are many prototype ICs on ZMR material, and some three-dimensional circuits have been reported.

The third SOI method considered is oxidized porous silicon, or FIPOS (Full Isolation by Porous Oxidized Silicon). This technique is more complicated than the other two but offers the potential for essentially defect-free active silicon. FIPOS is accomplished by first depositing epitaxial silicon on substrates with a heavily doped top layer. Openings are cut through the epi to the substrate, and anodization converts the doped layer to porous silicon. The porous silicon oxidizes readily, leaving the original high-quality silicon on top. The defect density in these materials is currently such that bipolar, as well as MOS, devices are being reported.

Special Issues on SOI

We have assembled the articles into two issues of the *IEEE Circuits and Devices Magazine*. There are three review papers, one on each of the principal SOI techniques and one article describing the design considerations for SOI material. In support of this work, there are several papers that report device performance for the different SOI methods. In the first issue, SIMOX and ZMR are reviewed with accompanying device articles. Also included in this issue is a paper describing recent material improvements for SOS. The second issue will review oxidized porous silicon and will include papers describing current and projected applications for the SOI technology; it will also include an overview of design considerations for SOI circuits.

A special feature of the first issue is an index of ongoing SOI development. We have tabulated responses to our advertisement for information on current work in an index, organized alphabetically by institution. A brief statement of the work and a point of contact are given.

In our first article, Dr. Hon-Wai Lam overviews the progress made with ion-implanted silicon as an SOI material. He has included a lengthy bibliography on the subject for further reading. An article by Dr. W. A. Krull and coworkers from Harris Corporation describes device performance using ion-implanted SOI and competing methods under study at Harris, and Dr. P. K. Vasudev gives an SOS status.

Dr. Bor-Yeu Tsaur reviews the status of recrystallized silicon as an SOI technique. Dr. Tsaur focuses primarily on materials formed with strip heaters, although recrystallization can be accomplished by using a variety of methods, i.e., lamps, lasers, electron beams, etc. A complementary article by Dr. M. Haond and coworkers gives device performance for materials fabricated using a lamp as the heating element.

As a preview to our second issue, Dr. T. Houston and coworkers describe the design issues for ICs that are uniquely associated with SOI materials. We then have Dr. Sylvia Tsao's article on the current status of oxidized porous silicon as an SOI technique. Device data from these materials, novel techniques for radiation hardening, and some potential applications to three-dimensional circuits will also be presented.

There are several other SOI methods under study, which are not addressed in these issues. Included among the techniques is the use of nitrides for isolation. In this case, ion implantation of nitrogen into silicon is again the most studied procedure. These materials have radiation-tolerant characteristics and some advantage in that fewer ions are needed to form the compound compared to oxides. In mentioning this, we note that our issues are not inclusive, but are intended to outline the principal activity areas for SOI.

Latchup in CMOS Technologies

Ronald R. Troutman

Abstract

This paper shows how a conceptually simple definition of a PNPN structure's blocking state is also a precise statement of when latchup occurs, a statement that leads to a concise (there are no fitting parameters), experimentally verified equation for switching current. This paper also categorizes the operational PNPN configurations for various triggering modes and shows how they reduce to two simply analyzed cases. It also discusses latchup avoidance techniques from the perspective of the new latchup criterion.

Introduction

Latchup is a key concern to bulk CMOS and stems from the parasitic bipolar transistors inherent to its structure. Under certain conditions, one of the many PNPN structures on a CMOS chip can be switched from its blocking state to a latched state. The resulting high current causes circuit malfunction and, often, a power supply or ground line to fuse. With proper process and layout design, however, bulk CMOS chips can be operated under relatively harsh conditions without ever encountering latchup; that is, none of the many PNPN structures ever leaves its blocking state even momentarily.

The concept of a blocking state is developed more fully by first introducing a new latchup criterion and then showing how this criterion leads to a precise definition of the blocking state in SAFE space. A second benefit of the new criterion is a simple equation for switching current, which enables a quantitative assessment of latchup hardness. Possible triggering techniques are categorized into three general groups in order to systematize application of the differential latchup criterion. Finally, techniques for achieving a latchup-free CMOS design and how they relate to the differential latchup criterion are discussed.

Differential Latchup Criterion

The parasitic PNPN structure inherent to bulk CMOS consists of an NPN and a PNP bipolar transistor connected so that one's output drives the other's input, as is shown in Fig. 1. Latchup can occur when the loop gain for the PNPN structure equals or exceeds unity. A frequently repeated criterion for latchup is that the sum of the common base current gains (or, equivalently, the product of the common emitter current gains) must equal or exceed unity [1]–[3], i.e.,

$$\alpha_{fn} + \alpha_{fp} \geq 1 \quad (1)$$

However, this condition is strictly valid only for the diode configuration, and only if the small-signal current gain values are used.[1]

The critical gain for the general PNPN tetrode configuration, in which each base/emitter junction is shunted by a bypass resistor, is substantially more complicated since it must account for the gain reduction from the bypass resistors and high-level injection effects. The new criterion can be written in the same form as the preceding one [4], namely,

$$\alpha^*_{fns} + \alpha^*_{fps} \geq 1 \quad (2)$$

where the effective small-signal alphas for forward operation are given by

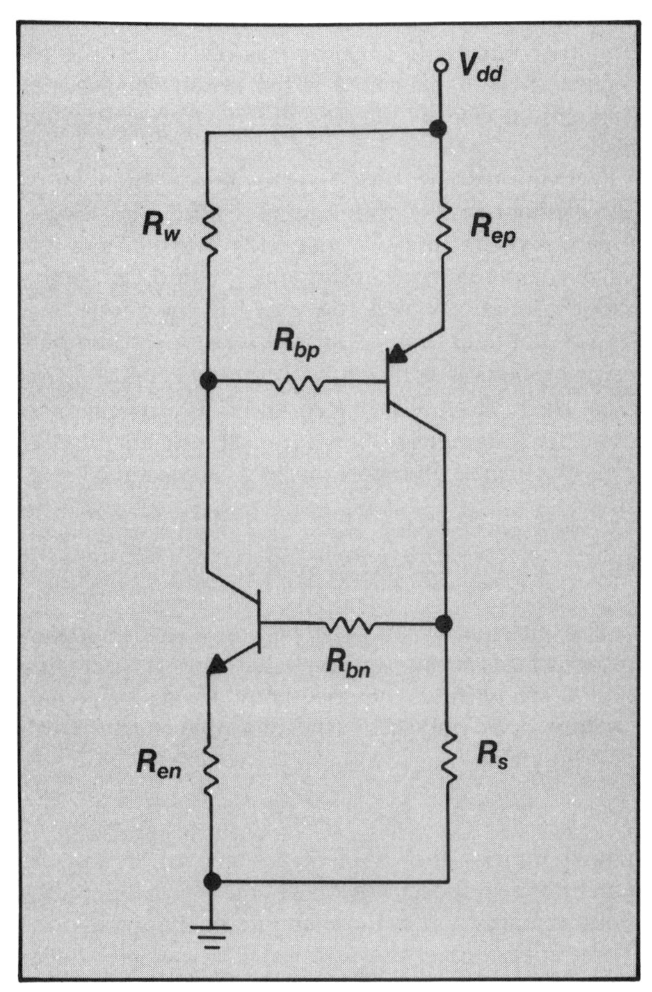

Fig. 1 Lumped-element equivalent circuit for a PNPN structure. Only the resistors influencing switching current are shown.

[1]The subscript f emphasizes current gain in the forward mode of transistor behavior. In the blocking state, the transistors operate strictly in the forward mode; in the latched state, at least one is in saturation and operates in the reverse mode as well. The subscript s denotes the small-signal value for the parameter.

Reprinted from *IEEE Circuits and Devices Magazine*, Vol. 3, No. 3, pp. 15-21, May 1987.

$$\alpha^*_{fns} = R_s\alpha_{fns}/[R_s + r_{en} + R_{en} + R_{bn}(1 - \alpha_{fns})] \quad (3)$$

and

$$\alpha^*_{fps} = R_w\alpha_{fps}/[R_w + r_{ep} + R_{ep} + R_{bp}(1 - \alpha_{fps})] \quad (4)$$

and the various lumped equivalent resistances for the interconnect network are defined in Fig. 1. The small-signal NPN emitter resistance in Eq. (3) is defined by $r_{en} = dV_{be}/dI_{en}$, which, for low-level injection, reduces to $r_{en} = V_t/I_{en}$. For room-temperature operation, the small-signal emitter resistance drops to 1 kΩ once the emitter current has been raised to just 25 μA. High-level injection increases r_{en} according to $r_{en} = V_t/(1 - H_n)I_{en}$, where the high-level injection factor H_n accounts for emitter current saturation as well as the Early effect [4]. Similar expressions can be written for the PNP small-signal emitter resistance r_{ep} in Eq. (4).

In many cases of interest, both transistors' emitter and base leg series resistance can be neglected compared to the small-signal emitter resistance. One then obtains a particularly useful form of the differential latchup criterion, namely,

$$R_s\alpha_{fns}/(R_s + r_{en}) + R_w\alpha_{fps}/(R_w + r_{ep}) \geq 1 \quad (5)$$

It is clear that until a parasitic transistor turns on so that its small-signal emitter resistance is reduced to a value comparable to its bypass resistance, the effective small-signal alpha is the actual small-signal value multiplied by the ratio of bypass to emitter resistance.

In many modern CMOS technologies, the sum of the common base current gains for the parasitic bipolars exceeds unity over a wide range of emitter current. Spontaneous latchup (as would occur for the diode configuration) is prevented in the triode and tetrode configuration because at least one of the parasitic bipolars is sufficiently shunted to hold α^*_{fns} (or α^*_{fps}) $\ll$ 1. The preceding condition is a precise and compact statement of how far each parasitic bipolar must be turned on before latchup can occur.

SAFE Space

The differential latchup criterion also provides a means of classifying turn-on behavior of the parasitic PNPN. Rewriting the inequality to define when latchup does not occur (the blocking state for the PNPN) yields

$$\alpha^*_{fns} + \alpha^*_{fps} < 1 \quad (6)$$

which defines the SAFE space shown in Fig. 2. Latchup cannot occur as long as the PNPN operating point remains within the triangularly shaped region. Three special cases are depicted. The dashed line labeled (1) is the locus of operating points in effective alpha space traced by a PNPN tetrode with exactly complementary NPN and PNP transistors as it turns on. The locus begins at the origin and ends at the switching boundary defined by $\alpha^*_{fns} + \alpha^*_{fps} = 1$. For the triode configuration in which the PNP is not bypassed, the locus labeled (2) begins at a point on the vertical axis equal to α_{fps} and moves horizontally to the switching boundary, where latchup occurs. Similarly, for the triode configuration in which the NPN is not bypassed, the locus labeled (3) begins at a point on the horizontal axis equal to α_{fns} and moves vertically to the switching boundary. The locus for any PNPN structure can be found by using Eqs. (3), (4), and (6) but, in the general case, is complicated by the nonlinear interdependence of r_{en} and r_{ep}.

Representative maps of SAFE space for the general tetrode configuration are shown in Fig. 3 for an N-well CMOS technology. The substrate bypass resistance R_w is held fixed at R_s = 1 kΩ, and the N-well resistance (bypassing the vertical PNP) is varied. Note that when $R_w \simeq 4R_s$, the tetrode's trajectory approaches that of a floating N-well triode configuration. Likewise, when $R_w \simeq R_s/4$, the tetrode's trajectory approaches that of a floating substrate triode configuration. In most cases of interest, the triode configuration is an excellent approximation, and it is easier to analyze since one of the small-signal emitter resistances is removed from the latchup criterion. Which triode configuration is the better approximation for the tetrode depends on the relative values of the bypass resistances and on how the PNPN structure is triggered. The section entitled "Triggering Taxonomy" provides more details on this choice.

Switching Current

The triode approximation to the general tetrode configuration is useful because it greatly simplifies

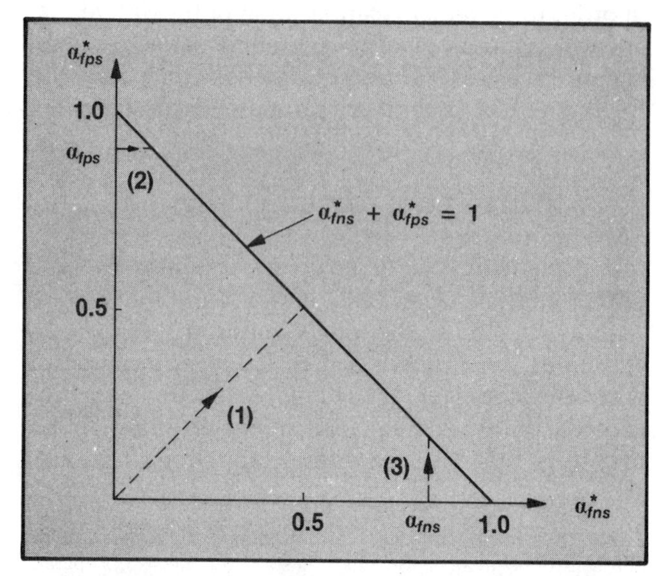

Fig. 2 SAFE space representation of the blocking state. Latchup does not occur as long as the operating point remains in the triangularly shaped region. Dashed lines show special case operating loci (see text). Copyright © 1986 Kluwer Academic Publishers; used with permission.

the description of general behavior in the vicinity of the switching point. For example, the switching current in N-well CMOS for the floating substrate triode when the PNP is in low-level injection (a good approximation here since its bypass resistor prevents it from being turned on hard) is given by [4]

$$I_s = I_{ep,s} + V_t[\ln(1 + I_{ep,s}/I_{sp})]/R_w \qquad (7)$$

where

$$I_{ep,s} = (1 - \alpha_{fns})V_t/(\alpha_{fns} + \alpha_{fps} - 1)R_w \qquad (8)$$

is the PNP emitter current flowing at the switching point. $V_t = kT/q$ is the thermal voltage, and I_{sp} is the saturation current for the PNP base/emitter junction.

Switching current increases as temperature increases and as saturation current and bypass resistance decrease. Reducing bypass resistance is the most effective way of raising switching current, and the section entitled "Latchup-Free Design" discusses how this is accomplished via process and layout design.

Figure 4 compares switching current calculated from the preceding triode equation with a numerical simulation of the general tetrode configuration. Even when one bypass resistor is only twice the other, the triode approximation for switching current is within 5 percent of that calculated for the tetrode configuration. Note that for the parameter values chosen, the worst error using the triode approximation is only 18 percent, since the floating N-well triode should be used when $R_w > R_s = 1\ \mathrm{k\Omega}$.

Switching currents calculated using the triode approximation are in excellent agreement with switching currents measured for bypass resistance values satisfying the triode approximation for bypass resistors. Figure 5 shows an example for the NPN-driven triode configuration (the NPN is not bypassed). Measurements were made with a current source in the NPN emitter, and in all cases, $R_s >> R_w$. The internal well resistance measured $R_{wi} = 285\ \Omega$, and various bypass resistance values were obtained by externally adding series resistance to the substrate and N-well contacts. To an excellent approximation over nearly three orders of magnitude, switching current for the floating substrate triode configuration increases linearly with the N-well bypass conductance. Because the emitter current at the switching point also increases with bypass conductance, the slope in Fig. 5 is somewhat greater than unity (actually, 1.05).

Design techniques that raise switching current also raise holding current. However, modeling holding current requires more information than modeling switching current. At the holding point, at least one of the parasitic transistors is operating in the reverse as well as forward mode, and both transistors are much more likely to exhibit high-level injection effects. In addition, current flow can be drastically altered from that at the switching point by conductivity modulation. Whereas the transistor gains

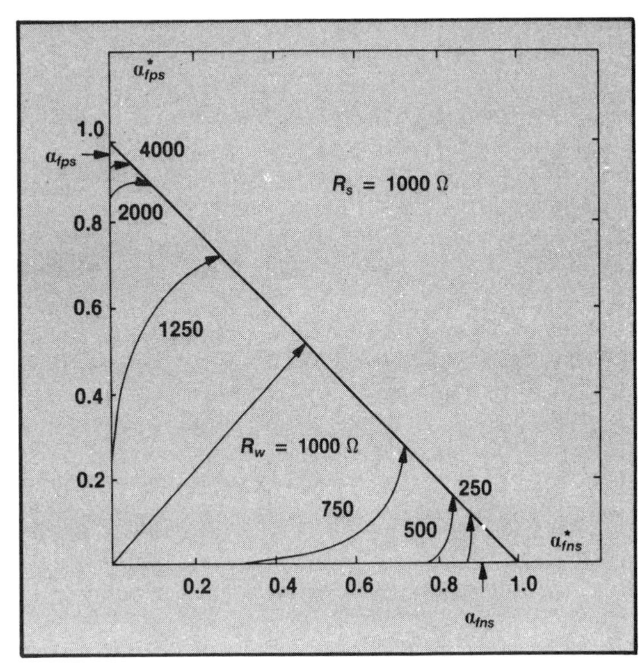

Fig. 3 Representative map of SAFE space. Increasing N-well/substrate current moves the operating point in SAFE space from the origin to the switching boundary. Modeled transistor parameters: $\alpha_{fns} = 0.90$, $\alpha_{fps} = 0.98$, $I_{sn} = I_{sp} = 4.5$ pA. Copyright © 1986 Kluwer Academic Publishers; used with permission.

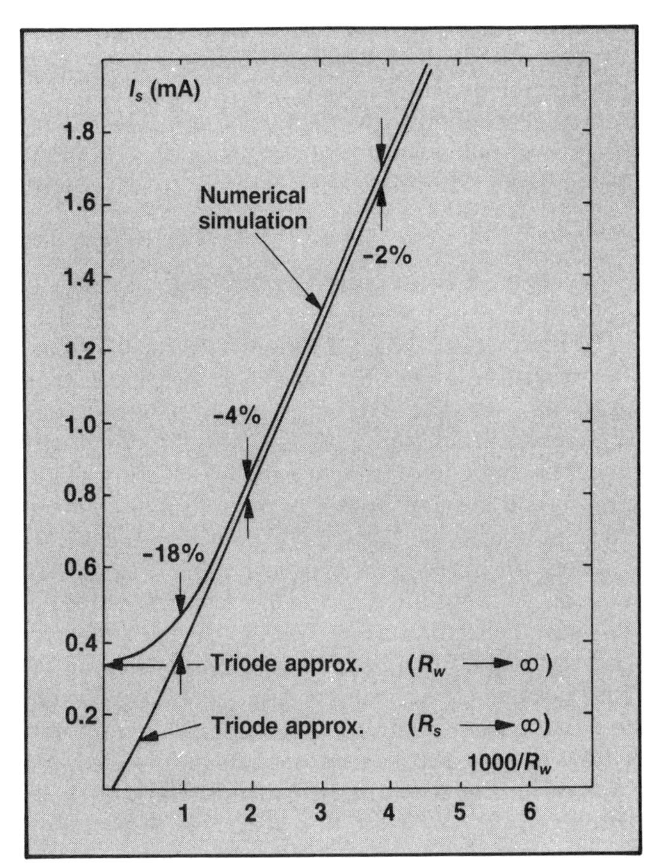

Fig. 4 A comparison of triode and tetrode switching currents for $R_s = 1\ \mathrm{k\Omega}$. The former is calculated from Eqs. (7) and (8); the latter is numerically simulated. Modeled transistor parameters: $\alpha_{fns} = 0.90$, $\alpha_{fps} = 0.98$, $I_{sn} = I_{sp} = 4.5$ pA. Copyright © 1986 Kluwer Academic Publishers; used with permission.

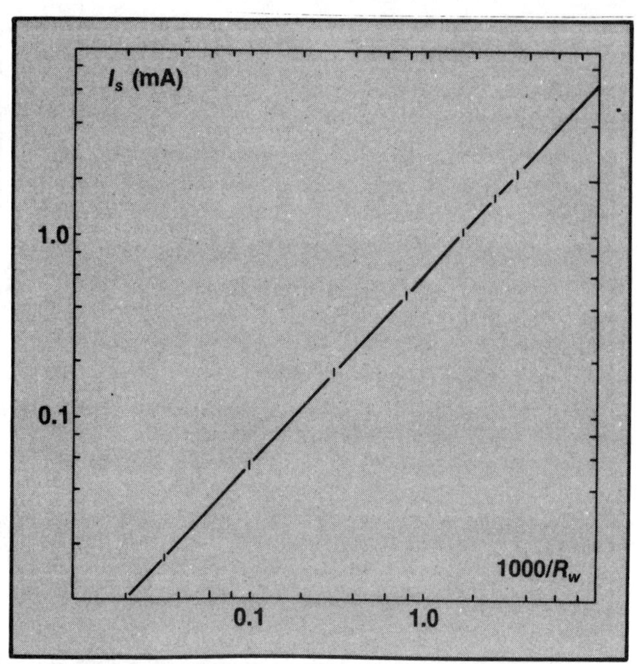

Fig. 5 Switching current for NPN-driven triode configuration. The vertical bars represent the range of three measurements; the solid curve is calculated from Eq. (7). Measured transistor parameters: $\alpha_{fns} = 0.88$, $\alpha_{fps} = 0.98$, $I_{sn} = 7.9\,fA$ (femto Amp), $I_{sp} = 0.33\,fA$. Copyright © 1986 Kluwer Academic Publishers; used with permission.

measured using standard three-terminal techniques usually are adequate for the switching current, they are definitely insufficient for describing holding current. Consequently, quantitative design assessments or theoretical/experimental design comparisons are easier for switching current than for holding current.

Triggering Taxonomy

To understand which is the operative PNPN configuration for a specific situation, it is important to know how the PNPN is being triggered. Both switching current and holding current (or holding voltage) generally depend on the operational PNPN configuration, and inherent to the configuration is the triggering mode.

There are three basic types of triggering. (1) An external stimulus turns on the first transistor, and the resulting collector current flowing through the second transistor's bypass resistance causes an emitter/base voltage. (2) An external stimulus causes current to flow through both bypass resistors. The transistor with the larger bypass resistor turns on first, helping to turn on the second transistor. (3) Current flows through one transistor due to some degradation (such as punchthrough). The resulting current flows through the second transistor's bypass resistor and then forces it on in turn. If large enough to complete a trajectory to the switching boundary, each of these stimuli can cause latchup.

The Table summarizes the various triggering mechanisms and the appropriate PNPN configurations for each. In type 1A, for example, a voltage on the output node of a CMOS driver exceeds V_{dd} by a sufficient amount, and for a sufficient length of time, to turn on the vertical PNP formed by the PMOS drain, N-well, and substrate. Its base current flows to the power supply, and its collector current flows through the lateral transistor's bypass resistor R_s. If the voltage drop across R_s is large enough to turn on the lateral NPN, two latchup scenarios are possible.

The first latchup scenario is called *output latchup* [5]. If the LNPN is turned on enough to satisfy the differential latchup criterion for the "floating N-well" triode configuration (R_w is connected to V_{dd} and does not bypass the emitter/base for the VPNP on the output node), this parasitic PNPN switches to the latched state. The trajectory through SAFE space for this case is indicated schematically by locus (2) in Fig. 2. In the latched state, a low impedance is presented to the output node. Depending on the output circuitry, the voltage on the output node may then drop to V_{dd} or below with the PNPN switching back to the blocking state. Latchup in this case is unsustained.

The second possible scenario is called *main latchup* [5]. If the LNPN is turned on enough to satisfy the differential latchup criterion for the tetrode configuration formed by a bypassed lateral NPN and the vertical PNP whose emitter is connected to V_{dd}, this parasitic PNPN switches to its latched state, and the power supply line is pulled to a low potential. In this case, latchup is generally sustained and can result in permanent damage.

Which of the preceding two possibilities occurs first depends on which PNPN configuration has the lower switching current. When the switching current for output latchup (whether sustained or unsustained) is lower, it is also possible for the increased current in the latched state of output latchup to trigger main latchup.

Bypass current in type-2 triggering, if large enough, turns on both transistors—the one with the larger bypass resistor turning on first. Latchup occurs when the differential latchup criterion is satisfied for the tetrode configuration.

In type-3 triggering, one of the transistors is shunted by a low-impedance path due to punchthrough, avalanche, or the channel current of a field FET (field-effect transistor). As long as this degradation persists, latchup occurs when the differential latchup criterion is met for a triode configuration. If the external stimulus causing such degradation is removed, the latchup may not be sustained. A sustained latchup occurs if the differential latchup criterion is met for the tetrode configuration.

Although the tetrode appears frequently in the table as the operational configuration, it can be approximated by the appropriate triode configuration when one bypass resistance is more than approxi-

Type	Triggering Mode	Operational Configuration
1A	I/O Node Overshoot	
	Output Latchup	Floating N-Well Triode
	Main Latchup	Tetrode
1B	I/O Node Undershoot	
	Output Latchup	Floating Substrate Triode
	Main Latchup	Tetrode
2A	Avalanching N-Well	Tetrode
2B	Photocurrent	Tetrode
2C	N-Well Displacement Current	Tetrode
3A	External Punchthrough	
	Unsustained	Floating Substrate Triode
	Sustained	Tetrode
3B	Internal Punchthrough	
	Unsustained	Floating N-Well Triode
	Sustained	Tetrode
3C	N-Channel Field FET	
	Unsustained	Floating Substrate Triode
	Sustained	Tetrode
3D	P-Channel Field FET	
	Unsustained	Floating N-Well Triode
	Sustained	Tetrode
3E	Avalanching P^+ Diffusion	
	Unsustained	Floating Substrate Triode
	Sustained	Tetrode
3F	Avalanching N^+ Diffusion	
	Unsustained	Floating N-Well Triode
	Sustained	Tetrode

Table Triggering taxonomy and operational PNPN configurations.

mately twice the other, as was discussed previously. For most cases of interest, the appropriate triode configuration can be used for analysis, which greatly simplifies the algebra. The advantage is a reduction of all triggering mechanisms to two simply analyzed configurations.

Latchup-Free Design

The best way to avoid latchup is to keep all PNPN structures in the blocking state. This ensures that a high impedance exists from the power supply rail to the ground at all times and prevents any PNPN switching. Even transient switching (unsustained latchup) can cause circuit problems and undesirable power dissipation.

Techniques for restricting parasitic PNPN operation to the blocking state are divided into two groups: layout guidelines and process design. The first applies to any CMOS technology and is of interest to device and circuit designers. The second is related to the process features incorporated into a CMOS technology by process engineers.

Guard structures are the most effective tool available to the circuit designer. A minority carrier guard is used to precollect minority carriers injected into the substrate before they reach a well comprising the base of a vertical bipolar [2], [6]. In doing so, it reduces the small-signal forward gain of the parasitic lateral transistor as measured to the collector incorporated into the PNPN structure. The minority carrier guard consists of a reverse-biased PN junction formed by a source/drain or well diffusion. It is more effective on epi-CMOS because of the reflecting boundary at the high/low junction and because of increased recombination in the highly doped substrate [7].

A majority carrier guard is used to reduce well or substrate sheet resistance locally, thus minimizing any ohmic drop from parasitic collector current [8]–[10]. As such, it reduces the bypass resistance and can be used on both lateral and vertical parasitic transistors. It consists of a source/drain diffusion of the same type as the background. Operation of a majority guard in the well is also enhanced on epi-CMOS because the elimination of substrate minority carrier collection at the bottom well junction means that majority carriers in the well appear only at its periphery, where they are more easily shunted away from parasitic emitters [4].

Multiple well contacts augment the low-bypass resistance effected by majority guards in the well. Some combination of the two should also be used on I/O circuits to ensure that the parasitic vertical bipolar is not turned on by I/O overshoots and undershoots. In addition to avoiding latchup, the goal here is to prevent large current spikes in these high-gain devices that otherwise would waste power.

A substrate contact ring should be mandatory for all chips. It minimizes lateral bypass resistance by distributing substrate majority carriers. On epi-CMOS, the contact ring can reduce lateral bypass resistance to below 1 Ω and is as effective as a backside substrate contact in eliminating latchup [11].

Butted source contacts can also be used to reduce bypass resistance for the source's parasitic emitter behavior [8]. Their effectiveness in the substrate is greater on epi-CMOS and is improved at reduced epi thicknesses [12]. They are, however, limited to FETs operating with grounded sources.

The second major technique for restricting PNPN operation to the blocking state is process design, which follows two major approaches: bipolar spoiling and bipolar decoupling. Early spoiling techniques attempted to reduce base minority carrier lifetime with either gold doping [13] or neutron irradiation [14], but this led to adverse device effects, such as high leakage current [2]. Later, internal gettering was introduced to lower lifetime in the substrate below the denuded zone, but this has little effect on parasitic lateral bipolars with small basewidth [15]. A retrograde doping profile in the well helps prevent latchup by reducing vertical transport [16]–[19], but care must be taken to remove injected carriers so that lateral transistor action to the well edge is not increased. Schottky source/drains on the PMOSFET have been used to degrade emitter injection efficiency, but, here, a fundamental trade-off arises between spoiling injection efficiency and maintaining decent FET behavior [20]–[22].

Bipolar decoupling has proved more effective and easier to implement than bipolar spoiling, and several methods have proved useful. A highly doped substrate beneath a lightly doped epitaxial layer very effectively shunts the lateral parasitic bipolar [2], [6], [12], [23], [24]. As discussed previously, this combination also improves the efficacy of minority carrier guards in the substrate and majority carrier guards in the well. A retrograde well can also be used to reduce the well's sheet resistance, although this is usually not as effective as including majority carrier guards in the well. Reverse bias on the substrate (or well) raises the bypass current needed to turn on the corresponding bipolar, but using on-chip generators to provide this bias necessitates careful guard designs to eliminate charge injected by the generator itself [25]. Trench isolation eliminates lateral current flow to and from the well and conceivably could allow a lithographically limited N^+/P^+ spacing [26]–[28]. Present sidewall inversion problems must be solved before diffusions can be butted against trench walls, however.

Bipolar decoupling provides a simple design procedure for avoiding latchup. (1) Decouple the vertical parasitic bipolar using epi-CMOS with a substrate contact ring to minimize lateral bypass resistance. (2) Decouple the lateral parasitic bipolar using majority carrier guards in the well or minority carrier guards in the substrate. Latchup hardness of such a design procedure can be rigorously tested using the differential latchup criterion. The data needed are (a) bypass resistance values (only the smaller is needed in the triode approximation), (b) small-signal alphas for both parasitic transistors, (c) base/emitter saturation currents for both transistors, and (d) temperature.

Whichever techniques are chosen to decouple the parasitic bipolars, they should prevent switching from the blocking state. Operating strictly in SAFE space guarantees a healthy, latchup-free CMOS technology.

References

[1] B. L. Gregory and B. D. Shafer, "Latchup in CMOS Integrated Circuits," *IEEE Trans. Nucl. Sci.*, vol. NS-20, pp. 293–299, Dec. 1973.

[2] A. Ochoa, Jr., W. Dawes, and D. Estreich, "Latchup Control in Integrated Circuits," *IEEE Trans. Nucl. Sci.*, vol. NS-26, pp. 5065–5068, Dec. 1979.

[3] D. B. Estreich, "The Physics and Modeling of Latchup and CMOS Integrated Circuits," Tech. Rept. G-201-9, Stanford Electronics Lab., Stanford Univ., Stanford, CA, Nov. 1980.

[4] R. R. Troutman, *Latchup in CMOS Technology: The Problem and Its Cure*, Boston, MA: Kluwer Academic Publishers, 1986.

[5] T. Iizuka and J. L. Moll, "A Figure of Merit for CMOS Latchup Tolerance," 1981 CMOS Workshop, San Francisco, CA, May 18, 1981.

[6] C. C. Huang, M. D. Hartranft, N. F. Pu, C. Yue, C. Rahn, J. Schrankler, G. D. Kirchner, F. L. Hampton, and T. E. Hendrickson, "Characterization of CMOS Latchup," *1982 IEDM Tech. Digest*, pp. 454–457, Dec. 1982.

[7] R. R. Troutman, "Epitaxial Layer Enhancement of n-Well Guard Rings for CMOS Circuits," *IEEE Elec. Dev. Letters*, vol. EDL-4, pp. 438–440, Dec. 1983.

[8] R. S. Payne, W. R. Grant, and W. J. Bertram, "Elimination of Latchup in Bulk CMOS," *1980 IEDM Tech. Digest*, pp. 248–251, Dec. 1980.

[9] R. D. Rung and H. Momose, "DC Holding and Dynamic Triggering Characteristics of Bulk CMOS Latchup," *IEEE Trans. Elec. Dev.*, vol. ED-30, pp. 1647–1655, Dec. 1983.

[10] E. Hamdy and A. Mohsen, "Characterization and Modeling of Transient Latchup in CHMOS Technology," *1983 IEDM Tech. Digest*, pp. 172–175, Dec. 1983.

[11] R. R. Troutman and M. J. Hargrove, "Transmission Line

Modeling of Substrate Resistance and CMOS Latchup," *IEEE Trans. Elec. Dev.*, vol. ED-33, pp. 945–954, July 1986.

[12] G. J. Hu and R. H. Bruce, "A CMOS Structure with High Latchup Holding Voltage," *IEEE Elec. Dev. Letters*, vol. EDL-5, pp. 211–214, June 1984.

[13] W. R. Dawes, Jr. and G. F. Derbenwick, "Prevention of CMOS Latchup in MOS Integrated Circuits," *IEEE Trans. Nucl. Sci.*, vol. NS-23, pp. 2027–2030, Dec. 1976.

[14] J. R. Adams and J. R. Sokel, "Neutron Irradiation for Prevention of Latchup in MOS Integrated Circuits," *IEEE Trans. Nucl. Sci.*, vol. NS-26, pp. 5069–5073, Dec. 1979.

[15] C. N. Anagnostopoulos, E. T. Nelson, J. P. Levine, K. Y. Wong, and N. Nichols, "Latchup and Image Crosstalk Suppression by Internal Gettering," *IEEE J. Solid-State Circuits*, vol. SC-19, pp. 91–97, Feb. 1984.

[16] D. B. Estreich, A. Ochoa, Jr., and R. W. Dutton, "An Analysis of Latch-up Prevention in CMOS ICs Using an Epitaxial-Buried Layer Process," *1978 IEDM Tech. Digest*, pp. 230–234, Dec. 1978.

[17] D. B. Estreich, "The Physics and Modeling of Latch-up and CMOS Integrated Circuits," Tech. Rept. G-201-9, Stanford Electronics Lab., Stanford Univ., Stanford, CA, Nov. 1980.

[18] R. D. Rung, C. J. Dell'Oca, and L. G. Walker, "A Retrograde P-well for Higher Density CMOS," *IEEE Trans. Elec. Dev.*, vol. ED-28, pp. 1115–1119, Dec. 1981.

[19] Y. Taur, W. H. Chang, and R. H. Dennard, "Characterization and Modeling of a Latchup-Free 1-μm CMOS Technology," *1984 IEDM Tech. Digest*, pp. 398–401, Dec. 1984.

[20] M. Sugino, L. A. Akers, and M. E. Rebeschini, "CMOS Latch-up Elimination Using Schottky Barrier PMOS," *1982 IEDM Tech. Digest*, pp. 462–465, Dec. 1982. See also, *IEEE Trans. Elec. Dev.*, vol. ED-30, pp. 110–118, Feb. 1983.

[21] C. J. Koeneke and W. T. Lynch, "Lightly Doped Schottky MOSFET," *1982 IEDM Tech. Digest*, pp. 466–469, Dec. 1982.

[22] E. Sangiori and S. Swirhun, "Trenched Schottky Barrier PMOS for Latchup Resistance," *IEEE Elec. Dev. Letters*, vol. EDL-5, pp. 293–295, Aug. 1984. See also, *1984 IEDM Tech. Digest*, pp. 402–405, Dec. 1984.

[23] D. Takacs, C. Werner, J. Harter, and U. Schwabe, "Surface Induced Latchup in VLSI CMOS Circuits," *1982 IEDM Tech. Digest*, pp. 458–461, Dec. 1982. See also, *IEEE Trans. Elec. Dev.*, vol. ED-31, pp. 279–286, Mar. 1984.

[24] A. G. Lewis, "Latchup Suppression in Fine-Dimension Shallow P-well CMOS Circuits," *IEEE Trans. Elec. Dev.*, vol. ED-31, pp. 1472–1481, Oct. 1984.

[25] R. Piro and F. Sporck, "Latchup-Free Substrate Bias Generators in CMOS," *IEEE Custom Integrated Circuits Conf. Digest*, pp. 524–527, May 1985.

[26] R. D. Rung, H. Momose, and Y. Nagakubo, "Deep Trench Isolated CMOS Devices," *1982 IEDM Tech. Digest*, pp. 237–240, Dec. 1982.

[27] K. M. Cham and S-Y. Chiang, "A Study of the Trench Surface Inversion Problem in the Trench CMOS Technology," *IEEE Elec. Dev. Letters*, vol. EDL-4, pp. 303–305, Sept. 1983.

[28] N. Endo, N. Kasai, A. Ishitani, and Y. Kurogi, "CMOS Technology Using SEG Isolation Techniques," *1982 IEDM Tech. Digest*, pp. 31–34, Dec. 1983.

Porous Silicon Techniques for SOI Structures

Sylvia S. Tsao

Abstract

Among the most promising techniques for producing silicon-on-insulator (SOI) substrates suitable for fabrication of high-performance devices are those based on the oxidation of porous silicon. Porous silicon has a unique set of material properties, which lends itself to a variety of different SOI fabrication techniques.

Introduction

Oxidized porous silicon is the basis for what is considered to be one of the more mature classes of silicon-on-insulator (SOI) fabrication techniques [1]. This paper reviews the SOI fabrication methods utilizing porous silicon, their current status, and their relative merits. Emphasis is placed not only on materials but also on processing and design issues. Only the case of full island isolation is considered here, although oxidized porous silicon has also been used for lateral dielectric isolation [2]–[5].

The mechanisms for porous Si formation have been reviewed [6], [7]. Briefly, porous Si is formed via an anodic electrochemical dissolution of Si in hydrofluoric acid, and P-type Si is more readily anodized than n-type Si. Due to its network of interconnected pores and enhanced surface area, porous Si is oxidized much faster than bulk Si. Depending on the Si doping and etching conditions, porous Si of various densities and pore microstructures can be formed. By proper control of these conditions, porous Si, which is 45 percent as dense as bulk Si, is formed and accommodates the volume increase upon its incorporation of oxygen during thermal oxidation. Because porous Si is formed in selected regions within the wafer and remains monocrystalline, it is attractive for SOI applications.

SOI Fabrication Methods Based on Porous Silicon

The various fabrication methods utilizing porous Si are classified into two main categories. The first is by selective formation of buried porous Si underneath islands, which will eventually contain the devices. It exploits the strong dependence of porous Si formation rates on Si dopant types and concentration. The second is by direct epitaxial Si deposition on a previously formed porous Si layer. This process relies on the fact that the remaining porous skeleton is a single crystal. In both cases, subsequent thermal oxidation is required to convert the porous Si into silicon dioxide.

Both methods have innate advantages and disadvantages. The buried layer techniques yield better quality device islands since these islands were part of the conventionally grown substrates. Their drawbacks stem from the lack of understanding and control of porous Si formation in the complicated geometries of device islands. The epitaxial techniques, on the other hand, while geometrically simpler, rely on an unconventional, albeit monocrystalline, seeding layer. Hence, device layers fabricated using the latter techniques have a higher density of defects. The various schemes of SOI fabrication using porous Si are outlined below.

Buried Porous-Silicon Formation

The most advanced devices to date have been fabricated using the FIPOS (full isolation by porous oxidized silicon) process, which originated from Nippon Telephone and Telegraph, Inc. (NTT) and was first reported in 1981 [8]. It relies on the ease of conversion of p-type Si surrounding n-type islands to porous Si. The process is shown schematically in Fig. 1. First, an Si_3N_4 film pattern is formed on p-type Si to protect device islands during porous Si oxidation. Boron ions are implanted into the unmasked regions to control the density of the porous Si surface layer. Proton ions are then implanted through the Si_3N_4 film to change the island regions to n-type. Porous Si is readily formed in the p-type regions surrounding the islands, which are converted into oxide following thermal oxidation. The Si_3N_4 mask is etched off selectively from the island surfaces prior to device fabrication. The FIPOS process has the advantage of being simple, requiring the

Reprinted from *IEEE Circuits and Devices Magazine,* Vol. 3, No. 6, pp. 3-7, November 1987.

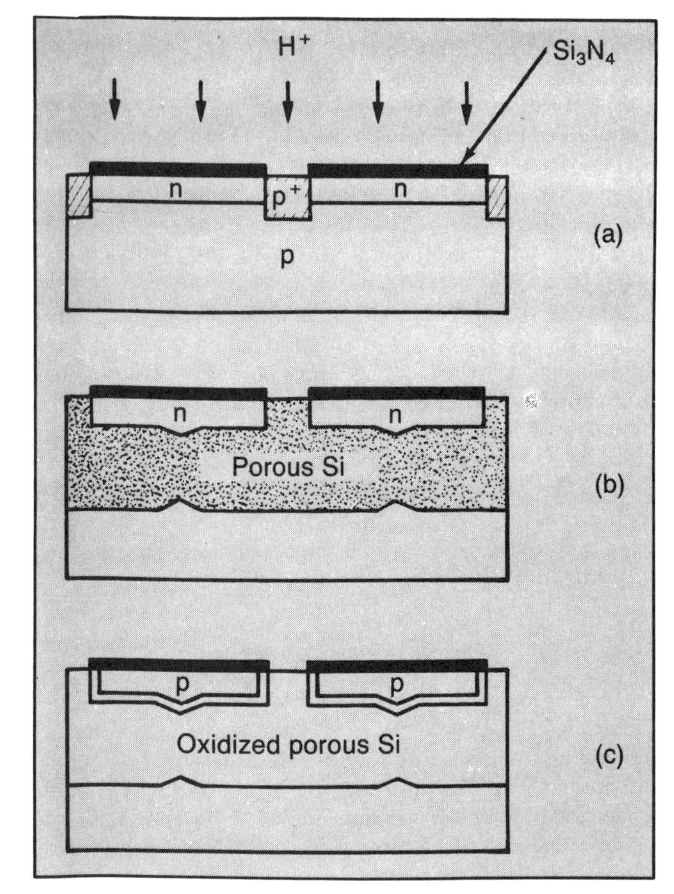

Fig. 1 Schematic of the original FIPOS process. (a) Convert islands from p-type to n-type by proton implantation, (b) form porous Si around n-type islands, and (c) oxidize porous Si [8]. A thin layer of dense thermal oxide is formed on the bottom and sides of the island.

addition of only two steps to the conventional LOCOS (local oxidation of silicon) selective oxidation method: proton implantation to convert p-type regions to n-type and porous Si formation. The resulting surface topography is suitable for LSI (large-scale integration) fabrication [8]. Its main disadvantage is that a thick, oxidized porous Si layer (PSL) is needed to isolate even small islands. For example, a 7-μm-thick layer is needed to isolate 8-μm-wide islands.

In 1984, Nesbit [9] at IBM showed that this drawback is eliminated by the addition of a p^+ layer, which is selectively anodized, between the n and p^- layers. The fabrication process used by Nesbit is shown schematically in Fig. 2. Trenches are etched through the n-type layer down to the p^+ layer to allow transport of the anodization reactants. The p^+ layer is preferentially anodized, resulting in island width-to-PSL thickness ratios of >50. This method retains some of the disadvantages of the original FIPOS method. Where two anodization fronts meet, a cusp in the silicon island is formed and a thin wisp of unanodized silicon remains underneath the middle of the island. The wisp of Si, ~100 Å thick [10], is easily oxidized and does not result in wafer warpage under proper oxidation conditions. However, the island thickness nonuniformity due to the cusp may be a problem during device fabrication, for example, if uniform backgate implant adjustments are required. Two conventional epitaxial depositions to deposit the p^+ and n-layers, or, boron ion implantation into a p^- substrate followed by n-type epitaxial Si deposition can be used to synthesize the $n/p^+/p^-$ structure. The latter was used by Nugent et al. [11] at Sperry Corporation for their version of this process, which they call POST (porous oxidized silicon with trench). Results similar to those in the $n/p^+/p^-$ structures can be obtained by preferential anodization of $p^-/p^+/p^-$ structures, since, near the wafer surface, the buried PSL formation occurs only in regions with net p-type dopant levels $>10^{15}/cm^3$ [10].

The remaining constraints of the modified FIPOS process have recently (1986) been circumvented by Benjamin et al. [12] at Royal Signals and Radar Establishment in England, by using an additional buried n-layer between the p-substrate and the anodizable p-layer underneath n-type islands, as shown in Fig. 3. The buried n-layer forces the current in the p-type layer to flow laterally rather than down into the substrate. Exit holes are provided in the buried n-layer for the return path of the current into the substrate. With this approach, uniformly thick islands and oxidized porous-Si layers using p-type porous Si have been formed. Since the buried n-layer was formed by proton implantation, it reverts back to p-type after high-temper-

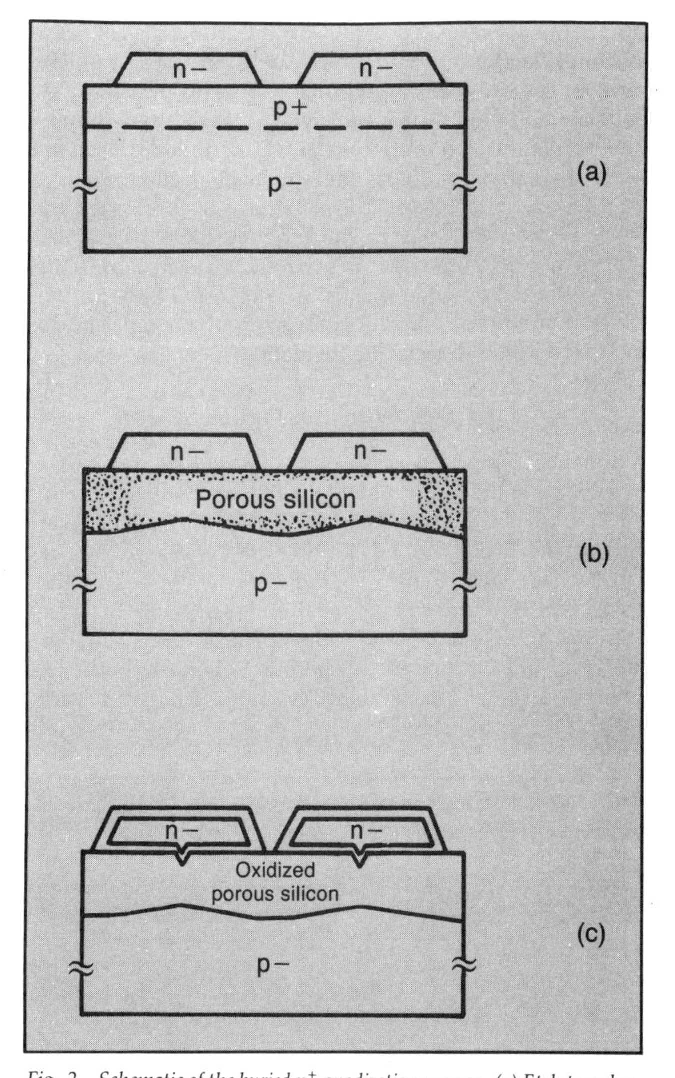

Fig. 2 Schematic of the buried p^+ anodization process. (a) Etch trenches through the n-layer down to the p^+ layer of the $n/p^+/p^-$ structure, (b) form porous Si selectively in the highly doped p-layer, and (c) oxidize the porous Si layer [9]. Note the thin layer of thermal oxide formed on the island surfaces.

ature annealing, thus avoiding occurrence of extra junctions. This method, however, requires an additional masking step to define the exit holes in the buried n-layer.

Selective formation of porous Si around fully dense islands can also be accomplished using n-type Si exclusively. Although n-type Si does not form porous Si as readily as p-type Si, anodization of increasingly higher doped n-type Si occurs at increasingly lower potentials. Holmstrom and Chi at GTE Laboratories were first to demonstrate selective anodization of a buried n^+ layer between an n^- island and an n^- substrate in 1983 [13]. The process is shown schematically in Fig. 4. Recently, this process has been refined by Zorinsky et al. at Texas Instruments and called the ISLANDS (Isolation by Self-Limiting Anodization of an N^+ Epitaxially Defined Sublayer) process [14]. With this method, the thickness of the island and the PSL are uniform and easily controlled by the n^+ doping profile. Another advantage of this method is the automatic endpoint on the island isolation. As soon as all of the easily anodized n^+ layer is converted to porous Si, the anodization current drops and etching stops due to the change in anodization potential threshold between the n^+ anodizable layer and the n^- island and substrate. For comparison, in the buried p^+ case, the island thickness is also easily controlled by the p^+ doping profile but the PSL thickness is not as readily controlled. Due to the ease of anodization of lightly doped p-type Si, even after completion of island isolation, porous Si will continue to form toward the substrate unless the anodization current is reduced to zero. A possible disadvantage of the buried n^+ process, however, may be the reported high dissolution rate of n-type porous Si in hydrofluoric acid [15]. This means that when wide islands are isolated, PSL with laterally nonuniform porosity is formed, which can lead to increased island defects after oxidation.

Epitaxial Deposition on Porous Silicon

Epitaxial deposition techniques on porous Si are attractive owing to the uniformity of the PSL and isolated island thicknesses. In these techniques, a uniform blanket layer of porous Si is first formed on the surface of the wafer. The PSL surface layer, which is single crystal, serves as the seeding layer for the epitaxy of a fully dense device Si layer. Trenches are etched through the epitaxial layer down to the PSL, and the underlying porous Si is later thermally oxidized through these trenches. With straightforward paths for current flow, the PSL is uniform and its characteristics are easily tailored. The main drawback to this method is that nonconventional low-temperature (<850°C) epitaxy is required to avoid sintering of the pores and to maintain their reactivity to oxidation. Low-temperature epitaxy techniques such as plasma chemical vapor deposition of SiH_4 (750°C) [16], molecular-beam epitaxy (MBE) (750°C) [17]–[21], and liquid-phase epitaxy [21] have been used. So far, the crystalline quality of Si overlayers fabricated with this technique have been inferior to those fabricated by the selective buried anodization methods. Residual defects such as microtwins and dislocations originating from the PSL/epitaxial Si interface are observed by cross-sectional transmission electron microscopy [17], [18]. However, reasonably good device characteristics have been obtained (discussed below) using these epitaxial techniques, since the bottom of the epitaxial Si, which is the most defective region of the island, is also oxidized during PSL oxidation.

Oxidation of Porous Si

Once porous Si has been formed by any of the techniques described above, the PSL must be oxidized without generating defects in the island. The oxidation treatments are optimized to yield oxidized porous Si (OPS) with electrical and physical properties similar to those of thermal oxides. The use of high-pressure oxidation has been re-

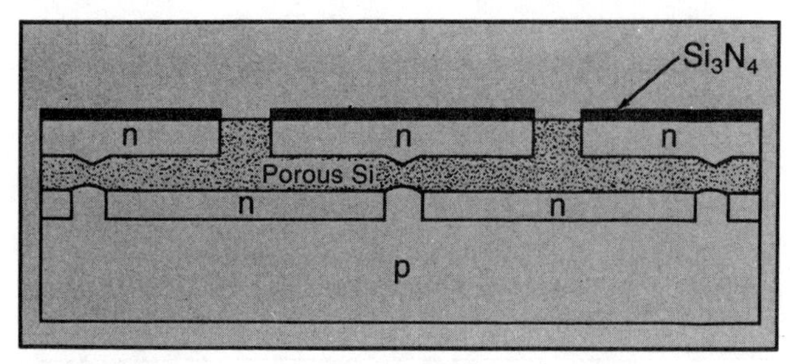

Fig. 3 Schematic cross section illustrating the wafer doping profile used by Benjamin et al. [12] to improve the control and uniformity of the FIPOS method. P-type regions surrounding n-type islands have been anodized to form porous Si. The buried n-layer forces the anodization current to flow laterally and, thus, enhances the uniformity of the buried porous-Si layer.

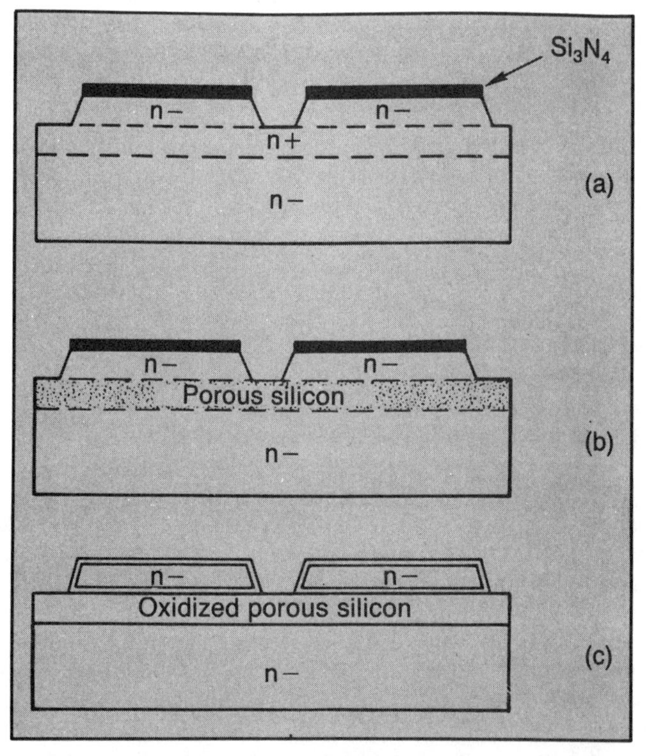

Fig. 4 Schematic of the buried n^+ anodization process. (a) Etch trenches through the n^- layer deep enough to expose the n^+ layer, (b) selectively anodize the n^+ layer, and (c) oxidize the buried porous-Si layer. As in Figs. 1 and 2, a thin layer of thermal oxide is formed on the exposed island surfaces. The protective Si_3N_4 layer during anodization is usually etched off before oxidation. Note the absence of a cusp underneath the middle of the island and the uniform thicknesses of the island and the oxidized porous-Si layer.

ported to eliminate formation of defects and reduce wafer warpage. For example, Nesbit found that when porous Si was oxidized at 800°C at a pressure of 10 atm, wafer warpage was reduced to <25 μm on 100-mm-diam wafers and no defects were observed in the isolated islands [9]. He also found that atmospheric oxidation always resulted in the formation of defects. Use of an initial low-temperature (≤ 450°C) treatment to stabilize the pores, followed by a higher temperature oxidation treatment, has been reported to help fully oxidize the PSL [14], [23]. Two-step treatments, with both employing high temperatures at atmospheric pressures, have also been found to reduce warpage and defects in the Si overlayer [23].

Actual oxidation of the PSL is accomplished in short times due to the short distance between pores and the relatively fast diffusivity of oxidant species in the PSL. However, the as-oxidized porous Si has a very high etch rate in hydrofluoric acid, by as much as an order of magnitude higher than that of thermal oxide. The porous Si oxide can be densified by annealing in steam ambients at sufficiently high temperatures to cause flow of the glass [22]. Otherwise, unless trenches are refilled with a slow-etching material, this can pose problems in device processing where hydrofluoric acid etches are used routinely.

Device Processing and Characteristics

Depending on applications, the required thicknesses of the isolated island and/or oxidized porous-Si layer will range from submicron to several microns thick. Thick buried dielectric layers are preferred for high-speed applications because of decreased wiring capacitance and for high-voltage applications. These are not easily obtained with the oxygen or nitrogen implantation (SIMOX/NI [separation by implanted oxygen or by nitrogen implantation]) techniques but are easily obtained with the oxidized porous-Si or zone-melt recrystallization (ZMR) techniques. The buried porous-Si formation techniques (such as those illustrated in Figs. 1, 2, and 4) can readily provide both varying buried dielectric and island thicknesses, depending on the initial doping profile of the substrate. Both the SIMOX/NI and ZMR techniques are not as readily adaptable to arbitrarily thick films. Both these techniques may require additional epitaxial deposition for film-thickness applications of a few microns—for example, as in some bipolar device applications. Note also that in these two techniques, the thickened device Si layer is not likely to be defect-free since the SOI epitaxial seeding layer provided by the implantation or recrystallization techniques is not defect-free. The porous-Si epitaxy techniques are more suitable for thin-film SOI applications than the buried porous-Si formation techniques, since low-temperature Si deposition rates are typically low, ~0.5–1 μm/hr with MBE.

In typical oxidation treatments, a 0.2-μm-thick thermal oxide is formed on the island surfaces, including the bottom interface between the island and the OPS layer, and the sidewalls of the island. In radiation-hardened CMOS applications wherein no trench refill will be used, thick oxide on the sidewalls may not be desirable because of increased oxide-trapped charge effects from the sidewall parasitic transistors.

The width of the island, which can be isolated by OPS, is one of the perceived limitations of the porous Si technique. At the present time, islands as wide as 325 μm have been successfully isolated [23]. If single or, at most, a few transistors are to be isolated by trenches anyway, then this magnitude of the islands size would not appear to be restrictive.

The most advanced devices have been fabricated by NTT using the original FIPOS process. Among these are high-speed 2-μm 16K-bit [24] and 1.5-μm 64K-bit [25] CMOS SRAMs (static random access memories). These had access times of 35 nsec and 20 nsec, respectively. Transistors of 2 μm with electron and hole mobilities of 750 cm^2/V-sec and 200 cm^2/V-sec, respectively; subthreshold slopes of 9 dec/V; and drain leakage current less than 10^{-11}A/8.5-μm channel width at $V_D = 5$ V have also been obtained by NTT [26]. OKI Electric has also used the FIPOS process to fabricate SOI/CMOS 1-μm 100-stage ring oscillators [27]. These exhibited speeds of 60 psec delay per stage at $V_D = 5$ V.

In addition, test transistors have been fabricated using the more recent modified FIPOS techniques and porous-Si epitaxy techniques. With the buried p$^+$ method, electron mobilities of ~500 cm^2/V-sec have been reported [9]. With the ISLANDS method, electron mobilities of 650–680 cm^2/V-sec, subthreshold slopes of 8–10 dec/V, and subthreshold leakage of 0.1-pA/μm gate width at $V_D = 5$ V have been obtained [14]. Transistors fabricated in MBE-deposited layers have shown good characteristics, too. Source-to-drain leakage currents were below 1 pA/20-μm channel width, and electron mobilities were ~730 cm^2/V-sec, at $V_D = 50$ mV [23].

Conclusion

Unique properties of porous Si have been exploited in creative ways to form SOI structures. There are two basic ways of isolating islands using porous Si. The first is by buried porous-Si formation; areas surrounding device islands are converted into porous Si by proper tailoring of the wafer dopant profile. The second is by epitaxy on porous Si; a uniform surface porous-Si layer is used as a seeding layer for low-temperature epitaxy of the device Si. Oxidation of the underlying porous Si layer, via trenches in the device Si, has been improved to the point that defect generation and wafer warpage are avoided. Advanced devices fabricated on the FIPOS material have shown that the porous silicon technology is among the front-runners for high-performance CMOS LSIs. The excellent crystalline quality of the device Si may make it the clear choice for SOI bipolar applications.

Acknowledgments

The author thanks T. R. Guilinger and M. J. Kelly for helpful discussions. This work was supported by the Department of Energy under Contract DE-AC04-76DP00789 and DNA.

References

[1] S. L. Partridge, "Silicon-on-Insulator Technology," *IEE Proc.*, vol. 133, p. 66, 1986.

[2] S. Nakajima, Y. Watanabe, T. Yokoyama, and K. Kato, "IPOS Scheme: A New Isolation Technique for Integrated Circuits," *Proc. 6th Conf. Solid State Dev.*, Tokyo, vol. 303, 1974.

[3] S. Nakajima and K. Kato, "An Isolation Technique for High Speed Bipolar Integrated Circuits," *Rev. Elect. Comm. Labs.*, vol. 25, p. 1039, 1977.

[4] T. Unagami and K. Imai, "An Isolation Technique Using Oxidized Porous Silicon," *Semiconductor Technologies*, vol. 8, p. 139, 1983.

[5] K. Imai and Y. Yoriume, "Application of IPOS Technique to MOS IC's," *Jap. J. Appl. Phys.*, vol. 18, p. 281, 1979.

[6] M. I. J. Beale, J. D. Benjamin, M. J. Uren, N. G. Chew, and A. G. Cullis, "An Experimental and Theoretical Study of the Formation and Microstructure of Porous Silicon," *J. Cryst. Growth*, vol. 73, p. 622, 1985.

[7] H. A. Tenhunen, "Formation of Anodic Porous Film on Silicon for Silicon-on-Insulator VLSI Structures," Ph.D. Thesis, Cornell University, 1986.

[8] K. Imai, "A New Dielectric Isolation Method Using Porous Silicon," *Solid-State Electron.*, vol. 24, p. 159, 1981.

[9] L. A. Nesbit, "Advances in Oxidized Porous Silicon for SOI," *Technical Digest IEDM*, p. 800, 1984.

[10] S. S. Tsao, D. R. Myers, T. R. Guilinger, M. J. Kelly, and A. K. Datye, "Selective Porous Silicon Formation in Buried p^+ Layers," to appear in *J. Appl. Phys.*

[11] S. Nugent, J. Seefeldt, M. Hanson, B. Bradford, D. Field, and M. Biswal, "Full Dielectric Isolation for VLSI Applications," *Electrochem. Soc. Ext. Abs.*, no. 271, 1986.

[12] J. D. Benjamin, J. M. Keen, A. G. Cullis, B. Innes, and N. G. Chew, "Large Area, Uniform Silicon-on-Insulator Using a Buried Layer of Oxidized Porous Silicon," *Appl. Phys. Lett.*, vol. 49, p. 716, 1986.

[13] R. P. Holmstrom and J. Y. Chi, "Complete Dielectric Isolation by Highly Selective and Self-Stopping Formation of Oxidized Porous Silicon," *Appl. Phys. Lett.*, vol. 42, p. 386, 1983.

[14] E. J. Zorinsky, D. B. Spratt, and R. L. Vinkus, "The ISLANDS Method—A Manufacturable Porous Silicon SOI Technology," *Technical Digest IEDM*, p. 431, 1986.

[15] R. Herino, K. Barla, G. Bomchil, and C. Bertrand, "Formation and Characterization of Porous Silicon Formed on Heavily Doped n-Silicon," paper presented at the 16th Electrochem. Soc. Meeting, Boston, 1986.

[16] H. Takai and T. Itoh, "Isolation of Silicon Film Grown on Porous Silicon Layers," *J. Elect. Mats.*, vol. 12, p. 973, 1983.

[17] T. L. Lin, S. C. Chen, Y. C. Kao, and K. L. Wang, "100-μm-Wide Silicon-on-Insulator Structures by Si Molecular Beam Epitaxy Growth on Porous Silicon," *Appl. Phys. Lett.*, vol. 48, p. 1793, 1986.

[18] M. I. J. Beale, N. G. Chew, A. G. Cullis, D. B. Garson, R. W. Hardeman, D. J. Robbins, and I. M. Young, "A Study of Silicon MBE on Porous Silicon," *J. Vac. Sci. Technol. B*, p. 732, 1985.

[19] F. d'Avitaya, K. Barla, R. Herino, and G. Bomchil, "Improvement of Silicon Epitaxy on Porous Silicon Substrates," *Proc. Electrochem. Soc.*, vol. 85, no. 7, p. 323, 1985.

[20] S. Konaka, M. Tabe, and T. Sakai, "A New Silicon-on-Insulator Structure Using a Silicon Molecular Beam Epitaxial Growth on Porous Silicon," *Appl. Phys. Lett.*, vol. 41, p. 86, 1982.

[21] H. Baumgart, R. C. Frye, F. Phillipp, and H. J. Leamy, "Dielectric Isolation Using Porous Silicon," *Mat. Res. Soc. Symp. Proc.*, vol. 33, p. 63, 1984.

[22] K. Barla, J. J. Yon, R. Herino, and G. Bomchil, "Oxide Formation from Porous Silicon Layers on p-Substrates," *Insulating Films on Semiconductors*, p. 53, 1986.

[23] T. L. Lin and K. L. Wang, "New Silicon-on-Insulator Technology Using a Two-Step Oxidation Technique," *Appl. Phys. Lett.*, vol. 49, p. 1104, 1986.

[24] T. Mano, T. Baba, H. Sawada, and K. Imai, "FIPOS CMOS 16K-bit Static RAM," *VLSI Technology*, p. 12, 1982.

[25] K. Ehara, H. Unno, and S. Muramoto, "1.5 μm FIPOS/CMOS VLSI Process with Low Wafer Warpage and Si Deposit-Defect-Free," *Electrochem. Soc. Ext. Abs.*, vol. 85, no. 2, p. 457, 1985.

[26] K. Imai and H. Unno, "FIPOS Technology and Its Application to LSI's," *IEEE Trans. Electron. Dev.*, vol. 31, p. 297, 1984.

[27] K. Anzai, F. Otio, M. Ohnishi, and H. Kitabayashi, "Fabrication of High Speed 1 micron FIPOS/CMOS," *Technical Digest IEDM*, p. 796, 1984.

Ferroelectric Materials For 64 Mb and 256 Mb DRAMs

Storing enough charge on the capacitors in the memory cells of ULSI DRAMs will require materials with much higher relative dielectric constants than those being used today. Some ferroelectric materials look promising.

by Laureen H. Parker and Al F. Tasch

Since its introduction in 1970, the solid-state dynamic random access memory (DRAM) has successfully used a simple charge-storage memory cell. Since 1972, the memory cell of choice has been a one-transistor (1-T) cell consisting of a single capacitor on which the charge is stored and one switching transistor to isolate the capacitor [1]. The continuing popularity of this cell is a direct result of its simplicity and small size.

As the number of memory cells has increased from 4 kilobits (kb) in the original 1-T design to the 1 Megabit (Mb) of current designs, the available area per memory cell has continually decreased to maintain acceptable die sizes. This reduction continues in the 4-Mb parts that are just beginning production and in the 16-Mb parts now in development. Further reduction will be essential in the 64-Mb and 256-Mb DRAMs which are now in the planning and early developmental stages.

The reduction in cell size required reducing the area of the planar storage capacitor area. This, in turn, made it necessary to increase the charge density on the capacitor to maintain adequate signal margins. Up through the 1 Mb DRAM, the required increase in storage-charge density has been achieved through improvements in process technology, reduction of the dielectric thickness, and innovations in the cell structure. Also, since the 256 kb, a Si_3N_4-SiO_2 sandwich-layer dielectric has often been used to increase both the effective relative dielectric constant and the dielectric reliability. Because of the great difficulty in continuing to increase the storage-charge density on a planar capacitor, the 4-Mb and 16-Mb parts are being designed with either a trench or a stacked capacitor. These geometries provide greater capacitive area, so the capacitor can store the total required charge even though it occupies less real estate on the die.

With optimum achievement in innovative 1-T cell design, soft-error protection, sense-amplifier sensitivity, and bit-line capacitance, capacitors that continue to use a nitride-oxide dielectric might possibly suffice for the 64-Mb DRAM [2]. However, nitride-oxide does not appear to be at all realistic for the 256 Mb. Therefore, if the simple charge-storage concept is to find continued use in ultra large scale integration (ULSI) DRAMs, we must find a new dielectric material that permits a greater charge density in the storage capacitor. In addition, the material must be compatible with ULSI processing.

Alternative dielectrics with higher dielectric constants have been sought over the years, but the search has not been so successful because the higher dielectric constants of the most promising materials—Ta_2O_5, Y_2O_3, and ZrO_2—have been offset by lower breakdown fields. The net result is that the storage-charge density has been no more than twice that of silicon dioxide. In this search, ferroelectric materials have been considered but never seriously pursued. However, the need for substantially higher charge-storage capacity in the dielectric has become so great that a serious examination of ferroelectrics is now in order.

We have performed an intensive literature search and analysis to assess whether it is feasible to use a ferroelectric dielectric in the storage capacitor of a 1-T memory cell.

This work was supported in part by the Semiconductor Research Corp. and a grant from Texas Instruments, Inc.

Reprinted from *IEEE Circuits and Devices Magazine*, Vol. 6, No. 1, pp. 17-26, January 1990.

When the general properties and reliability issues of ferroelectrics that are applicable to ULSI DRAMs are combined with the projected electrical requirements for 64-Mb and 256-Mb DRAMs, we can determine the required material characteristics of a ferroelectric dielectric. We can then identify those specific materials (from among those analyzed) that hold the greatest promise for use in ULSI DRAMs, and we do that at the end of this article.

Ferroelectric Properties*

Ferroelectric materials exhibit a number of unique and interesting properties— both physical and electrical. The defining property of a ferroelectric material is that it possesses spontaneous polarization that can be reversed by an applied electric field.

Ferroelectrics are a subgroup of the pyroelectric materials, which are in turn a subgroup of the piezoelectric materials. Ferroelectrics therefore possess both piezoelectric and pyroelectric properties, in addition to their unique ferroelectric properties. These materials have a characteristic temperature—the transition temperature—at which the material makes a structural phase change from a polar phase (ferroelectric) to a non-polar phase, typically called the paraelectric phase.

Crystal Structure

All of the materials reviewed in this article—except for KNO_3—possess the perovskite crystal structure described by the general chemical formula ABO_3. The *A* element is a large cation situated at the corners of the unit cell and the *B* element is a smaller cation located at the body center. The oxygen atoms are positioned at the face centers (Fig. 1).

In the ferroelectric phase, the perovskite structure usually assumes one of three structural formations: tetragonal, orthorhombic or rhombohedral. In the tetragonal symmetry, a cubic cell stretches along one side (the "c-axis") and shrinks along the other two sides ("a-axes") forming a rectangular prism. The spontaneous polarization aligns itself parallel to the longest side. The orthorhombic structure is formed by stretching a cube along a face diagonal along which the spontaneous polarization aligns itself. In a rhombohedral structure a cube is stretched along a body diagonal and the spontaneous polarization aligns in the direction of the stretched body diagonal. In the paraelectric phase, the perovskite structure has cubic symmetry—it is neither stretched nor distorted.

The Hysteresis Curve

One of the best-known features of ferroelectric materials is the response of the polarization *P* to external electric fields *E*, which is often referred to as simply the hysteresis loop or hysteresis curve (Fig. 2a). The displacement charge density D is related to the polarization by

$$D = \epsilon_0 E + P (C/cm^2) \quad (1)$$

where ϵ_0 is the permittivity of free space. Since $\epsilon_0 E << P$ in most ferroelectrics, the relationship of D to E — the D-E loop — is nearly identical to the P-E loop (Fig. 2b).

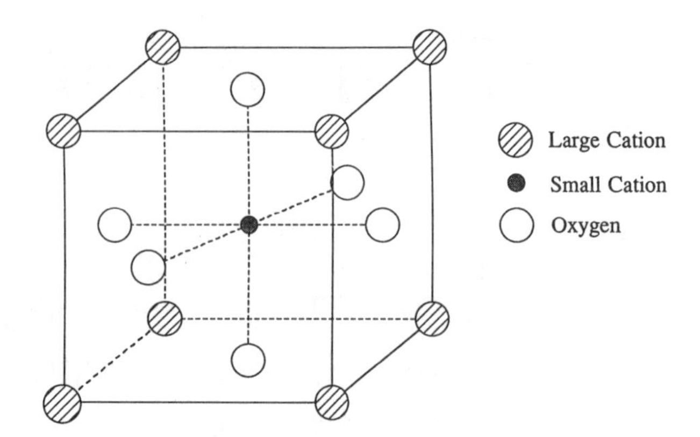

Fig. 1 Most of the interesting ferroelectric candidate materials for the dielectric in ULSI charge-storage memory cells have a perovskite structure. An example is barium titanate ($BaTiO_3$), where Ba^{2+} is the A cation and Ti^{4+} is the B cation.

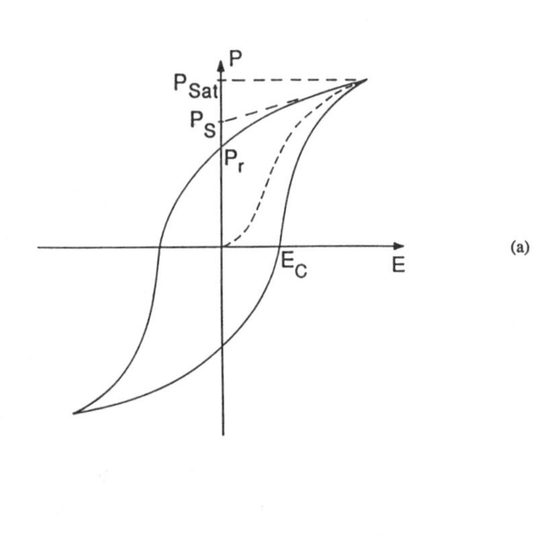

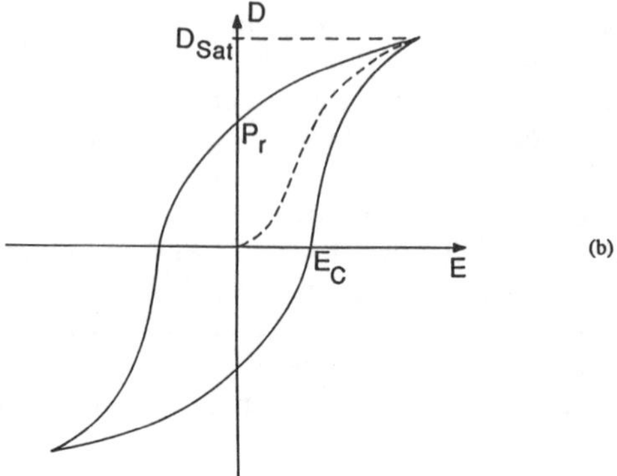

Fig. 2 Curves showing typical relationships between a) polarization vs. electrical field, and b) displacement charge vs. electric field for a ferroelectric dielectric are similar.

*There isn't enough published data on the frequency dependence of the relative dielectric constant for us to draw any meaningful conclusions here. Frequency dependance must be evaluated experimentally to determine its impact in DRAM applications.

As a sample is cooled from the paraelectric phase through the transition temperature, regions of aligned electric dipoles called domains form to produce the spontaneous polarization. If an electric field is applied, the domains that are closest to being parallel to the field grow while the others shrink. The macroscopic polarization with an external field applied is composed of the aligned spontaneous polarization, as well as electronic and additional ionic polarization generated by the external field. At a characteristic field, a maximum alignment of the spontaneous polarization occurs and the hysteresis curve saturates at P_{Sat} because additional electronic and ionic polarization produced by an increase in the field is typically quite small compared to the spontaneous polarization. However, there are some sputtered thin films in which the polarization does not appear to saturate [3–7]. For the sake of generality, then, we give the name "maximum polarization" P_{max} – to the polarization that corresponds to the maximum applied field (E_{max}).

When the electric field is removed, the ionic and electronic polarization decrease to zero. The remanent – remaining – polarization P_r is the spontaneous polarization that remains aligned with the previously applied field. It can assume both positive and negative values.

In the hysteresis curve, the magnitude for the reverse electric field which decreases the net polarization to zero is called the coercive field E_C. At this point the net polarization reverses polarity if the reverse applied field is increased further.

The relative dielectric constant K, which is nearly proportional to the slope, $\epsilon_0(K\text{-}1)$, of the P-E curve, is a nonlinear function of the electric field and possesses hysteresis [8–10]. The constant K can be nearly infinite at the coercive field and nearly zero at the saturation field. Because of the non-linearity and hysteresis, published values of relative dielectric constants cannot be used to calculate capacitor charge densities for a given applied electric field. A different approach is needed, and we will describe one later.

Temperature Dependence

The spontaneous polarization P_s is a function of temperature (Fig. 3). At a characteristic temperature T_0 the material changes from the ferroelectric phase, in which spontaneous polarization exists, to the paraelectric phase, in which P_S equals zero. Typically, the ferroelectric phase exists at temperatures below T_0, and the paraelectric phase exists above T_0. In a few cases, this relationship is reversed [11]. And in the case of Rochelle salt, among others, there are two transition temperatures, and the ferroelectric phase is bounded by two paraelectric phases. The materials we will consider all have the first characteristic.

The relative dielectric constant K is not only a function of the electric field; it is also a function of temperature. In the ferroelectric phase, the dielectric constant increases as the temperature increases. At the transition temperature, it becomes anomalously large (Fig. 4). Above T_0, K often exhibits Curie-Weiss behavior:

$$K = \frac{C}{T - T_C}$$

where C is the Curie constant and T_C is the Curie temperature*.

Leakage Current

In a 1-T DRAM cell, charge is lost from the storage capacitor over a period of time through various leakage mechanisms. This means the contents of the memory cell must be refreshed periodically. A tolerance for charge leakage is built into each DRAM design, but information can be lost from the cell if the leakage current is excessive. Thus, the

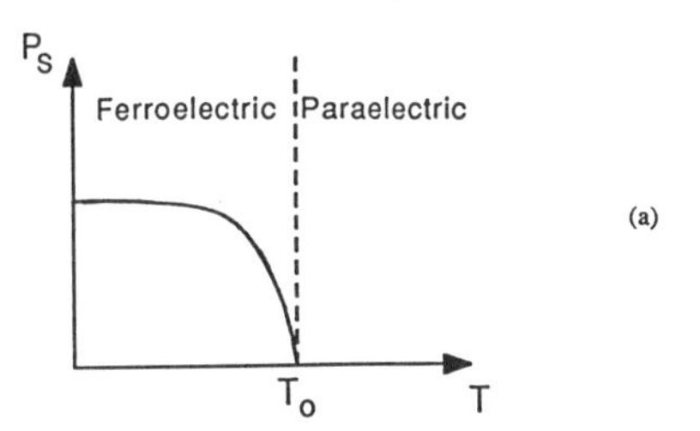

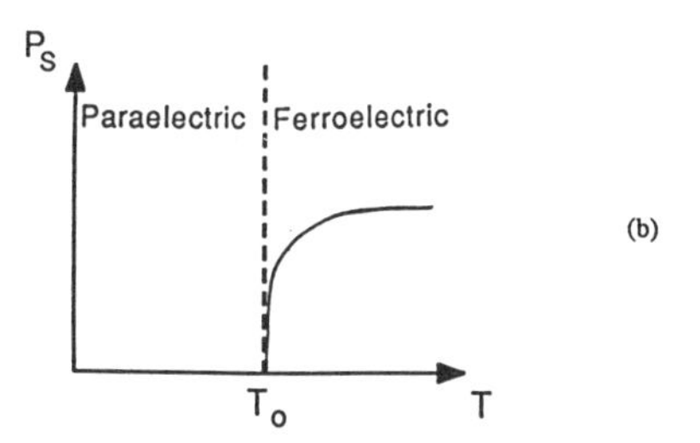

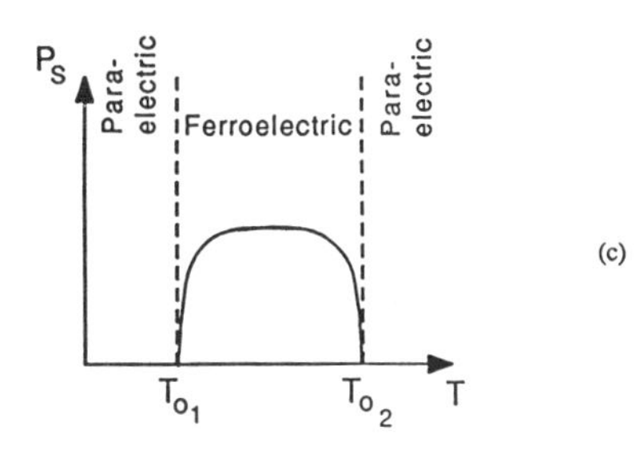

Fig. 3 There are the three types of ferroelectric-paraelectric phase transitions. All of our candidate materials have the transition illustrated in (a).

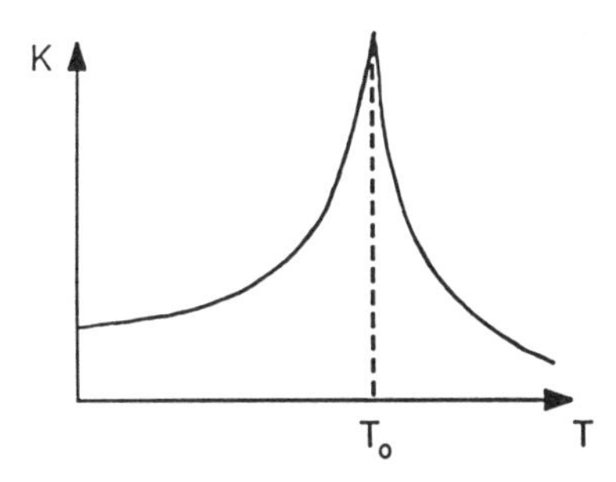

Fig. 4 The relative dielectric constant rises to anomalously high values near the phase-transition temperature, T_o.

leakage-current characteristics of the storage capacitor dielectric are very important in a 1-T DRAM cell. Unfortunately, there is very little data in the published literature regarding leakage-current values or leakage mechanisms of thin ferroelectric films.

However, it is known that in sputtered barium titanate the leakage current decreases as the crystallinity of the film decreases [12]. In $BaTiO_3$ amorphous films, the current is at least two orders of magnitude less than that in polycrystalline films. If this holds true for other materials, then amorphous films would be preferable to polycrystalline films from a leakage-current point of view. It is also generally believed that the leakage current increases as the film thickness decreases. This could be a potential problem in the thin films that will be required for the 64-Mb and 256-Mb DRAMs. On the positive side, leakage current in some materials can be reduced by adding impurities. Lanthanum, for instance, acts as a compensating donor in $PbTiO_3$, which is normally p-type conductive [13–15].

Dielectric Breakdown

The electric field at which the dielectric breaks down E_{Br} is another important parameter in the characterization of DRAM capacitor dielectrics. The operating field across the capacitor must be designed to be less than the breakdown field if the capacitor is to operate reliably. There is little data in the published literature regarding the mechanisms and characteristics of dielectric breakdown fields in thin-film ferroelectric materials. Until more is known, we will assume that an acceptable operating field is approximately the same fraction of the breakdown field as is currently used in DRAM products — approximately one third of the breakdown field. In ferroelectric capacitors, this relationship may be different and needs to be determined experimentally.

Limited data indicates that in sputtered films of barium titanate the breakdown field decreases as the crystallinity of the film increases [12]. If the film is polycrystalline, the breakdown field is much smaller than if the film is amorphous. This seems to indicate that the breakdown field will decrease as the film thickness decreases because there are fewer grain boundaries and thus greater crystallinity (unless the film is amorphous).

On the other hand, some studies of ceramic samples indicate that thermal mechanisms contribute to the dielectric breakdown, and that thinner samples exhibit larger breakdown fields because there is less internal heating to contribute to the breakdown [16,17]. Thin — 100-600 nm — KNO_3 films have also exhibited a breakdown field that increased with decreasing film thickness [18].

Reliability

As with any capacitor dielectric, there are many physical properties of a ferroelectric dielectric that can change with time, including the dielectric constant, remanent and maximum polarization, coercive field, shape of the hysteresis loop, dielectric breakdown field, loss tangent, and leakage current. There are three time-dependent mechanisms that can affect these parameters in ferroelectric materials. The first — aging — refers to the changes that occur under static conditions, either with or without dc bias. The second mechanism — fatigue — refers to the changes that occur through continuous polarization reversals due to the application of an electric field. This is sometimes called "field-induced aging" or "excitation aging." The last mechanism is time-dependent dielectric breakdown (TDDB), in which breakdown of the dielectric occurs after a period of time under electric field-stress and, in some cases, under electrical and temperature stress.

Aging

In an aged sample, the remanent polarization, maximum polarization [19–21], coercive field [19–23], and loss tangent [24–26] are usually reduced. In some samples, the maximum polarization does not change over time [21]. Aging occurs only in the ferroelectric phase of the material [27].

There are two dominant theories regarding the cause of aging: (1) domain relaxation, and (2) the development of a space-charge field. Both mechanisms can clamp the domains of an aged sample so that the domains do not move and grow as they do in a fresh sample [26]. This reduces their ability to contribute to the macroscopic polarization. The first mechanism, reorientation and clamping of the domain walls, is due to lattice strain at the domain walls and grain boundaries [28]. Over time, these strains are relieved through the reorientation of the domains [26–27, 29–32]. Materials with large internal strains have been shown to age faster than those with smaller internal strains. As these samples aged, the internal strain was reduced [28]. Materials with very small grains ages more slowly because there is less room within a grain to accommodate domain splitting and reorientation [33–34].

The second aging mechanism is the movement and pinning of charge [35–36] at either the domain walls, grain boundaries, or dielectric-electrode interfaces. This mechanism has been found to have a larger effect in samples with smaller grain sizes [37–38]. In the ferroelectric phase with zero external bias, this charge — which could be due to dopants, other impurities, or vacancies — aligns itself with the remanent polarization and creates an internal electric field in the direction opposite to the electric field associated with P_r.

When an external field is applied to reverse the polarization, the space charge field opposes the applied field. This effectively increases the magnitude of the applied field E_C required to induce polarization reversal. Once the polarization is reversed and the applied field is removed, the space-charge field causes part of the remanent polarization to switch back to its original direction, thus reducing P_r. Another effect of the space charge is that it can prevent the domains from growing, which is required for polarization reversal. This reduces the maximum polarization. Space-charge effects are also referred to as "waiting-time effects" because the build-up of space charge is dependent on the time between polarization reversals [39–40].

If the material is in the paraelectric phase, aging due to reorientation should not be an issue. However, if the material is under constant dc bias (as it is in a DRAM), then free charge could drift to the grain boundaries and set up an internal space-charge field. It may be possible to control space charge by a careful choice of material composition and processing.

Fatigue

The second time-dependent mechanism that affects the electrical parameters of a ferroelectric dielectric is fatigue, which occurs when the polarization is continuously reversed. The material's response to fatigue is very similar to the aging response: the remanent and maximum polarization decrease, the coercive field may increase [28,41,42] or decrease [42,43], the hysteresis loop loses its squareness [43,44], and microcracks often appear in the material [41,43,45]. If the polarization is never reversed, fatigue either does not occur or is reduced to an insignificant level.

The mechanisms that cause fatigue are less well understood than the aging mechanisms and are not appreciably treated in the literature. As in aging, internal strain may be a fundamental cause for the change in electrical properties. During polarization reversals, there is an even greater strain within a sample because of the continuous structural changes. Some of the strain is relieved through reorientation and pinning of the domains, and large strains may be relieved through spontaneous microcracking [43]. These microcracks can cause a permanent change in the electrical parameters and might well produce catastrophic failure of the dielectric. As the grain size decreases or as the cell distortion decreases due to compositional variations, the microcracking has been found to decrease [46–48]. The electrode material has also been suggested as a possible influence in fatigue, but there is a lack of concensus in the literature regarding which electrode materials affect fatigue.

Time Dependent Dielectric Breakdown

Under sufficient electrical stress, the dielectric material will break down over a time period that is characteristic of the intrinsic material, the procedures and quality of the processing, and the electrode material [49–51]. Although TDDB must be characterized in order to guarantee sufficient reliability for commercial application, we have not been able to find any published information on TDDB of thin-film ferroelectric materials.

Charge Storage Capacity in a 1-T DRAM Cell

In normal operation, the voltage on one plate of the capacitor in a one-transistor DRAM cell is held constant while the voltage on the other plate can be either a positive (or negative) voltage or zero. These two different voltage levels correspond to the logical "1" and "0". The charge-storage capacity of the cell Q_C is equal to the difference in the charge stored on the electrodes for the two different voltage levels and is given by

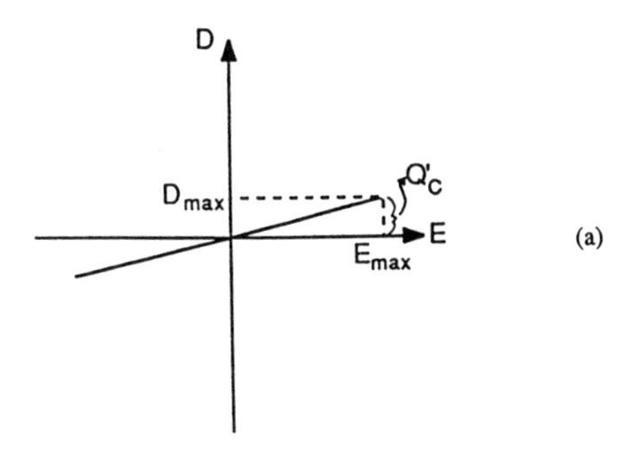

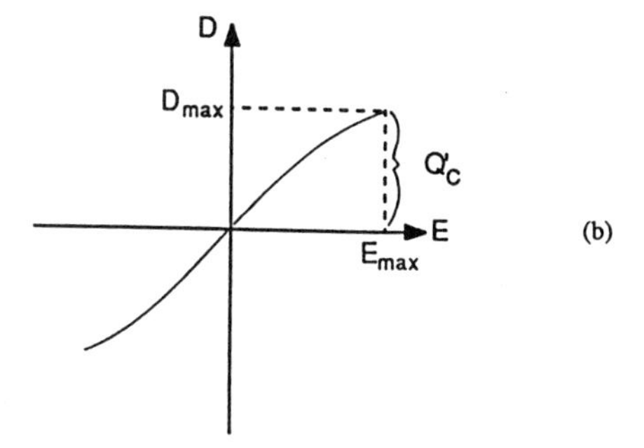

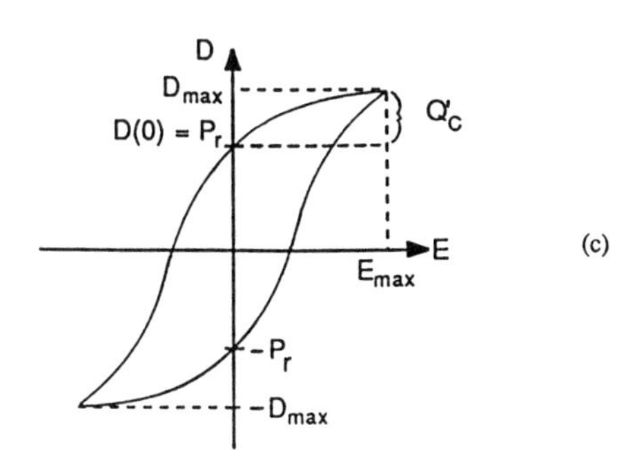

Fig. 5 The storage-charge density for ferroelectric capacitors (c) must be calculated differently than the storage-charge densities for linear (a) and non-linear capacitors (b).

$$Q_C = [D(E) - D(0)]A_{Cap} = \Delta D\ A_{Cap}(C) \qquad (3)$$

where D is the displacement charge and A_{Cap} is the area of the capacitor.

The storage-charge density $Q'_C = Q_C/A_{Cap}$ can be found as a function of the applied field and the polarization by combining Equations 1 and 3.

$$Q'_C = \epsilon_0 \Delta E + \Delta P \qquad (C/cm^2) \qquad (4)$$

For a linear or non-linear capacitor, Q'_C is equal to the displacement charge at the maximum applied field (Fig. 5). If the D-E curve is linear – as it is in SiO_2, for example – and if the dielectric constant and the applied operating field are known, then the storage-charge density can be determined by

$$Q'_C = \epsilon_0 KE \qquad (C/cm^2) \qquad (5)$$

(non-ferroelectric dielectric only)

However, if the D-E curve is nonlinear, (Eq.) 5 cannot be used because the slope, e_0K, is not constant. Then, Q'_C must be determined from a plot of D vs. E. Since the D-E curve of paraelectric phase materials can be either linear or non-linear, one cannot automatically apply (Eq.) 5 to the paraelectric phase.

For a dielectric material in the ferroelectric phase, the storage-charge density is less then the maximum displacement charge due to the remanent polarization at zero field. Thus, it is not equal to the displacement at the maximum field, nor can it be calculated from (Eq.) 5. The storage-charge density can only be found by subtracting P_r from D_{max} (Fig. 5). In most ferroelectrics P is approximately equal to D, so $P_{max} - P_r$ is very nearly equal to Q'_C.

Projected Requirements for Electrical Parameters

In this section and the following one, a number of ferroelectric materials are analyzed for their potential use as the dielectric in 64 Mb and 256 Mb DRAM capacitors. This analysis is based on projected requirements for the electrical properties of 64 Mb and 256 Mb DRAMs. Preferred material characteristics will be discussed in the next section.

The authors have performed an in-depth analysis of the issues and technology requirements for 64-Mb and 256-Mb DRAM memory cells (2). This analysis produced projected ranges for the required memory-cell area, ratio of cell area to storage-capacitor area, total storage charge, storage-charge density, and leakage current. Many of these parameters are relevant to the current discussion (Table 1).

Using currently available technology, the 4- and 16-Mb DRAMs require trench or stacked capacitors in order to achieve the required total storage charge in the small memory cells. The range for the trench capacitor specifies the minimum level of improvement required in the storage-charge density for 64- and 256-Mb DRAMs. The range for the planar capacitor specifies the required storage-charge density if the process limitations of the new dielectric necessitate a planar capacitor.

Preferred Material Characteristics

After careful consideration of the characteristics of ferroelectric materials and the reliability issues (including aging, fatigue, and microcracking), we have identified several general characteristics that appear desirable. Because of the large variations in material characteristics that take place at T_0, it appears desirable to use a material whose transition temperature is located outside the normal temperature range of DRAM operation — 0-70°C ambient and 0-100°C on chip. In short, the selected material should always be in one phase — ferroelectric or paraelectric — without substantial variation of key parameters over the operating temperature range.

In order to minimize problems with aging, fatigue and microcracking, it appears that paraelectric or nearly cubic ferroelectric materials [33] are more desirable than ferroelectric materials with large distortions — a large c/a ratio in a tetragonal structure, for example. Although a large distortion in the cell structure hinders domain relaxation and thus inhibits aging, it increases problems with fatigue and microcracking [48]. Taking these considerations together, large-distortion materials are undesirable.

Aging caused by domain relaxation can be reduced by optimizing the process to produce films with minimal initial strain or by adding dopants such as lanthanum in $(Pb,La)TiO_3$ and $(Pb,La)(Zr,Ti)O_3$ [13]. The dopant is thought to increase the domain-wall mobility, which results in lower initial strain and, consequently, reduced aging [52]. However, even if aging in a ferroelectric, composition is reduced by lowering the initial internal strain, fatigue and microcracking may still be a problem unless the polarity of the polarization is never switched. It may also be desirable to add dopants to reduce the conductivity and thus the leakage current [13,52].

Small-grain films may be preferable to large-grain films because the former reduce microcracking and aging due to domain relaxation [46]. However, any internal space-charge fields are increased because the distance between the space-charge regions at the grain boundaries is reduced [37]. Since

Table 1:

Parameter	64 Mb	256 Mb	Comments
Memory cell area, A_C (μm^2)	0.7 - 1.0	0.25 - 0.34	
Storage capacitor area, A_S (μm^2)	2 - 6	0.7 - 2	Trench
	0.2 - 0.5	0.07 - 0.17	Planar
Storage charge density, Q'_C ($\mu C/cm^2$)			
• Successful SOI	1 - 5	2 - 11	Trench
	10 - 50	20 - 115	Planar
• Partially successful SOI	2 - 7	4 - 17	Trench
	20 - 75	45 - 170	Planar
Leakage current density, $J_{L,max}$ ($\mu A/cm^2$)	2 - 25	3 - 35	Trench
	20 - 250	30 - 360	Planar
Dielectric thickness, t_d (nm)	10 - 200	10 - 200	

the effects of internal space charges are not observed in all samples, the advantages of small-grain films may be greater than the disadvantages.

Achieving uniform composition of the films can be difficult. If the relative dielectric constant of nonstoichiometric regions is much smaller than the relative dielectric constant in the surrounding material, then the total displacement charge will be reduced. Thus, processing parameters should be optimized to produce maximum uniformity of composition. In addition, the capacitor electrodes must consist of a material that will not diffuse into or react with the dielectric material. The electrode material must also match the thermal expansion of the dielectric material and adhere to it, as well as be compatible with ULSI processing. Furthermore, the electrodes should either be metal or possess metal-like conductivity in order to exploit the maximum capacitance of the structure. That is, even if an electrode is heavily doped silicon (or polysilicon), there will always be a small depletion-layer capacitance in the silicon. Since this is in series with the large dielectric capacitance, the total capacitance is significantly reduced.

Potential Ferroelectric Materials for 1-T DRAM Capacitors

Out of the very large number of ferroelectric materials that could have potential for use as the capacitor dielectric in ULSI 1-T DRAM capacitors, we selected the following for analysis because experimental thin-film data was available: $BaTiO_3$, $PbTiO_3$, $Pb(Zr,Ti)O_3$, $(Pb,La)TiO_3$, $(Pb,La)(Zr,Ti)O_3$, $SrTiO_3$, KNO_3, amorphous $BaTiO_3$, amorphous $PbTiO_3$, and amorphous $LiNbO_3$. Since the characteristics of these materials are dependent on processing, the initial selection of materials for analysis was based on whether the processing was closely related to that which is anticipated for ULSI DRAMs. Thus, except for two cases, deposited films were selected over ceramic materials. Two ceramic compounds – $Pb(Mg,Nb)O_3$ and $Pb(Mg,Nb)O_3$:$PbTiO_3$ (PMN and PMN:PT) – are included in the analysis because reports show that PMN:PT (90:10) does not age, which is most important for ULSI DRAMs. The drawback of the ceramic materials is that one cannot be certain how their electrical characteristics will change when the material is deposited as a thin film. Only two deposited materials, $Bi_4Ti_3O_{12}$ and $Pb_{0.92}Bi_{0.07}La_{0.01}(Fe_{0.405}Nb_{0.325}Zr_{0.27})O_3$, are not included in the following analysis because of insufficient published data. They may deserve further investigation.

In the analysis, several handicaps were consistently encountered. One of the major drawbacks of the published data is that the film thicknesses are much larger than what will be required for ULSI DRAMs. Extrapolation of the data to thinner films is difficult because information on the variation of the parameters with thickness is scarce. The spontaneous polarization and remanent polarization are known to decrease with decreasing thickness, but information has not been located on how the maximum polarization varies with thickness. It is also expected that leakage currents of polycrystalline films will be higher for thinner samples because of their greater crystallinity, but the extent of the increase is unknown [12]. Of the films reviewed, over half of the film thicknesses were greater than 1 μm, and only in rare instances was the thickness <200 nm. In these cases the data was too limited to develop any real understanding of the characteristics of the thinner films. Data for leakage currents and dielectric-breakdown fields are scarce, and information on TDDB is non-existent. Furthermore, the variations in the processing, the electrode material, and the film thickness make it difficult to correlate data among different publications.

In spite of the scarcity of information and correlation between experiments, we have attempted to draw conclusions from the published experimental data. Because the available data on breakdown fields and leakage current is marginal and is not representative of thinner films, the following review will cover only estimated storage-charge densities and notable features of these materials. Even though the published storage-charge densities are expected to change for thinner films, a comparison of the materials with one another should indicate those that have a higher probability of being suitable dielectrics for ULSI DRAMs.

The films with the largest reported storage charge densities ($\triangle D$) are $PbTiO_3$ where $\triangle P$ ($\triangle P$ approximately equals $\triangle D$) ranges from 10 to 30 $\mu C/cm^2$ [53-59]. The large tetragonality of this material is thought to reduce aging caused by domain relaxation, but it is also a major cause of microcracking. If small grain sizes can be produced in thin films, it might be possible to reduce microcracking and fatigue.

Very large storage-charge densities of 6-25 $\mu C/cm^2$ have been reported for $Pb(Zr,Ti)O_3$ (PZT) [60-69]. For the ferroelectric phase of $(Pb,La)TiO_3$ (PLT), $\triangle P$ is 10-16 $\mu C/cm^2$ [70-73], and $(Pb,La)(Zr,Ti)O_3$ (PLZT) storage-charge densities are 5-17 $\mu C/cm^2$ for ferroelectric compositions [73-78]. All of these ferroelectric-phase compositions are of questionable reliability. Although the addition of lanthanum has been shown to reduce aging and increase the resistivity of the film, compositions of PLZA with large concentrations of lanthanum still have problems with aging and fatigue. Films with small grain size may exhibit reduced aging and fatigue effects, but experimental evaluation will be required to determine the extent of any reduction.

Unless aging and fatigue can be minimized, – perhaps through optimized processing, compositional purity, and the use of a single-polarity electric field in actual DRAM operation, paraelectric-phase compositions appear to offer the most promising combination of features. At the top of the list is paraelectric $(Pb,La)TiO_3$ with a dielectric constant in the range of 1000-3000 [73]. Based on the experimental data, 12 $\mu C/cm^2$ may be possible with these compositions. Since they are paraelectric, fatigue and aging from domain relaxation should not be problems. The dielectric constants for paraelectric PLZT compositions – 1450-9200 – are much larger than those for PLT, but the data is for ceramic samples [79]. How this will change in deposited thin films is not known.

For $BaTiO_3$ sputtered onto Pt or Pt-Rh substrates, the storage-charge densities of 0.3-5 $\mu C/cm^2$ are at the low end of the ULSI DRAM requirement range, but it may be possible to increase these charge densities with larger applied fields since the polarizations showed no indications of saturating at the applied fields that were reported [3,4,5,12,80]. The structure of sputtered barium titanate is nearly cubic instead of the normal tetragonal structure seen in the single-crystal or ceramic form [81–82]. This reduces the remanent polarization [5], should prevent microcracks, and should reduce aging and fatigue. If the charge densities can be increased by increasing the applied field, these features might make barium titanate attractive as far as reliability is concerned. $BaTiO_3$ films sputtered onto silicon substrates in an argon-oxygen atmosphere are in general too conductive for ULSI DRAM applications [6,7,83–85]. Leakage currents for films sputtered in pure argon are much better, but charge-density data is not available [86].

Although ceramic $Pb(Mg,Nb)O_3$:$PbTiO_3$ (PMN:PT) does not age [23,87–89], the transition temperature is in the middle of the operating temperature range for commercial DRAMs, and this may cause reliability problems. In addition, PMN:PT has exhibited significant fatigue problems [90]. However, without the $PbTiO_3$, $Pb(Mg,Nb)O_3$ (PMN) is paraelectric above −20°C and it has a minimum dielectric constant of 4000 in ceramic materials [91–92] in the temperature range of 0-100°C. The estimated charge density is 12 $\mu C/cm_2$, but experimental data is required to determine if this value can be maintained in deposited thin films.

The storage-charge density for $SrTiO_3$ is estimated to be around 8 $\mu C/cm^2$ for an applied field of 500 kV/cm [93]. But the maximum applied field may be limited to the neighborhood of 70 kV/cm because the breakdown field decreases rapidly with increasing ambient temperature. This would reduce the storage-charge density below the requirements for ULSI DRAMs. There also appears to be a compatibility problem with most electrode materials.

We could not locate information on charge densities for KNO_3 [94]. However, excessive fatigue, waiting-time effects, and water sensitivity during processing make KNO_3 an unsuitable choice for DRAM application [39].

Although only limited data exists on amorphous materials, the processing advantages of these films make them attractive. They can be deposited at very low temperatures, and it should be easier to repeatedly deposit a high-quality amorphous film than a polycrystalline film. The reproducibility should be much greater since the maximum and remanent polarization in polycrystalline films depend on the orientation of the grains. Furthermore, there should be no aging nor fatigue because there is no crystalline structure. For amorphous $BaTiO_3$, the dielectric constant ranges from 12 to 80 depending on film thickness and perhaps depending on electrode material [12,95–97]. If K is in the range of 30–65 for films <200 nm and the applied field is at least 1 MV/cm, then the available storage-charge density would be 2.5-6 $\mu C/cm^2$. This is at the low end of the range required for ULSI DRAMs, but the processing and reliability advantages may make amorphous $BaTiO_3$ a better candidate than polycrystalline films.

There is insufficient data for an analysis of amorphous $PbTiO_3$ [98–100]. The dielectric constants reported for amorphous $LiNbO_3$ are 50–90, but because breakdown-field data were not reported in these studies, charge densities cannot be estimated [100–102].

Conclusions

We have studied the feasibility of using ferroelectric materials as the capacitor dielectric in the one-transistor memory cells for 64-Mb and 256-Mb DRAMs. It is significant that the storage-charge density of a ferroelectric capacitor is not equal to the displacement charge at the maximum applied field. Rather, it is equal to the difference between the displacement at maximum and minimum applied fields, ΔD. In the paraelectric phase, ΔD is equal to the maximum displacement charge, D_{max}.

An extensive literature search led us to published experimental data for a number of ferroelectric compounds, which we have analyzed. We discovered that, although the number of publications concerning ferroelectrics is large, data relevant to the DRAM application is scarce. For example, there is very limited data on thin films less than 1 μm, paraelectric phase marerial, breakdown fields, or leakage current. The information on aging and fatigue is for ceramic materials only.

Characteristics that appear to be desirable in ferroelectric films include: compositions that do not change phase in the normal DRAM operating temperature range; paraelectric-phase compositions; amorphous films; films with small grains, small unit-cell distortion, and low internal strain; and metal or metal-like conductive electrodes that do not react with the ferroelectric material during fabrication of the integrated circuit or over the lifetime of the part. Also, it may be desirable to use a single-polarity electric field in the DRAM operation. This should reduce and possibly eliminate fatigue effects.

Compared to the relatively thick samples on which the studies in the literature were performed, leakage currents may be higher and breakdown fields may be lower for thinner films that will be used in ULSI DRAM applications. The breakdown field may be higher in ULSI DRAMs, however, because the very small capacitor area produces a lower probability that a potential breakdown path exists. Which feature will dominate is difficult to judge without additional information.

Of the materials examined, those that appear to hold the greatest promise for 64-Mb and 256-Mb DRAMs are the paraelectric phase compositions of PLT and PLZT. PMN is also attractive because it is paraelectric, but it may not be able to achieve the required charge densities when deposited as a thin film.

Amorphous $BaTiO_3$ may have lower charge densities than the paraelectric PLT, PLZT, or PMN; however, easier processing and lower leakage currents may make it a better choice overall. Small-grain polycrystalline $BaTiO_3$ may possess reliability characteristics similar to paraelectric films, and may achieve charge densities similar to amorphous $BaTiO_3$; however, leakage currents may be larger than in the amorphous films.

Ferroelectric-phase $PbTiO_3$, PZT, PLT, and PLZT have large storage-charge densities; however, reliability issues initially make them less desirable. $SrTiO_3$ and KNO_3 do not seem sufficiently attractive overall for use in ULSI DRAMs, and there is insufficient data for analysis of amorphous

$PbTiO_3$ and amorphous $LiNbO_3$.

We feel that several of the analyzed films have sufficiently large storage-charge densities to justify experimental evaluation. It must be determined if sufficient storage change densities for ULSI DRAM application can be achieved in very thin films—less than 200 nm—and breakdown fields, leakage currents, and reliability characteristics must be investigated. We also recommend that both amorphous and polycrystalline films be investigated at the same time to determine the trade-offs of the different film structures. While many questions remain regarding the viability of ferroelectrics in DRAM applications, the variety of available materials and their appealing prospects for greatly improved storage-charge densities certainly merit detailed experimental examination.

References

Note: A list of these references containing complete entries, rather than the abbreviated ones presented here, is available by circling Number 50 on the Reader Service Card.

[1] A. Rosenblatt, *Electronics* p. 67, Aug. 30, 1971.
[2] A. F. Tasch and L. H. Parker, *Proc. of the IEEE*, Vol. 77, No. 3, pp. 374–88, 1989.
[3] Y. Shintani et al., *Jap. J. Appl. Phys*, Vol. 17, No. 3, pp. 573–74, 1978.
[4] T. Nagatomo et al., *Ferroelectrics*, Vol. 37, pp. 681–84, 1981.
[5] T. Nagatomo and O. Omoto, *Jap. J. Appl. Phys.*, Vol. 26 (Suppl. 26–2), pp. 11–14, 1987.
[6] K. Sreenivas and A. Mansingh, *ISAE*, Bethlehem, Pa., pp. 602–5, June 8–11, 1986.
[7] V. S. Dharmadhikari and W. W. Grannemann, *J. Appl. Phys.*, Vol. 53, No. 12, pp. 8988–92, 1982.
[8] Jack C. Burfoot, "Ferroelectrics: An Introduction to the Physical Principles," D. Van Nostrand, Ltd., London, 1967.
[9] Ennio Fatuzzo and Walter J. Merz, "Ferroelectricity," North-Holland, Amsterdam, (1967).
[10] Helen D. Megaw, "Ferroelectricity in Crystals," Methuen, London, 1957.
[11] M. E. Lines and A. M. Glass, "Principles and Applications of Ferroelectrics and Related Materials," Oxford U. P., 1977.
[12] I. H. Pratt and S. Firestone, *J. Vac. Sci. Tech.*, Vol. 8, pp. 256–60, 1971.
[13] F. Kulcsar, *J. Am. Ceram. Soc.*, Vol. 42, No. 7, pp. 343–49, July 1959.
[14] J. de Launay and P. L. Smith, *Naval Res. Lab. Report 7172*, Oct. 15, 1970.
[15] R. Gerson and H. Jaffe, *J. Phys. Chem. Solids*, Vol. 24, pp. 979–84 1963.
[16] R. Gerson, and T. C. Marshall, *J. Appl. Phys.*, Vol. 30, No. 11, pp. 1650–53, Nov. 1959.
[17] B.-C. Shin, and H.-G. Kim, Ferroelectrics, Vol. 77, pp. 161–66, 1988.
[18] F. Scott et al., *ISAF*, Bethlehem, Pa., pp. 569–71, June 8–11, 1986.
[19] M. McQuarrie, *J. Appl. Phys.*, Vol. 24, No. 10, pp. 1334–35, 1953.
[20] Z. Pajak and J. Stankowski, *Acta Phys. Polon.*, pp. 1144–46, 1958.
[21] K. Carl and K. H. Hardtl, *Ferroelectrics*, Vol. 17, pp. 473–86, 1978.
[22] G. H. Jonker, *J. Am. Ceram. Soc.*, Vol. 55, No. 1, pp. 57–58, Jan. 1972.
[23] W. Pan et al., *J. Mat. Sci. Ltrs.*, Vol. 5, pp. 647–49, 1986.
[24] W. A. Schulze and J. V. Biggers, *Ferroelectrics*, Vol. 9, pp. 203–7 1975.
[25] K. W. Plessner, *Proc. Phys. Soc.*, Vol. 69 B, pp. 1261–68, 1956.
[26] S. Ikegami and I. Ueda, *J. Phys. Soc. Jap.*, Vol. 22, No. 3, pp. 725–34, March 1967.
[27] R. C. Bradt and G. S. Ansell, *J. Am. Ceram. Soc.*, Vol. 52, No. 4, pp. 192–99 (Apr. 1969).
[28] G. W. Taylor, *J. Appl. Phys.*, Vol. 38, No. 12, pp. 4697–706, Nov. 1967.
[29] A. Cohen et al., *J. Am. Ceram. Soc.*, Vol. 53, No. 7, pp. 396–98, July 1970.
[30] G. Arlt, *Ferroelectrics*, Vol. 76, pp. 451–58, 1987.
[31] P. V. Lambeck and G. H. Jonker, *J. Phys. Chem. Solids*, Vol. 47, No. 5, pp. 453–61 (1986).
[32] K. Tsuzuki, *Jap. J. Appl. Phys.*, Vol. 24 (Suppl. 24–3), pp. 126–29, 1985.
[33] M. Kuwabara et al., *J. Am. Ceram. Soc.*, Vol. 71, No. 2, pp. C110–12, Feb. 1988.
[34] C. A. Miller, *Brit. J. Appl. Phys.*, Vol. 18, pp. 1689–97, (1967).
[35] M. Takahashi, *Jap. J. Appl. Phys.*, Vol. 9, No. 10, pp. 1236–46, Oct. 1970.
[36] S. Takahashi, *Jap. J. Appl. Phys.*, Vol. 20, No. 1, pp. 95–101, Jan. 1981.
[37] K. Okazaki and K. Nagata, *J. Am. Ceram. Soc.*, Vol. 56, pp. 82–86, 1973.
[38] H. Neumann and G. Arlt, *Ferroelectrics*, Vol. 76, pp. 303–10, 1987.
[39] J. F. Scott and B. Pouligny, *J. Appl. Phys.*, Vol. 64, No. 3, pp. 1547–51, Aug. 1, 1988.
[40] J. F. Scott et al., *J. Appl. Phys.*, Vol. 62, No. 11, pp. 4510–13, Dec. 1, 1987.
[41] W. R. Salaneck, *Ferroelectrics*, Vol. 4, pp. 97–101, 1972.
[42] D. B. Fraser and J. R. Maldonado, *J. Appl. Phys.*, Vol. 41, No. 5, pp. 2172–76, Apr. 1970.
[43] W. C. Stewart and L. S. Cosentino, *Ferroelectrics*, Vol. 1, pp. 149–67, 1970.
[44] J. R. Anderson et al., *J. Appl. Phys.*, Vol. 26, pp. 1387–88, 1955.
[45] V. G. Gavrilyachenko et al., *Sov. Phys. Sol. State*, Vol 12, No. 5, pp. 1203–4, Nov. 1970.
[46] R. W. Rice and R. C. Pohanka, *J. Amer. Ceramic Soc.*, Vol 62, No. 11–12, pp. 559–63, Nov.-Dec. 1979.
[47] S. S. Chiang et al., *Comm. Amer. Ceramic Soc*, pp. C141–43, Oct. 1981.
[48] H. T. Chung and H. G. Kim, *Ferroelectrics*, Vol. 76, pp. 327–33, 1987.
[49] T. N. Nguyen et al., *Proc. of Symp. on Rel. of Semiconductor Dev. and Interconnection and Multilevel Metallization, Interconnection, and Contact Tech.*, Vol. 89-6, pp. 185–206, Oct. 9–14, 1989.
[50] P. Hiergeist et al., *IEEE Trans. Elect. Dev.*, Vol. 36, No. 5, pp. 913–19 May 1989.
[51] I.-C. Chen et al., *IEEE Trans. Elect. Dev.*, Vol. ED-32, No. 2, pp. 413–22, Feb. 1985.
[52] R. Gerson, *J. Appl. Phys.*, Vol. 31, No. 1, pp. 188–94, Jan. 1960.
[53] K. Iijima et al., *J. Appl. Phys.*, Vol. 60, No. 1, pp. 361–67, July 1, 1986.
[54] Y. Matsui et al., *Appl. Phys. A.*, Vol. 28, pp. 161–66, 1982.
[55] Y. Matsui et al., *Jap. J. Appl. Phys.*, Vol. 20, Suppl. 20-4, pp. 23–26, 1981.
[56] Y. Matsui et al., *J. Appl. Phys.*, Vol. 52, No. 8, pp. 5107–11, Aug. 1981.
[57] M. Okuyama, and Y. Hamakawa, *Ferroelectrics*, Vol. 63, pp. 243–52, 1985.
[58] M. Okuyama et al., *Jap. J. Appl. Phys.*, Vol 18, No. 8, pp. 1633–34, 1979.
[59] N. Shohata et al., *ISAF*, Bethlehem, PA., pp. 580–84, June 8–11, 1986.
[60] M. Adachi et al., *Jap. J. Appl. Phys.*, Vol. 26, No. 4, pp. 550–53, Apr. 1987.
[61] R. N. Castellano, and L. G. Feinstein, *J. Appl. Phys.*, Vol. 50, No. 6, p. 4406–11, June 1979.
[62] A. Croteau et al., *Jap. J. Appl. Phys.*, Vol. 26 (Suppl. 26-2), pp. 18–21, 1987.
[63] A. Croteau and M. Sayer, *ISAF*, Bethlehem, PA., pp. 606–9.

June 8–11, 1986.
[64] S. B. Krupanidhi et al., *J. Appl. Phys.*, Vol. 54, No. 11, pp. 6601–9, Nov. 1983.
[65] A. Okada, *J. Appl. Phys.*, Vol. 49, No. 8, pp. 4495–99, Aug. 1978.
[66] A. Okada, *J. Appl. Phys.*, Vol. 48, No. 7, pp. 2905–9, July 1977.
[67] Y. Shintani and O. Tada, *J. Appl. Phys.*, Vol. 41, No. 6, pp. 2376–80, 1970.
[68] E. V. Sviridov et al., *Sov. Phys. Tech. Phys.*, Vol. 30, No. 5, pp. 576–77, May 1985.
[69] K. Sreenivas and M. Sayer, *J. Appl. Phys.*, Vol. 64, No. 3, pp. 1484–93, Aug. 1988.
[70] K. Iijima et al., *J. Appl. Phys.*, Vol. 60, No. 8, pp. 2914–19, Oct. 15, 1986.
[71] H. Adachi et al., *Appl. Phys. Lett.*, Vol. 42, No. 10, pp. 867–68, May 15, 1983.
[72] H. Adachi et al., *Jap. J. Appl. Phys.*, Vol. 22 (Suppl. 22–2), pp. 11–13 1983.
[73] H. Adachi et al., *J. Appl. Phys.*, Vol. 60, No. 2, pp. 736–41, July 15, 1986.
[74] Y. Hamakawa et al., *IEDM*, Washington, D.C., pp. 294–297, Dec. 5–7, 1977.
[75] Y. Higuma et al., *Jap. J. Appl. Phys.*, Suppl. 17–1, pp. 209–14, 1978.
[76] M. Ishida et al., *J. Appl. Phys.*, Vol. 48, No. 3, pp. 951–53, Mar. 1977.
[77] T. Nakagawa et al., *Jap. J. Appl. Phys.*, Vol. 18, No. 5, pp. 897–902 May 1979.
[78] K. Tanaka et al., *Jap. J. Appl. Phys.*, Vol. 15, No. 7, pp. 1381–82, 1976.
[79] G. H. Haertling and C. E. Land, *J. Am. Ceram. Soc.*, vol. 54, No. 1, pp. 1–11, Jan. 1971.
[80] R. Vu Huy Dat and C. Baumberger, *Physica Status Solidi.*, Vol. 22, pp. K67–K70, 1967.
[81] G. Arlt et al., *J. Appl. Phys.*, Vol. 58, No. 4, pp. 1619–25, 1985.
[82] A. S. Shaikh et al., *ISAF*, Bethlehem, PA., pp. 126–29, June 8–11, 1986.
[83] J. K. Panitz and C.-C. Hu, *Ferroelectrics*, Vol. 27, pp. 161–64, 1980.
[84] J. K. Panitz, *J. Vac. Sci. Tech.*, Vol. 16, No. 2, pp. 315–18, 1979.
[85] A. Mansingh et al., *ISAF*, Bethlehem, PA., pp. 576–79, June 8–11, 1986.
[86] C. A. T. Salama and E. Siciunas, *J. Vac. Sci. Tech,.*, Vol. 9, No. 1., pp. 91–96, 1972.
[87] L. E. Cross, *Ferroelectrics*, Vol. 76, pp. 241–67, 1987.
[88] W. Pan et al., *J. Am. Ceram. Soc.*, Vol. 71, No. 1, pp. C17–C19, Jan. 1988.
[89] W. Pan et al., *ISAF*, Bethlehem, PA., pp. 645–47, June 8–11, 1986.
[90] Dr. Wuyi Pan, private communication.
[91] A. J. Gorton et al., ISAF, Bethlehem, PA., pp. 150–52, June 8–11, 1986.
[92] S. L. Swartz et al., *J. Am. Ceram. Soc.*, Vol. 67, No. 5, pp. 311–15, May 1984.
[93] W. B. Pennebaker, *IBM J. Res. Develop.*, Vol. 15, pp. 686–95, Nov. 1969.
[94] C. Araujo et al., *Appl. Phys. Lett.*, Vol. 48, No. 21, pp. 1439–40, 1986.
[95] D. J. McClure and J. R. Crowe, *J. Vac. Sci. Tech.*, Vol. 16, No. 2, pp. 311–14, Mar.-Apr. 1979.
[96] J. C. Olson and D. F. Stevison, *Ferroelectrics*, Vol. 37, pp. 685–87, 1981.
[97] K. Sreenivas and A. Mansingh, *J. Appl. Phys.*, Vol. 63, No. 11, pp. 4475–81, Dec. 1, 1987.
[98] M. Kitabatake and K. Wasa, *Jap. J. Appl. Phys.*, Vol. 24 (Suppl. 24–3), pp. 33–35, 1985.
[99] M. Kitabatake et al., *J. Non-Cryst. Sol.*, Vol. 53, pp. 1–10, 1982.
[100] M. Kitabatake et al., *Jap. J. Appl. Phys.*, Vol. 22 (Suppl. 22–2), pp. 31–34, 1983.
[101] T. Mitsuyu and K. Wasa, *Jap. J. Appl. Phys.*, Vol. 20, No. 1, pp. L48–50, Jan. 1981.
[102] M. Kitabatake et al., *J. Appl. Phys.*, Vol. 56, No. 6, pp. 1780–84, Sept. 15. 1984.

CHAPTER 2

INTEGRATED CIRCUIT MANUFACTURING AND NOVEL CIRCUIT DESIGN

Manufacturing-Based Simulation: An Overview

Stephen W. Director

Abstract

Integrated circuit designers have become increasingly interested in predicting the quality, in terms of yield as well as performance, of a design prior to manufacture. Toward this end, manufacturing-based methods have evolved. This paper reviews this class of techniques and their applications.

Introduction

As device sizes continue to decrease, the sensitivity of a circuit's performance to the inevitable variations in the manufacturing process increases. This sensitivity is translated into low manufacturing yields. (Manufacturing yield is the percentage of manufactured chips that meet all of the specified performance constraints.) Because design times and production costs are increasing, designs that will result in low yields should be identified as such before they are committed to production. Thus, it is becoming increasingly important for the circuit designer to be able to predict the effects of statistical variations inherent in the manufacturing process on circuit performance in order to take any necessary corrective action during the design process. With this goal in mind, a new class of simulators—known as *manufacturing-based simulators*—has been developed. With these simulators, it is possible to predict the yield of a design given the layout of the design and a description of the fabrication process. Typically, the description of the fabrication process includes a list of the processing steps, the values of the process controls (e.g., diffusion times and temperatures), and the characterization of the disturbances that occur in the process.

At the present time, manufacturing-based simulators fall into one of two categories: those aimed at predicting *functional yield* and those aimed at predicting *parametric yield.* Functional yield is defined as the percentage of manufactured circuits that are functionally correct, i.e., they perform the correct logic function or have all of the correct state transitions. Parametric yield is the percentage of manufactured circuits that have parametric performance (e.g., speed or power consumption) within specified limits. The overall manufacturing yield can be thought of then as the percentage of manufactured circuits that are functionally correct and have satisfactory parametric performance. To first order, manufacturing yield is therefore the product of functional yield and parametric yield.

Given the capability of predicting manufacturing yield, it is possible to perform a number of interesting tasks, such as evaluating the effects of redundancy in a design, optimizing process design to maximize yield in terms of device performance, maximizing parametric circuit yield, etc. This paper reviews some of the advances that have been made in the area of manufacturing-based simulation as well as some interesting applications of these methods.

This paper begins with an overview of the VLSI fabrication process in order to define some terms and identify the primary causes of yield loss. The sources of disturbances in the manufacturing process are discussed next. To first order [1], we will see that functional yield is dependent on the distribution of local disturbances in the process, such as spot defects on a mask and oxide pinholes, while parametric performance is dependent on global disturbances in the process, such as mask misalignment and oxide growth rates. Further evaluation of manufacturing yield is based upon the evaluation of functional and parametric yield. Methods for evaluating functional yield are presented, and applications of these methods are discussed. Applications of parametric yield prediction are presented. It is shown how parametric performance simulation forms the basis for an improved worst-case analysis methodology. Lastly, a brief conclusion is given.

VLSI Fabrication Process

In this section, we review the steps involved in integrated circuit (IC) manufacture and thereby introduce some terminology. The IC manufacturing process involves a sequence of basic processing steps. To facilitate process observability and process control, one or more processing steps are usually separated by an evaluation or inspection and a selection step. An evaluation step is a measurement, or sequence of measurements, called *in-line measurements.*

If one or more in-line measurements fall outside of some range defined by a set of *selection thresholds*, the wafer is considered defective and is discarded. We refer to a processing step followed by an evaluation and selection step as a *manufacturing operation*.

The outcome of the ith manufacturing operation depends on three major factors: the process controlling parameters, or *control* denoted by the vector C_i; the geometry of the fabricated IC, or *layout* denoted by the vector L_i; and some randomly changing environmental factors, or *disturbances* denoted by the vector of random variables D_i, that are inherent in any manufacturing process.

A typical IC fabrication process involves many manufacturing operations followed by a sequence of final tests. In general, two types of tests are made: functional and parametric. Functional tests are used to detect errors in the function performed by IC chips. For digital circuits, functional tests involve the application of binary testing sequences to chip inputs, and the results are compared against expected output binary vectors. For analog circuits, functional evaluation depends on the type of chip, but usually consists of some frequency- or dynamic-response measurements. Parametric tests are used to detect basic discrepancies between the performance (in terms of power, critical signal levels, and other DC parameters) of the IC under test and desired performance. Such tests are a measure of quality of the chip's performance.

Reprinted from *IEEE Circuits and Devices Magazine*, Vol. 3, No. 5, pp. 3-9, September 1987.

Process Disturbances and Circuit Faults

All physical manufacturing processes are affected by inherent variations, or disturbances. In order to reduce the effects of such disturbances, we need to understand their cause and the effect they have on IC performance. (For a thorough discussion of these issues, see [1].) Physical causes of disturbances in an IC process include:

- Human errors and equipment failures.
- Instabilities in the process conditions (e.g., turbulent flow of gases used for diffusion and oxidation).
- Material instabilities (e.g., variations in the physical parameters of the chemical compounds and other materials used in the manufacturing process).
- Substrate inhomogeneities (e.g., point defects, dislocations, and surface imperfections).
- Photolithographic mask imperfections.

In general, disturbances in the process can be viewed as affecting either the geometry of an IC device and/or the electrical characteristics of an IC device.

The actual geometry of an IC device is determined by the two-dimensional layout parameters and the physics of the manufacturing process. Ideally, the edges of the layout pattern define the boundaries between various regions in an IC, and various manufacturing process steps determine the depth, or thickness, of a region. In an actual IC, this is not the case. Deformations in geometry can be caused by global disturbances, such as mask misalignment, over- and underetching, and lateral diffusion, as well as by local disturbances, such as spot defects and oxide pinholes. Figure 1 illustrates the difference between an "ideal" IC device and an actual device.

The electrical characteristics of an actual IC are determined by the three-dimensional impurity distributions in the conducting and semiconducting regions, and charge distribution in the insulating layers. In an ideal IC, a number of simplifying assumptions are typically made that allow for analysis of a design prior to manufacturing. It is important to note that these simplifications introduce a systematic error in the analysis and cause the description of the electrical properties of an IC used in the design process to differ from physical reality. However, there also exist many phenomena that are random in nature and cause deformations of an IC's electrical characteristics. Some of these are global in nature, such as the thickness of a gate oxide due to temperature fluctuations during growth and interlevel dielectric thickness due to randomness in curing temperature. Others are local in nature, such as local impurity precipitates or silicides that affect the percentage of electrically active impurities in silicon.

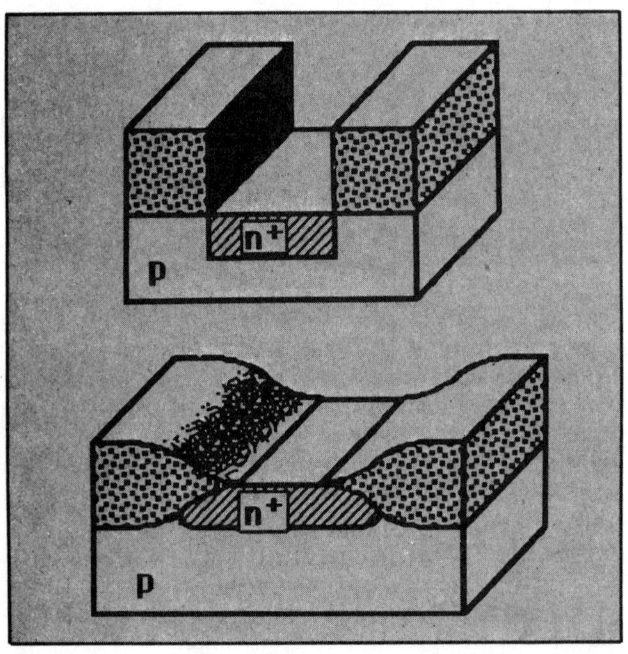

Fig. 1 Ideal and actual IC devices.

In general, each process disturbance may cause a number of different changes in IC performance. For instance, a spot defect introduced in a lithography step may cause a short between two adjacent conducting paths, a break in the path, or may only alter effective device dimensions. These changes in behavior are called *faults*. It is convenient to classify faults as either *structural faults*, which are changes in the topology of the circuit, or *performance faults*, in which the IC topology remains unchanged; however, circuit performance (e.g., speed or dissipated power) falls outside of some set of allowable tolerance limits. Performance faults can be further divided into two classes: *soft-performance faults* and *hard-performance faults*. If an IC is functionally correct (e.g., in a digital circuit, all state transitions occur in the proper order and the output pattern is the correct response to an input pattern vector) but some performance measure (e.g., signal delay) lies marginally outside of the specified range, we say a soft-performance fault has occurred. If an IC does not function properly (e.g., some state transitions do not occur) or some performance is orders of magnitude away from the desired value, we say a hard-performance fault has occurred. In general, structural faults and hard-performance faults are referred to as *functional faults* in that the basic function of an IC is incorrect, while soft-performance faults are often merely called *performance faults*.

Structural faults require some further elaboration. While some deformations will not alter the structure of an IC under no bias, or very small bias conditions, for other values of DC bias, a short between two conducting paths can occur. On the other hand, some geometrical variations may introduce new parasitic elements (such as overlap capacitances), which do not necessarily cause an IC to be defective. Therefore, structural faults are best defined in terms of changes in the equivalent DC circuit diagram under worst-case bias conditions.

Measures of Process Efficiency

Because of the inherent fluctuations in an IC fabrication process, and the attendant variations in circuit performance, not all of the chips that are manufactured will actually meet all of the design specifications. By being able to *predict yield* during the design process, it may be possible to adjust the design itself to improve yield. In order to predict yield during the design process, we need to be able to determine the probability that a single IC chip will not be rejected during fabrication. To evaluate this probability, it is convenient to introduce the concept of an *acceptability region*.

Assume that a manufacturing process has n steps. Then let the vector $C_i = (C_1^T, C_2^T, \ldots, C_n^T)^T$ denote all of the process control parameters, vector $L = (L_1^T, L_2^T, \ldots, L_n^T)^T$ denote all of the layout parameters, and $D = (D_1^T, D_2^T, \ldots, D_n^T)^T$ denote all of the process disturbances.

Also, let $\mathbf{S_D}$ denote the Euclidean space that contains all of the process disturbances, D. For given values of L and C, denoted by L^0 and C^0, in the space of process disturbances, there exists a set of disturbances $\mathbf{A}_D(C^0, L^0) \subset \mathbf{S_D}$, which will not cause the performance of an IC to be unacceptable. We call this set of disturbances the region of acceptability. Design yield can, therefore, be formally expressed as

$$Y = \int_{\mathbf{A}_D(C^0, L^0)} f_D(\delta)\, d\delta \tag{1}$$

where $f_D(\delta)$ is the joint probability function, which describes D. Figure 2 illustrates the relationship that yield has to the particular values of C^0 and L^0. In Fig. 2(a), a typical joint probability density function (jpdf) for two disturbances is shown, and Fig. 2(b) represents the equiprobability level contours associated with this jpdf projected onto the "space" of process disturbances. Figures 2(c) and 2(d) illustrate low- and high-yield placements of the acceptability region within the disturbance space. In the low-yield placement, the disturbances that are most likely to occur fall outside of $\mathbf{A}_D$, while they fall inside $\mathbf{A}_D$ in the high-yield placement.

The design yield is difficult to evaluate, because, in general, it is not possible to obtain an explicit expression for $\mathbf{A}_D(C^0, L^0)$. However, through an analysis of the basic properties of the process disturbances, we can develop techniques for evaluation of design yield. Toward this end, it is convenient to decompose $\mathbf{A}_D$ into a functional acceptability region and a parametric performance acceptability region:

$$\mathbf{A}_D = \mathbf{A}_D^F \cup \mathbf{A}_D^P \tag{2}$$

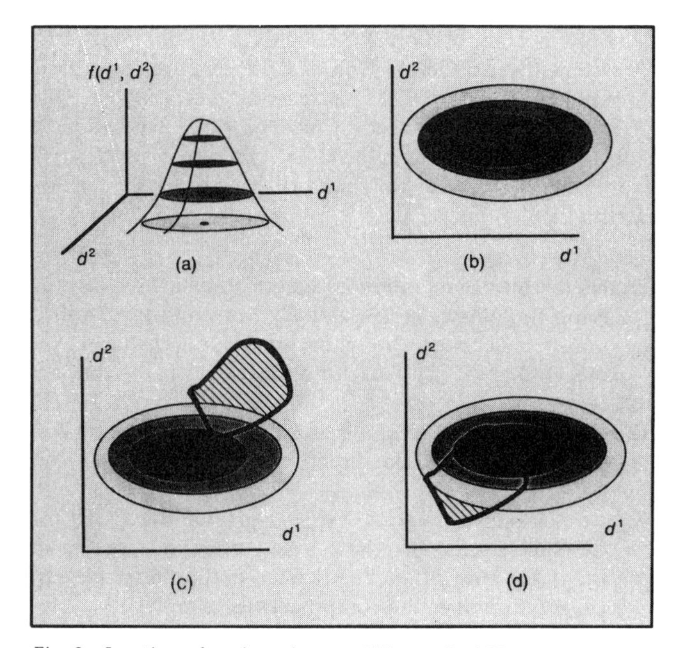

Fig. 2 Location of region of acceptability and yield: (a) probability density function, (b) level contours, (c) low-yield placement, and (d) high-yield placement.

where $\mathbf{A}_D^F$ is the disturbance set that does not cause the IC to have incorrect functionality and $\mathbf{A}_D^P$ is the disturbance set that does not cause poor performance. (We note that existence of such a decomposition is an assumption that is valid only to first order.) Thus, we can define the *functional yield* as

$$Y_{\text{FUN}} = \int_{\mathbf{A}_D^F(C^0, L^0)} f_D(\delta)\, d\delta \tag{3}$$

where, as before, $f_D(\delta)$ is the jpdf describing D, and the *parametric yield* is defined as

$$Y_{\text{PAR}} = \int_{\mathbf{A}_D^P(C^0, L^0)} f_D(\delta)\, d\delta \tag{4}$$

which are respectively probabilities that the IC is correct functionally and its performance characteristics are within some desired limits.

Decomposition of the design yield into parametric and functional components provides an opportunity for significant simplifications in evaluating design yield. This becomes evident if we note that, since $\mathbf{A}_D^F(C^0, L^0)$ contains those process disturbances that do not disturb functional correctness of the IC, then $(1 - Y_{\text{FUN}})$ can be estimated by searching for the complement of $\mathbf{A}_D^F(C^0, L^0)$, i.e., determining the set of process disturbances that causes structural failure and/or hard-performance faults only. Thus, in estimating $(1 - Y_{\text{FUN}})$, we need consider only those defects that cause structural changes in the connectivity of an IC, i.e., defects due to local disturbances. We do not need to evaluate IC performance itself, because process disturbances that cause structural changes in the connectivity of an IC can be found by analyzing the layout of the IC and the statistical characterization of the local defects only. In particular, to determine $\mathbf{A}_D^F(C^0, L^0)$, we need investigate only the *critical area* of a layout, i.e., that area of an IC that is vulnerable to the occurrence of a point or spot defect. We review some of the techniques developed for this purpose in the next section.

Computation of parametric yield is also simplified because of decomposition, because single spot or point defects that do not change circuit connectivity usually do not affect the quality of IC performance. Therefore, local defects can be ignored when computing parametric yield and we need consider only the effects of global disturbances, such as mask misalignment and lateral edge displacements. A method for computing parametric yield is reviewed in a subsequent section.

Functional Yield Prediction

As indicated previously, degradation of functional yield is due to process disturbances that cause structural failures due to changes in the DC connectivity of a circuit under worst-case bias conditions. There are many methods available for simulating the yield losses due to structural failures. Most of these are based on simple statistics, e.g., binomial or Poisson distributions, which have limited simulation accuracy because they cannot account for defect characteristics, such as size or clustering or specific IC layout information. Improved yield simulation methods were developed to account for the effects of defect clustering by incorporating compound and negative binomial statistics [2], [3]. However, these statistical techniques still failed to consider the particulars of the IC layout.

In order to address IC layout information, Maly [4] proposed a method for functional yield prediction based upon the concept of virtual layout. The virtual layout has the same statistical features as the nominal IC layout but is arranged in a more computationally efficient form. More specifically, this method assumes that functional yield is equal to the probability of the event that point defects combined with line registration errors, i.e., line registration errors themselves, do not cause shorts or openings in the fabricated IC. To calculate this probability, Maly employs the concept of *critical area*. The center of a defect is located in a critical area when the defect causes a short or open in an IC layer.

In the general case, computation of the critical area is complex and cannot be explicitly evaluated as a function of defect radii and layout dimensions. However, it is possible to compute the critical area by means of a sequence of layout resizing operations or by using a Monte Carlo technique [5] to facilitate computation of the critical area in terms of the defect radius. More specifically, the virtual layout has the same statistical features as the nominal IC layout but is arranged in a form that allows for easy calculation of the critical area as a function of the defect radius.

In order to overcome the difficulty the virtual layout method has with accounting for defect clustering, a Monte Carlo bench method was developed. This method was implemented in a simulator called VLASIC [6]. VLASIC (see Fig. 3) employs a Monte Carlo approach of generating, placing, and analyzing defects on a chip layout. After the defects have been placed on the layout, a series of fault analysis procedures is used to determine what, if any, circuit faults have occurred. The resulting faults are passed through a filtering phase that ignores those faults that do not affect the functional yield. The resulting output is a chip sample containing a list of the circuit faults, if any, that have occurred on the chip during the simulated fabrication. The sample is then passed to application postprocessors.

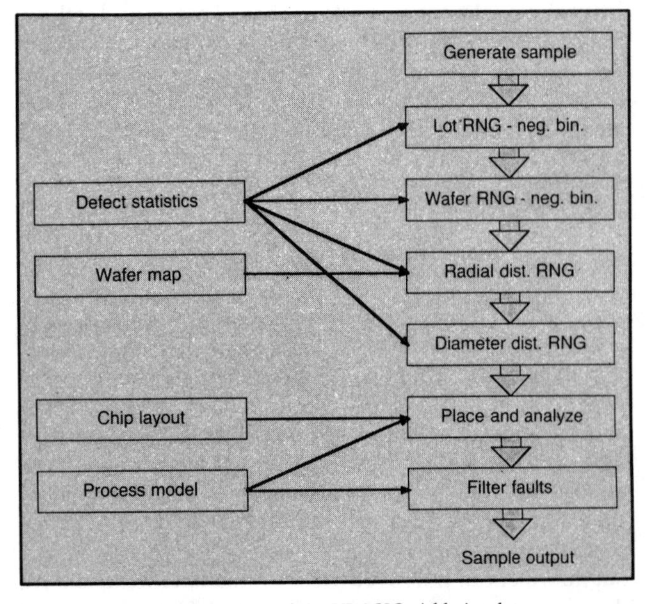

Fig. 3 Structure of the VLASIC yield simulator.

The defect random number generators (RNGs) in VLASIC are organized in a hierarchical manner, with separate generators for the spatial and size distributions. These generators are controlled by parameters such as the mean and variance of the defect density. The random number generators output a list of defect types, locations within the die sample, and defect diameters. For each defect, a series of fault analysis procedures is called to examine the layout geometry in the neighborhood of the defect in order to determine if any circuit faults have occurred. A separate fault analysis procedure is used for each possible circuit fault type, such as a short or open circuit. The result of the fault analysis is a list of the raw circuit faults caused by the defects.

The raw circuit faults pass through a filtering and combination phase. Those faults that do not cause a DC change to the circuit topology are ignored. Some faults are combined together into a composite fault, such as using short-circuit faults to hook up the terminals of a new device. The operation of both the fault analysis and filtering phases is guided by defect models. The models specify what circuit faults can be caused by each defect type, what layers interact with the defect, and how layers are electrically connected together.

The resulting output is a chip sample containing a list of the circuit fault groups that have occurred on it during the simulated fabrication. A *fault group* is a list of circuit faults caused by a single defect. Each fault in the fault group specifies what type of fault it was (e.g., short, open), the type of defect causing the fault (e.g., extra metal), and the details of the fault, such as what nets have been shorted together. The chip sample fault lists are summarized to record the frequency of each unique fault combination and are then passed to postprocessors when simulation is complete.

Both of the approaches described previously for functional yield prediction fail to account for the effects of global disturbances. Furthermore, the approach implemented in VLASIC is computationally expensive. As an attempt to overcome these difficulties, an analytically based approach to simulate functional yield has recently been proposed [5]. This approach can account for most of the global and local effects and allows hierarchical calculation of yield, e.g., from cell and chip, to wafer level. Three key steps of this method are:

(1) Hierarchical generation of defect statistics by considering the effects of the globally nonuniform distribution of defects on the wafer level, local clustering on the chip level, and uniform distribution on the cell level.
(2) Analytic calculation of the probabilities of failure (POFs) due to global and combined effects for simple layout patterns.
(3) Derivations of analytic expressions for the POFs for functional cells, or chips, based upon the results of Steps (1) and (2). A hierarchical method can be employed to reduce the computational complexity.

This approach is still in its early stages of development, although the results have been promising.

Applications of Functional Yield Prediction

The output of a program such as VLASIC can be used for a number of tasks. For example, a redundancy analysis postprocessor has been developed, which can be used to calculate yield in the presence of redundancy. A set of heuristics is used to convert circuit faults into cell failures. For example, a short circuit causes all cells containing the shorted nets to fail, while an open device causes only the cell containing it to fail. The heuristics use signal strengths to determine whether one net can overpower another. The postprocessor then uses the knowledge of which cells have spares, which cells are spares, which cells have no spares, the constraints on spare swapping, and the spare selection algorithm to determine whether all failing cells can be replaced by spares.

Another application for the output of a program such as VLASIC is the development of improved testing procedures, such as that known as inductive fault analysis (IFA) [7]. The four major steps of the IFA procedure are: (1) generation and placement, on the circuit layout, of physical defects using statistical data obtained from the fabrication process; (2) extraction of primitive (geometric) faults caused by these defects; (3) abstraction of these primitive faults to the transistor, logic, or even the functional level; and (4) classification of fault types and ranking of faults based on the likelihood of occurrence. Hence, given the layout of an integrated circuit, a customized and very accurate fault model and an associated ranked fault list can be automatically generated, which take into account the technology, layout, and process characteristics.

Parametric Yield Prediction

The central tool developed for aiding in the prediction of the degradation of parametric performance, which results primarily from global variations in the manufacturing process, is the IC process/device simulator FABRICS II [8]-[10]. The models implemented in FABRICS II allow for the accurate simulation of typical semiconductor devices manufactured using a variety of fabrication processes (e.g., NMOS, CMOS, or bipolar). FABRICS II employs efficient numerical and analytical models, which are solutions of the partial differential equations that describe each fabrication step, under a set of restricted or simplifying conditions that have been found to yield reasonable results. The input to the process simulator, part of FABRICS II, consists of *process parameters*—such as times and temperatures of diffusion steps, and doses and energies of ions in implantation—and *process disturbances*—such as diffusivities of impurity atoms and misalignments between photolithographic masks. To account for local (intradie) and global (interdie) variations in device parameters, process disturbances are modeled by hierarchically defined RNGs at levels that correspond to natural divisions, i.e., at the lot, wafer, chip, and device levels. Since the process disturbances cannot be obtained from direct measurements, the probability density functions describing process disturbances are identified through in-line and test structure measurements [11].

The device simulator in FABRICS II uses the physical parameters generated by the process simulator, in combination with the layout of the devices, to produce device model parameters. The model parameters generated by FABRICS II are compatible with the device models implemented in circuit simulators such as in SPICE.

By coupling FABRICS II with a circuit simulator, as illustrated in Fig. 4, it is possible to generate histograms of various performances that are typical of what would be measured in a real fabrication line. Parametric yield can then be easily calculated [12].

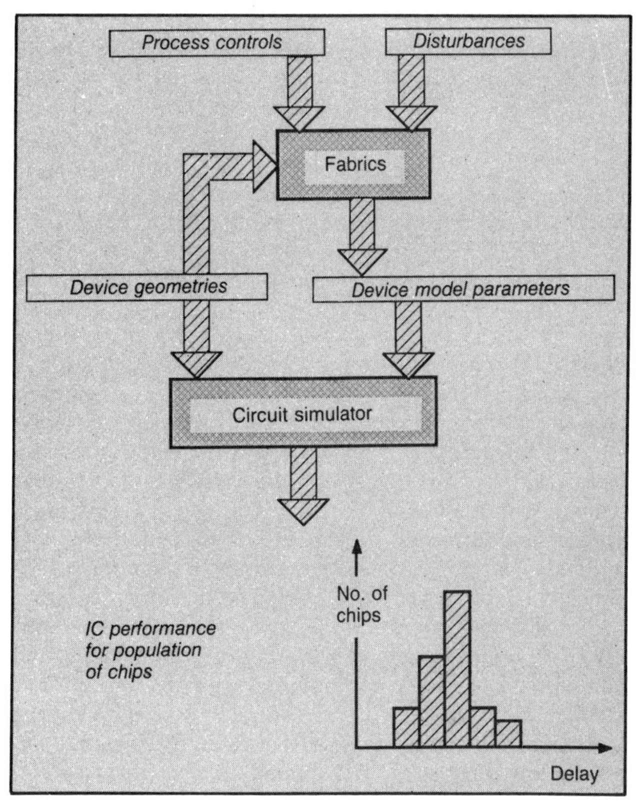

Fig. 4 System for generating histograms of parametric circuit performance.

Applications of Parametric Yield Prediction

FABRICS II, combined with a circuit simulator, can be employed for a number of tasks ranging from design verification and optimization to process diagnosis and statistical process control.

Figure 5 illustrates the role of such a simulator in optimization tasks. In this figure, the disturbances, D, are assumed to be characterized in terms of statistical moments. Process yield optimization is achieved by specifying a test structure with a fixed set of device dimensions, L, and a desired device behavior. The goal then is to adjust the process controls to maximize the number of devices produced whose performance falls within acceptable limits of behavior [13]. Circuit yield optimization assumes that the process control parameters are fixed and that the layout L is adjusted to maximize the number of circuits whose performance falls within acceptable limits [14]. Both of these optimization problems are costly due to the process, device, and circuit simulation required as well as the high dimensionality of the problem. However, recent work has resulted in approaches that can realistically attack these problems.

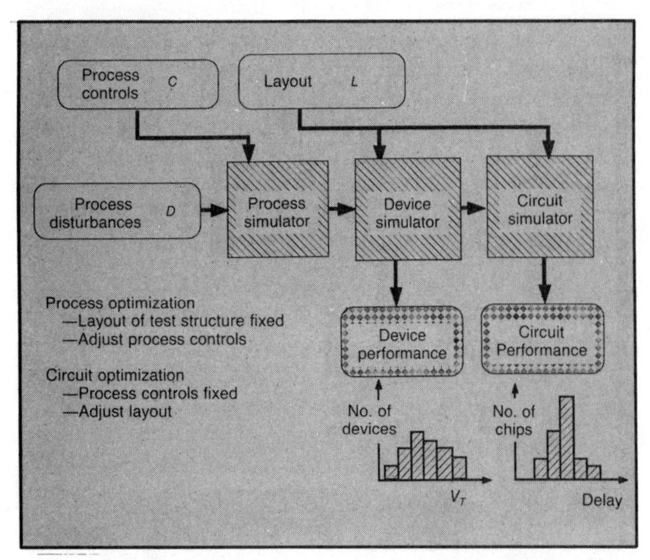

Fig. 5 Process and circuit optimization to maximize parametric yield.

The capability provided by FABRICS II also provides the basis for an IC fabrication line diagnostic system called PROD [15]. The operation of PROD can be divided into two phases: *learning* and *fault identification*. In the learning phase, fault simulation experiments are performed, in which single faults or combinations of faults are artificially introduced to FABRICS II and the corresponding probability distributions of the circuit performances are obtained via combined process, device, and circuit simulations. To guide the learning process in a meaningful manner, a sensitivity analysis is performed. This analysis provides information about which process disturbances have the strongest effect upon the circuit performances, allowing added emphasis to be placed on these disturbances during fault simulation. To further reduce the fault simulation cost, a multilayer regression tool [16] has been implemented to extract the regression polynomial model from the sequence of simulation steps. The sequence of simulations can then be replaced with this regression model. Efficient transformations of these fault simulation results are then performed to compress the data and extract the relevant pattern features. The learning phase continues until a number of the most significant fault combinations have been simulated and stored in a library. At this point, PROD is ready to be used for the identification of actual faults occurring in the IC fabrication process. To identify faults in an operating fabrication line, the same performances as those chosen during the simulation stage are extracted from the circuit, and pattern recognition algorithms are used to compare against those in the library. Once the ''closest'' pattern of performance failure has been found the fault causing the failure has been identified.

Worst-Case Analysis

In order to guarantee satisfactory performance, a design is typically verified by the designer under what is defined as worst-case conditions. This process is traditionally performed by adjusting certain electrical parameters of the IC devices, such as threshold voltages and transconductances for MOSFETs, to their extreme limits. In reality, however, these parameters are statistically dependent (correlated) random variables with a multilevel structure of variance (intradie and interdie). In traditional approaches, the correlation coefficients between device parameters are not taken into account and, therefore, IC performances are estimated for some *unrealistic* combinations of the device parameters. Hence, the results of such as an analysis are usually too pessimistic. A better worst-case estimate can be achieved if analysis was performed in terms of a truly independent set of parameters. Since the process disturbances used in FABRICS II are statistically independent, we can use them for the worst-case analysis.

In this alternate approach [14], the selection of the significant worst-case parameters is performed based upon the *sensitivities* of the performances to process disturbances. The sensitivities are estimated by perturbations using data obtained from FABRICS II, coupled with a circuit simulator. Since sensitivities are *local* estimates of this dependence, a more accurate method of approximating the relationship over a wide range of changes in process disturbances is to build nonlinear regression models relating performances to the process disturbances. Such models have been successfully built, and it was found that these dependencies are *monotonic* over a wide range of process disturbances. Therefore, sensitivities can be reliably estimated by *large* perturbations and the worst-case combinations of significant process disturbances can be obtained for each IC performance. Then *realistic* worst-case sets of device parameters may be generated by FABRICS II if the significant process disturbances are changed by, for example, one standard deviation from their identified values in the ''worst-case direction.'' Because of the independence of the process disturbances, it is possible to estimate the *probability* of occurrence of this case, which is valuable information for IC designers. Furthermore, if the device models are defined in such a way as to be independent of device dimensions, then the designer may alter the IC layout to improve performance. If, however, some changes in the fabrication process parameters are necessary, the worst-case device models have to be evaluated once again. Observe that such an evaluation is computationally inexpensive because, in the proposed approach, Monte Carlo simulations are not required.

Conclusions

In this paper, we have reviewed some of the simulation methods developed for the purpose of predicting how the random effects inherent in any integrated circuit (IC) fabrication process affect performance, and, hence, yield.

In summary, the ability to predict manufacturing yield prior to the start of an actual manufacturing process is crucial in determining the quality of an IC design. While there has been considerable progress in the development of tools for aiding in the prediction of manufacturing yield, additional tools are still needed.

Acknowledgments

The research presented herein was sponsored by the Semiconductor Research Corporation. Much of the work described in this paper has been under development at

Carnegie-Mellon University for a number of years. The author would like to acknowledge the valuable contributions of A. J. Strojwas, W. Maly, J. Shen, S. R. Nassif, C. J. Spanos, D. Giannopoulos, M. Trick, P. Odryna, K. K. Low, P. Kager, H. Walker, R. Razdan, I. Chen, P. K. Mozumder, C. R. Shyamsundar, and J. Ferguson.

References

[1] W. Maly, A. J. Strojwas, and S. W. Director, "VLSI Yield Prediction and Estimation: A Unified Framework," *IEEE Trans. Computer-Aided Design of Integrated Circuits and Systems*, vol. CAD-5, no. 1, pp. 114–130, Jan. 1986.

[2] C. H. Stapper, Jr., "Modeling of Integrated Circuit Defect Sensitivities," *IBM J. Res. Dev.*, vol. 27, no. 6, pp. 549-557, Nov. 1983.

[3] R. M. Warner, Jr., "Applying a Composite Model to the IC Yield Problem," *IEEE J. Solid-State Circuits*, vol. SC-9, no. 3, pp. 86–95, June 1974.

[4] W. Maly, "Modeling of Lithography Related Yield Losses for CAD of VLSI Circuits," *IEEE Trans. Computer-Aided Design of Integrated Circuits and Systems*, vol. CAD-4, no. 3, pp. 166–177, July 1985.

[5] I. Chen and A. J. Strojwas, "Realistic Yield Simulation for IC Structural Failures," *IEEE Int. Conf. Computer-Aided Design (ICCAD) Digest of Technical Papers*, pp. 220–223, Nov. 1986.

[6] D. M. H. Walker, "Yield Simulation for Integrated Circuits," Ph.D. thesis, Computer Science Department, Carnegie-Mellon University, Pittsburgh, PA, July 1986.

[7] J. P. Shen, W. Maly, and F. J. Ferguson, "Inductive Fault Analysis of MOS Integrated Circuits," *IEEE Design and Test of Computers*, vol. 2, no. 6, pp. 13–26, Dec. 1985.

[8] P. Kager and A. J. Strojwas, "PI/C: Process Interpreter/Compiler," *IEEE Int. Conf. Computer-Aided Design (ICCAD) Digest of Technical Papers*, pp. 321-323, Nov. 1985.

[9] K. K. Low and S. W. Director, "PED: A Graphical Process Editor," *Proc. IEEE Int. Symp. Circuits and Systems*, pp. 560–566, May 1986.

[10] S. R. Nassif, A. J. Strojwas, and S. W. Director, "FABRICS II: A Statistically Based IC Fabrication Process Simulator," vol. CAD-3, no. 1, pp. 40–46, Jan. 1984.

[11] C. J. Spanos and S. W. Director, "Parameter Extraction for Statistical IC Process Characterization," *IEEE Trans. Computer-Aided Design of Integrated Circuits and Systems*, vol. CAD-5, no. 1, pp. 66–78, Jan. 1986.

[12] A. J. Strojwas, "The CMU-CAM System," *IEEE Design and Test of Computers*, vol. 3, no. 1, pp. 35-44, Feb. 1986.

[13] D. J. Giannopoulos and S. W. Director, "IC Fabrication Process Optimization," *IEEE Int. Conf. Computer-Aided Design (ICCAD) Digest of Technical Papers*, pp. 164–166, Nov. 1984.

[14] A. J. Strojwas, S. R. Nassif, and S. W. Director, "Optimal Design of VLSI Minicells Using a Statistical Process Simulator," *Proc. IEEE Int. Symp. Circuits and Systems*, Jan. 1983.

[15] P. Odryna and A. J. Strojwas, "PROD: A VLSI Fault Diagnosis System," *IEEE Design and Test of Computers*, vol. 2, no. 6, pp. 27-35, Dec. 1985.

[16] A. J. Strojwas and S. W. Director, "A Pattern Recognition Based Method for IC Failure Analysis," *IEEE Trans. Computer-Aided Design of Integrated Circuits and Systems*, vol. CAD-4, no. 1, pp. 76–92, Jan. 1985.

Micro-Automating Semiconductor Fabrication

Automating very-small-scale mechanical processes has confounded the best efforts of engineers, but now there's progress

by Ilene J. Busch-Vishniac

Microelectronic fabrication depends upon chemical, optical, and mechanical processes. Chemical processes, such as etching a feature; and optical processes, such as exposing a mask, are well-suited to automation. But small-scale mechanical processes resist automation because of friction.

Friction poses difficulties on all scales, but its significance grows as device size decreases because of the accompanying increase in the ratio of surface area to volume. This is one reason why most of the processes used in microelectronic fabrication rely on purely optical and chemical techniques. Unfortunately, there are no purely optical or chemical techniques for moving one object relative to another, and such motions are at the heart of many important manufacturing processes. These motions require mechanical approaches, but mechanical approaches lack precision on a small scale. This drawback poses a significant obstacle to automation.

Automation in Miniature

What we need is practical micro-automation; i.e., the automatic control of processes requiring relative motion at micron or sub-micron accuracies. In microelectronics manufacturing, micro-automation might play a hand in material transport, electrical probing, and mechanical probing. Many of the problems associated with micro-automation, such as the need for smart sensors and robust control systems, are shared with conventional macroautomation. Micro-automation, however, is distinctly different in two important ways. First, although friction can be significant in macro-automation, it is one of the dominant forces in micro-automation. Second, large-scale automatic control of manufacturing usually involves machines that are designed to handle high payloads and apply large forces. Micro-automation is much more likely to be directed at delivering well-controlled, delicate forces and transporting small payloads. These differences suggest that methods appropriate in the macro realm may not be appropriate in the micro realm.

Applying Micro- Sensors and Actuators

A recent National Science Foundation report on the emerging field of microelectromechanical systems discussed the potential applications of micro-actuators and micro-sensors, which are key components of a micro-automation system [1]. The application areas included biotechnology, optics, materials handling, equipment miniaturization, and microelectronics.

In biotechnology, one can imagine a smart pill capable of taking measurements, and even of performing surgery. In optics, high-speed alignment of optical elements with micro-automation could lead to high-speed multimode optical switching. In materials handling, the ability to carry small payloads to positions specified within a micron might permit the repair of masks used in semiconductor fabrication. Micron-scale automatic controls could also reduce dramatically the size of some equipment. A reduction of space provides tangible benefits for clean-room facilities, where, for example, costs are directly tied to the required volume.

Applications of micro-automation to semiconductor manufacturing are particularly appealing. Mechanical processes currently pose significant obstacles to automatic fabrication, and advances are sorely needed. But an automated approach would simultaneously result in cost reductions and product quality improvements.

Positioning

In my opinion, the three most attractive applications for microautomation are positioning, electrical probing, and mechanical probing [2]. Many of the processing steps in semiconductor fabrication require high-speed transport of materials over large ranges with high precision. Such positioning tasks span several scales, and include mask alignment, hybrid-circuit manufacturing, and wafer transport. Mask alignment, for example, continues to grow more difficult. Feature sizes in microelectronics are shrinking, so wafer-positioning errors must be controlled ever more vigorously during mask alignment. At the same time, wafer dimensions and circuit sizes are increasing, and thus wafer nonplanarity is becoming a greater problem. A total

Reprinted from *IEEE Circuits and Devices Magazine,* Vol. 7, No. 4, pp. 32-37, July 1991.

positioning tolerance of 100 nm over an entire 8-inch wafer requires a positioning accuracy of roughly 5 x 10^{-7}, where accuracy is defined as the ratio of the required resolution to the workspace dimension. Typical positioning approaches used on large scales cannot provide this accuracy. One unconventional approach that has been successfully demonstrated in scanning tunneling microscopes is magnetic levitation.

Mask alignment is an example of one task that requires extreme positioning. At the other extreme is the transport of wafers between process stations, which has much in common with an airplane ride. The speed and smoothness of the ride are critical, but positioning accuracy is important only at the path endpoints.

Traditional approaches to the transport problem use robots or conveyor belts, but both of these present difficulties. Robots generally involve jointed parts that rub during operation, so they must be specially built to prevent particles from entering the clean-room environment. Further, robots are often difficult to reconfigure for fabrication-line changes. Conveyor belts tend to be dirty, but they are much less costly than robots. Unconventional approaches using air tracks, magnetic levitation, or electrostatic levitation are likely to offer significant improvements in the near future.

Electrical Probing

The desire for improved quality control in semiconductor fabrication has prompted increased analytical probing of the interiors of VLSI circuits. Analytical probe stations currently on the market are of two types: mechanical probers that rely on manual or semi-automatic translation of contact probes; and noncontacting probe stations that determine voltages and currents optically.

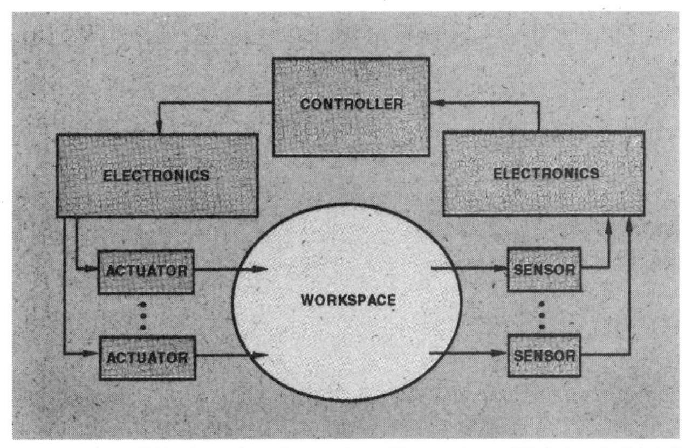

1. System architecture of a micro-automation system. At this level of detail, the architecture looks much like that of a macro-automation system.

The mechanical probe stations are difficult and time-consuming to use, and they may destroy the device under test because the probes are often on long lever arms that cause skating as they make contact. Noncontact probe stations can pinpoint the location of a failure with great speed and accuracy, but they cannot determine the nature of a fault since they cannot inject a signal. They are typically an order of magnitude more expensive than their contacting counterparts. A new approach to analytical probing is needed.

Analytical probe stations require positioning with at least three degrees-of-freedom. With electronic feature sizes moving below 0.5 μm, resolution of roughly 0.1 μm is required. (In a workspace of roughly 1 cm^2, we require an accuracy of roughly 10^{-5}.) Another consideration that makes this application challenging is the desire to place multiple probes on a circuit simultaneously. Multiprobe testing requires the automatic control of multiple objects such that interference and collisions are prevented. A further issue is the electrical load presented by the probe itself. To meet the needs of broadband operation, restrictions must be placed on the permitted materials and geometries. Taken in combination, the requirements of automated analytical contact probing provide a challenge that can only be answered well through micro-automation.

Mechanical Probing

Another application for micro-automation in semiconductor fabrication is mechanical quality-control testing. The bonds between circuits and their packages, for example, are often viewed as the place where most nondesign failures occur. Researchers have developed a number of ways to look at bonds, including laser/infrared signatures, acoustic microscopy, and mechanical stress tests.

Mechanical stress tests directly measure the mechanical integrity of a bond, but current techniques suffer from three serious defects. First, the tests are usually destructive, so they can only be used to provide statistical information on a fabrication lot. Second, even nondestructive pull or shear testers cannot perform their tests quickly. As a result, these tests must be executed off-line on a small number of devices, instead of on all bonds of all devices. Third, the lack of position and force control in the testing equipment ultimately yields binary information on whether a bond is good or bad. Ideally, one would prefer to know how good or how bad a bond is, so that various fabrication methods can be objectively compared.

To improve mechanical shear and pull measurements, test all bonds of all microelectronic devices nondestructively and quickly. For example, apply a known shear force to a bond, and then monitor the deflection. Such a quality control tester would require automated positioning in at least three degrees of freedom. For a 400-lead device with 4-mil lead separation, the required accuracy would be about 10^{-3}. If one of these devices is to be tested every two seconds, testing of at least 200 bonds per second would be required. While these requirements would be difficult to achieve on large scales, they should be reasonably easy to satisfy on small scales using low- mass testers with high acceleration. Thus, the task is ideal for micro-automation. Table 1 summarizes the attributes of the applications discussed above.

System Architecture

At the level of detail shown in a schematic of a general micro-automation system, there is no real distinction between a micro- and a macro-automation system (Fig. 1). Both generally require that the process be controlled and monitored. Control is through one or more actuators that affect the object on which work is to be performed. The quality of the work is monitored through sensors on or near the workpiece, and the desired dynamic performance is usually compared to the achieved response for error compensation via some real-time, closed-loop control strategy. This strategy can be implemented in either hardware or software, connecting an electronics interface between the controller and the sensors and actuators.

Of the four main system elements shown in Fig. 1, it is the sensors and actuators that divide micro-automation from macro-automation. These are also the most challenging elements to construct at any scale.

For reasonable sensing, one needs a device capable of measuring the state of a system with minimal impact on that state. This generally mandates a low-power transducer, but micro-automation must confront additional problems: the added weight of a sensor, and the need to make measurements in the micron or submicron range.

Appropriate actuation requires a device capable of imposing a state on a system with the least possible sensitivity to the load. This generally results in a high-power transducer, but actuator power needs in micro-automation systems are modest and must be kept low to avoid thermal dissipation problems. In addition, the need for extremely accurate actuation supersedes the importance of load sensitivity.

The State of the Art in Transducers

Micro-sensors, i.e., devices capable of measuring micron or submicron motions, were developed before micro-actuators. They come in a variety of forms, many of which are microfabricated and are therefore quite small. Also available are optical sensors such as fiber-optic devices, photodiodes, and phototransistors, and they are suitable for many systems. There are also a host of sensing elements that have been fabricated in silicon [3]. Generally, these solid-state devices are capacitive (relying on the motion of one silicon membrane that serves as the plate of a capacitor); or resistive (using strain-induced resistivity changes in a Wheatstone bridge embedded into a thin membrane to produce an output).

Microfabrication—using semiconductor-processing techniques to produce micron-scale mechanisms—is used for micro-actuators as well as for micro-sensors, but it is not a requirement. The defining characteristic of a micro-actuator is that it be capable of producing micron or submicron motions, regardless of the micro-actuator's own size. (There are some differences of opinion concerning these definitions. The term "micro-actuator," for example, is often used to mean "microfabricated actuator." We will keep with the definition given above.)

Micro-actuators have been studied intensely over the past few years. They are generally more difficult to design and construct than their sensor counterparts, because actuation requires more motion than does sensing. The excursion of a loudspeaker cone, for example, is on the order of millimeters, while a microphone membrane rarely moves more than a micron. Much of the effort in micro-actuation has been expended on electrostatic approaches to silicon microfabricated motors [4]. Other micro-actuators have used hydraulics, shape-memory alloys, piezoelectrics, and magnetics.

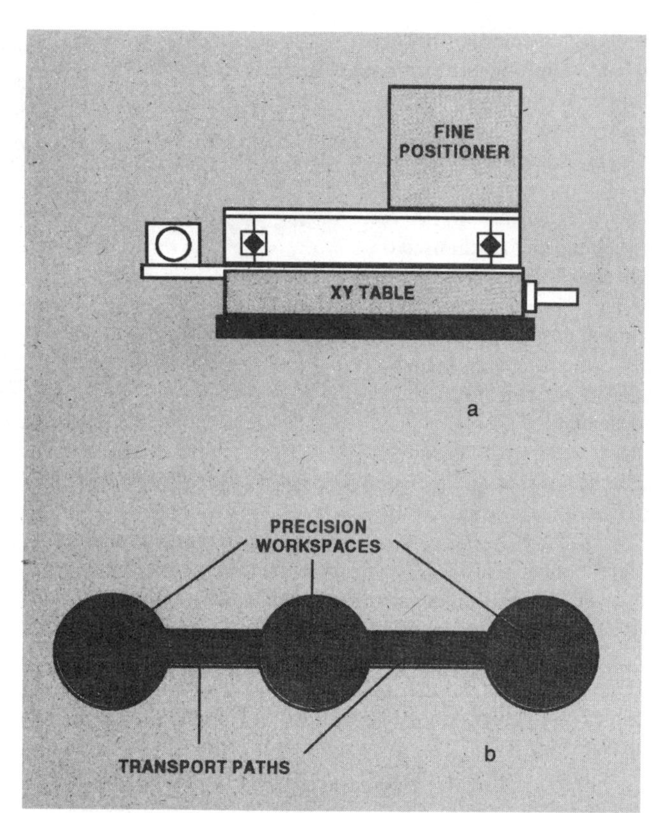

2. For applications of micro-manipulation that require high-precision positioning over a large workspace, it is often useful to couple a fine manipulator to a coarse manipulator (a). When high-resolution motion is needed only at specific locations in a large workspace, as on semiconductor fab lines, precision workspaces can be linked with pathways along which the motion is less precise (b).

Designing Systems

Actuators for micro-automation are in one of two basic forms: a fine manipulator, or a multi-fingered hand. A hand's flexibility is valuable when workpieces are of unknown shape, the environment is unconstrained, and only low-accuracy information is required. (They are well suited to exploration and part-identification tasks.) But in the highly constrained micro-automation environment, we are working with a limited set of well-characterized workpieces and require high accuracy. Under these circumstances, it is best that the actuator move as a rigid body. In addition, a fine manipulator is preferred to a multifingered hand [5]. Most of the work in micro-actuation has indeed concentrated on fine manipulation rather than development of anthropomorphic devices.

Another key issue in micro-automation system design is the geometry of the workspace over which high-resolution motion is required. Some applications, such as mask alignment of a large silicon wafer, require micron-scale resolution throughout a sizable workspace. Other applications, such as automation of wafer transport on a fabrication line, require high resolution only at well-defined sites within a large workspace.

For applications requiring high-precision positioning over a large workspace, it is often advantageous to couple a fine manipulator to a coarse manipulator (Fig. 2a). This removes the pressure to maintain extraordinarily high accuracy in the fine manipulator over a very wide range. The trade-off is that the approach requires a more complex control scheme. The amount of added complexity is determined by the amount of mechanical coupling between the coarse and fine manipulators.

For applications such as fabrication lines that require high-resolution motion only at specific locations in a large workspace, it is reasonable to design precision workspaces that are linked via pathways along which the motion is less precise (Fig. 2b). Dividing the environment into precision workspaces and transport paths requires a way of linking the two divisions. This scheme results in more complex hardware and a

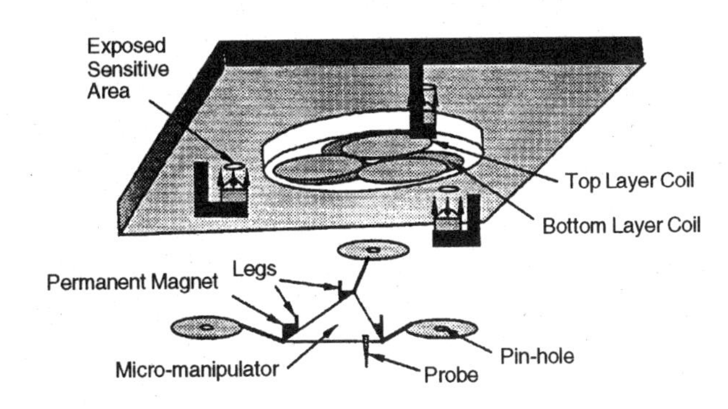

3. The University of Texas micro-manipulator uses magnetic levitation and feedback from optical sensors. It can be positioned to within 0.1 μm.

more complex control strategy, but we gain two substantial advantages. First, the workspace's size can be expanded tremendously without loss of resolution. Second, the design is very flexible because the two divisions of the workspace can be interconnected in many ways. Although it is not necessary for the transport paths and the precision workspaces to operate with the same technology, using a single approach, such as air bearings, eases the integration problem.

The Magnetic Attraction

When our work began in micro-automation at the University of Texas at Austin about five years ago, we chose magnetics technology for driving the micro-automation system. Magnetic and electrostatic approaches to microfabricated actuators have been compared widely. Three main arguments favor electrostatic approaches. First, with structural gaps on the order of a micron, the energy densities of electric fields are comparable to those of magnetic fields. Second, electrostatic devices are more efficient at very small scales. Third, only electrostatic approaches permit full integration of the electronics with the micro-actuator. These arguments are certainly correct in the context of their original assumptions: that actuators would be monolithic, microfabricated, and inexpensive. It is not at all clear, however, that these assumptions apply to the actual micro-automation systems that will be built.

The defining characteristic of micro-actuation is the ability to achieve micron-scale positioning with high speed and accuracy. We mentioned previously that the actuator itself need not be of micron scale itself. The simple micrometer, among many other devices, is capable of resolutions far smaller than the scale length of its own features. And there is no *a priori* reason for micro-actuators to be monolithic in construction.

From this perspective, magnetics deserve a fresh look. The maximum energy density of an electric field is roughly comparable to that of a magnetic field from iron-based devices. However, superconducting magnets could produce a magnetic field two orders of magnitude greater than iron-based magnets.

Proponents of electrostatic approaches to microfabricated actuators have cited their superior efficiency, an assessment based on comparisons between voltage- and current-driven systems. But for very small devices, efficiency is dominated by surface forces such as friction, not by electrical processes. Thus, the efficiency of microfabricated actuators should not depend on whether they are magnetically or electrostatically driven. In any case, the question of micro-actuator efficiency may not be important. Efficiency is only important when a device is power-starved or heat-dissipation limited. Neither is the case for the micro-actuators discussed in the literature to date.

The ability to integrate electronics with micro-actuators and micro-sensors is an interesting issue. If one uses a monolithic micron-scale electrostatic actuator or sensor, then failure to integrate the electronics can result in serious problems, because the parasitic capacitance can easily exceed the device capacitance. For larger-scale micro-actuators, this capacitance consideration is not a problem, so the ability to integrate electronics is not crucial.

Magnetic approaches yield some special benefits for semiconductor fabrication. First, the presence of high electric fields is intolerable in many tasks, such as electrical probing. Second, some process steps in semiconductor fabrication are wet rather than dry, and magnetic devices can be operated in wet environments. Third, a magnetically driven system is more compatible with clean-room operation since electric fields attract dust.

Demonstrating a Precision Workspace

We have constructed precision workspaces for demonstrating a micro-automation system (Fig. 3). The system with which we are working has four main components: precision workspaces, means of expanding the working range through transport paths or a coarse positioner, sensors, and controlling electronics. The novel portions of the system are the precision workspaces and the working-range expanders.

In the precision workspace, magnets are mounted onto a rigid-body manipulator that is suspended below a set of six air-core solenoids. Currents running through the solenoids induce forces and moments on the manipulator, which responds by moving. Permanent magnets are mounted on the manipulator so that no tether to the manipulator is necessary. The system is capable of operation in a free-floating magnetic levitation, but such operation requires real-time feedback control. We therefore use magnetic levitation with endpoint friction for stabilization. In operation, the manipulator is permitted to drop momentarily, then a set of currents is supplied to the solenoids, which drives the manipulator laterally and back up to the solid surface. (The motion between two points is similar to the motion of a grasshopper, only upside down.) Using this system, we have established motion in four degrees-of-freedom: two translations in a plane parallel to the solenoid plane, an in-plane rotation, and an out-of-plane rotation. We are able to achieve

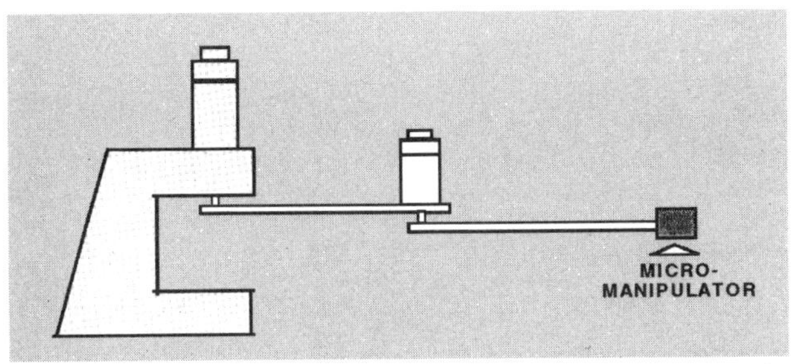

4. A micro-manipulator can be fastened to the endpoint of a robot for precision manipulation over a large workspace.

resolutions of 0.1 μm over a workspace range of roughly 400 μm x 400 μm. This resolution is the limit of our sensing system, which relies on monitoring the light shining through pinholes in the manipulator onto lateral-effect photodiodes fixed on the same structure as the driving solenoids.

This particular precision workspace is far from the optimum solution for semiconductor fabrication. Because of the presence of friction, operating the system generates particles, and the system is less precise than it might be otherwise. Miniature jeweled bearings, free-floating levitation based on real-time feedback control, superconducting levitation, and supporting leaf springs, all show great potential for eliminating friction.

Coarse/Fine Positioner

The working range of micro--actuators is generally quite small. To achieve submicron resolutions everywhere in a large workspace, we coupled the precision workspace with a two-link articulated robot (Fig. 4). The robot can position objects with an accuracy of about 100 μm, and the fine manipulator positions objects within the roughly 200 μm x 200 μm space defined by the robot endpoint. The micro-actuator serves as a tool at the robot endpoint, much like the fingers that many robots have at their endpoints. The weight of the fine manipulator itself is less than 10^{-3} multiplied by the weight of the robot arm, so for all practical purposes there is no dynamic coupling between the coarse and fine positioners. This permits them to be controlled independently.

We have used the automated coarse/fine positioners to track a very fine line. The line is a 25-μm cut made in a copper foil. The positioner tracked the line over an 8-cm length using a standard semiconductor electrical probe mounted at one of the vertices of the fine manipulator. Given the probe size and cut width, our clearance for probing was ±5 μm.

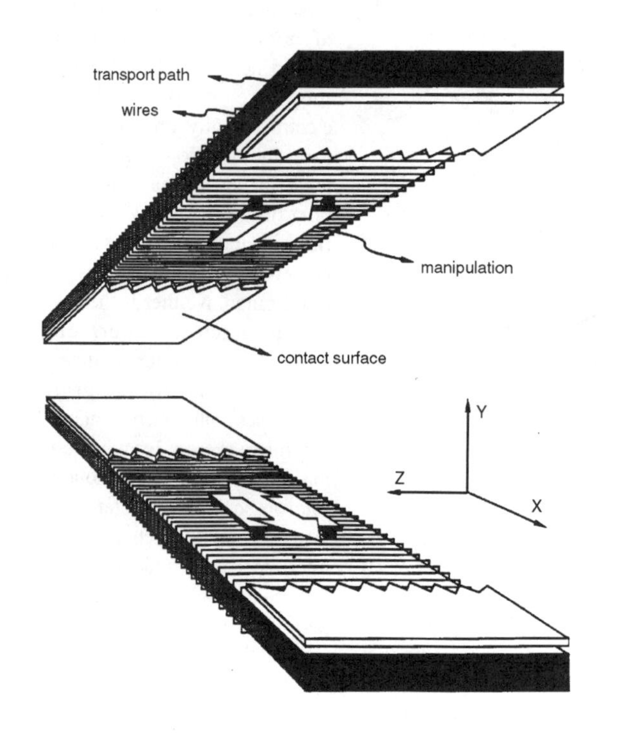

5. A micro-manipulator can also be transported rapidly along a transport path by magnetic levitation at speeds up to 30 cm/sec.

Transport Path

For micro-automation applications (such as the transport of wafers along a fabrication line) that require high resolution only in limited sections of a large workspace, we have developed a linear transport path capable of moving the fine manipulator from one precision-demanding area to another (Fig. 5). It consists of an aluminum block wrapped with a set of six repeating wires. The manipulator either rides above the path or below it, and is propelled by switching currents in the wires. This path stabilizes the manipulator's motion through the occasional use of friction, but the up-and-down motion of the manipulator is small enough to produce a very smooth ride. For our unloaded manipulator, we can generate linear speeds in excess of 30 cm/s, which is far greater than current transport speeds for wafers along a fabrication line.

The results with this first-generation path are encouraging, but the path is not suitable for our target application because of friction. A second-generation path, which uses a free-floating magnetic levitation, has been designed and constructed but has not yet been tested.

A number of crucial steps are still required before this micro-automation system will be suitable for use. We must demonstrate that the transport path can be linked with a

Table 1: Application Attributes

Application	Degrees of Freedom	Accuracy	Speed?	Scale Length
Positioning: Mask Alignment	5-6	$5(10^{-7})$	Yes	20 cm
Positioning: Wafer Transport	1	N/A	Yes	meters
Analytical Probing	3 min.	10^{-3}	No	2-3 cm
Mechanical Probing	3 min.	10^{-5}	Yes	2-3 cm

precision workspace without intervention by other hardware or by humans. Once this is achieved, the vision of an automated micro-fabrication line (in which material is loaded at end and processed parts are removed at the other end) will be possible. As we progress in the demonstration of the hardware necessary for micro--automation, we will turn our attention more to the software issues. This will permit the development of workspaces relying on free-floating magnetic levitation.

The Future of Micro-Automation

A growing proportion of all manufactured devices relies on features that are of micron scale. In order for this segment of our industry to continue development, it is absolutely necessary that we find a means to automate all of the processes involved in fabrication. The main obstacle to automation at very small scales has been, and continues to be, the dominance of friction. The research currently under way, particularly in the U.S., Germany, and Japan, are producing new micro-sensors and micro-actuators. These are the devices that hold the key to micro-automation.

Most current research efforts emphasize the transducer technology itself, rather than applications to a specific task. As the technology matures, basic research will make room for applied research, and micro-automation will be a clear priority. While it is difficult to predict when the emphasis on micro-automation systems will take hold, Japan's MITI recently announced that it is targeting the development of micro-sensors and micro-actuators, and this activity should serve to stimulate the field dramatically.

Acknowledgments

I am pleased to acknowledge the assistance of the members of the Micro-Automation Group of the Mechanical Engineering Department of The University of Texas. The magnetic-levitation-based, micro-automation-system work has been supported by Texas Instruments, the Semiconductor Research Corp., the Bosque Foundation of Central Texas, and the National Science Foundation. **CD**

Ilene J. Busch-Vishniac is an Associate Professor of Mechanical Engineering at The University of Texas at Austin. She is the 1987 recipient of the Lindsay Award of the Acoustical Society of America, and was named an NSF Presidential Young Investigator in 1985.

References

1. K. Gabriel, J. Jarvis, and W. Trimmer (eds.), "Small Machines, Large Opportunities: A Report on the Emerging Field of Microdynamics," National Science Foundation, 1988.
2. I. J. Busch-Vishniac, "Applications of Magnetic Levitation-Based Micro-Automation in Semiconductor Manufacturing," *IEEE Trans. Semiconductor Fab.*, **T-SMN 3**, p. 109-115, 1990.
3. B. J. Korites, *Microsensors*, Kern, Rockland, MA, 1987.
4. S. D. Senturia, "Microsensors vs. ICs: A Study in Contrasts," *Circuits and Devices*, **6**, p. 20-38, 1990.
5. R. L. Hollis, A. P. Allan, and S. Salcudean, "A Six Degree-of-Freedom Magnetically Levitated Variable Compliance Fine Motion Wrist," 4th ISIR, Santa Cruz, CA, 1987.

Robotics Applications in Electronics Manufacturing

John H. Powers, Jr.

Robot Basics

To appreciate how robotics is being applied in electronics manufacturing, we should first address some fundamental considerations such as what a robot really is, why we use them, and where we use them. The term *robot* means different things to different people and has been fantasized in science fiction for years. To the manufacturing industry, however, robots are a new form of automated tools that can be programmed to perform a wide variety of tasks using humanlike capabilities. The Robot Institute of America has developed a formal definition for the industry: "a robot is a reprogrammable, multifunctional manipulator designed to move material, parts, tools, or specialized devices through a variety of tasks." There are many things that are called robots being used in industry today that differ substantially in their capabilities; they do not all fit that definition. The most elementary "robots" have very limited flexibility and are often used for specific, simple, repetitive tasks such as loading and unloading a tool. More sophisticated robots are capable of complex motion and sensing functions that permit them to perform much more difficult tasks, such as assembly operations.

Robots are used in manufacturing for a variety of reasons, most of which can be related to a few fundamental objectives. For robots to be productive, they should, in some way, help manufacturing to improve its costs, quality, or ability to meet schedules. In most cases, particularly in relatively simple applications, a robot is justified economically by reducing the direct labor cost associated with a particular operation. More advanced applications usually take advantage of a robot's capabilities to do things that are difficult or impossible for humans, such as handling large amounts of data, precision movement, or integrated process control. Other considerations that may be involved in robotics applications include performance of tasks that are potentially hazardous to humans or where, for technological reasons, human contact is not desirable. There are many such applications to be found in the electronics industry in particular.

Robots can be used in a wide variety of manufacturing applications and in many different types of manufacturing operations. It might be helpful to think of three major types of manufacturing environments, each with its own peculiar characteristics and needs. *Process* manufacturing involves a continuous flow of materials through a series of sequential operations that transform these materials into a product. In this environment, it is essential to have process, equipment, and material control at every step. Perhaps the best example of this type of environment in the electronics industry is semiconductor manufacturing. Other examples could be the manufacture of ceramic substrates or magnetic disks. *Part* manufacturing, on the other hand, is a fabrication process that usually involves discrete operations on batches of objects to create parts or subassemblies. Examples of this would, of course, include electrical and electronic parts and component assemblies. Finally, *product* manufacturing can be thought of as the final process that assembles parts and subassemblies into a functional end product, such as a machine or piece of equipment. In electronics, this of course includes such products as instruments, home entertainment equipment, communications equipment, and computers.

What Are Typical Applications Today?

Robots are already being used in a wide variety of electronics manufacturing applications. Following are some examples from process, part, and product manufacturing operations in IBM. Although these are IBM examples, they are not by their nature unique to IBM or even to the data-processing industry, but serve as illustrations of the variety of types of applications that exist in electronics manufacturing today. Hopefully, these examples demonstrate that robots can be economically and productively employed in the many different simple, as well as complex, tasks found in our industry.

Process Manufacturing

There are obviously many process operations in electronics manufacturing; so there should be ample opportunity to find ways to use robots to reduce costs, improve quality, or provide control. Perhaps the type of application that is the most pervasive in a process environment is materials handling. Note that the type of robots that are often used in such applications are elementary, limited-function devices. Here are several examples from three different types of processes that can be found in our industry: semiconductor, packaging, and disk manufacturing.

Figure 1 shows a wafer handler in an automated dipping station. The handler is comprised of a pair of grippers that can be programmed to move individual

Reprinted from IEEE Components, Hybrids, and Manufacturing Technology Society Newsletter.

Reprinted from *IEEE Circuits and Devices Magazine*, Vol. 1, No. 1, pp. 63-66, January 1985.

wafers through a series of etching and cleaning steps, whose sequence and times can be varied as required by the specific process of the devices involved.

Figure 2 shows a common anthropomorphic-type robot being used to load and unload greensheets of multilayer ceramic substrates at a laser-scanning inspection station.

Figure 3 illustrates another type of pick-and-place robot being used to handle aluminum substrates in the fabrication of magnetic disks.

Part Manufacturing

The production of electronic parts and assemblies has many applications that lend themselves to automation. In the past, such operations were often very labor intensive; today, to be cost-competitive and to provide the kinds of control required by the technologies involved, robots are being employed to perform some of these tasks. A common area for this type of manufacturing in the electronics industry is circuit packaging, that is, printed circuit card and board assemblies. Here are two examples from different stages of a board assembly process.

Figure 4 shows a two-arm robot being used to solder tiny twisted pairs of wire to the back of a large board for overflow and engineering change connections.

Figure 5 illustrates a robotic system designed to insert electronic components onto printed circuit cards, automatically.

Product Manufacturing

The manufacture of finished electronic products or machines usually involves a lot of assembly operations that have traditionally been manual. Here, too, the robot is finding a place, because it is capable of being programmed to perform a variety of complex tasks in a batch manufacturing environment. This is just one such example.

Figure 6 is a picture of a robot performing the final assembly of an entire data-processing terminal.

These are just some of the numerous applications where robots can be employed in electronics manufacturing today. Although they vary in their complexity, the challenge in many of these applications is the development of the data-handling system and the supporting fixturing and tools.

What Are the Trends for the Future?

Even though robots have already been widely used in manufacturing applications, they are far from reaching their full potential. As the technology advances and industry gains more experience and success in using them, significant changes and improvements will be realized in terms of robotics capabilities and applications. Following are some general observations and examples of these trends.

Features and Functions

Most of the robots in manufacturing today are relatively simple, with limited functions such as loading and unloading tools. Recent developments in robotics and computer technology have made substantially more complex tasks possible. Advances in both hardware and software are occurring rapidly, and only time is necessary for industry to take advantage of them in actual production applications. The major thrust of these advances is to provide additional features and functions in robotics systems that will permit them to be used in a wider variety of tasks, particularly those that are more complex and typically performed manually today. Some of the key areas of activity to achieve this follow.

Fig. 1 Wafer Handler.

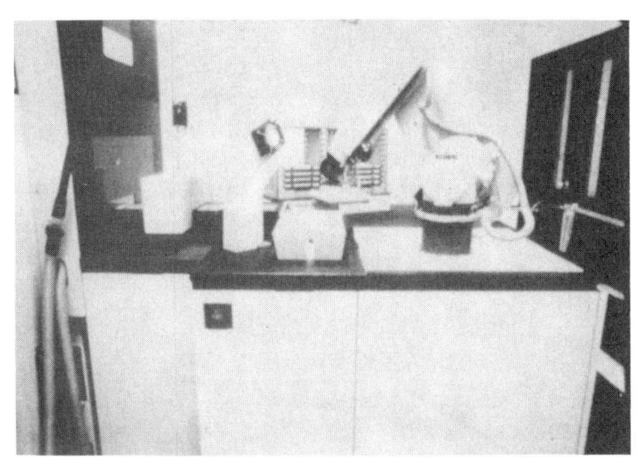

Fig. 2 Greensheet Inspection Loader.

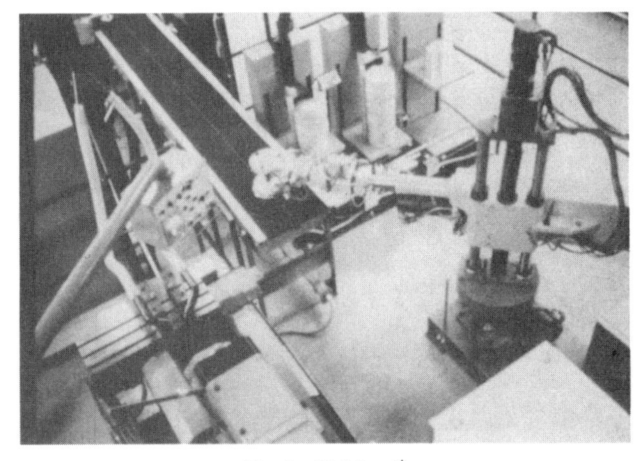

Fig. 3 Disk Loader.

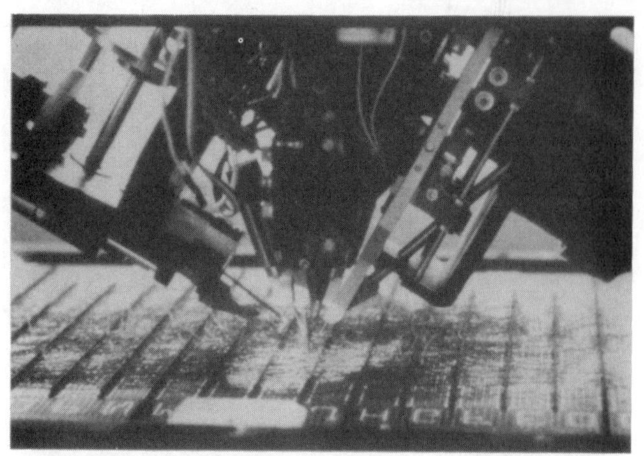

Fig. 4 Wire Bonder.

Fig. 5 Component Inserter (multiple exposu

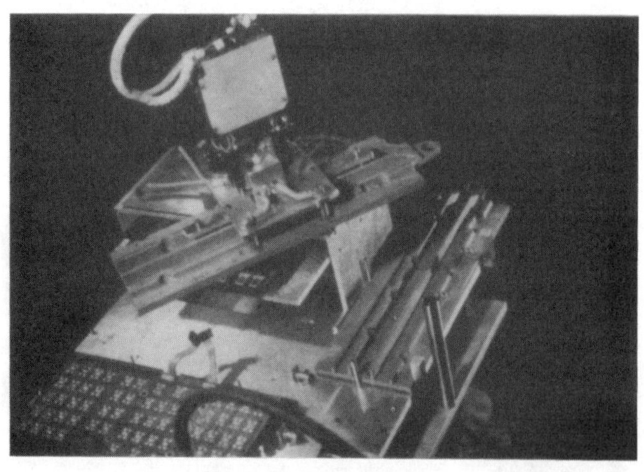

Fig. 6 Terminal Assembly.

Precision: To perform complex tasks, robots will have to be able to move in very precise, repeatable paths and to specific points with great accuracy. This will require advances in programming and control systems, as well as in the mechanical design of the robots themselves. It should be noted that a high degree of absolute accuracy may not always be necessary if the robot's position can be calibrated to assure repeatability. Some robots today have demonstrated a repeatability of less than 0.001 inch.

Sensors: Most robots today are programmed to make specific moves from point to point, but capabilities are being developed to add sensing functions that will permit robots to adjust to their environment and to situations in their tasks. Tactile sensing in terms of proximity, pressure, and force can be achieved using transducers and strain gages. Sensors have also been developed to respond to a variety of stimuli outside the visible spectrum such as thermal, ultrasonic, and x-ray. These functions, to be used successfully on robots, involve not only advances in the sensor elements themselves but also in the instrumentation and data-processing systems that control them.

Vision: One of the human senses that we often take for granted, but which most robots today lack, is vision. The ability to "see" objects that you are working on and adjust for differences in size or shape can be a significant advantage in a complex manufacturing task. Image recognition systems are now being developed and applied to a limited extent, which can be tied to robotics systems to give them the capability to select, retrieve, or avoid objects. Most current systems are relatively simplistic in that the vision capability is usually two dimensional with little range of size, shape, or contrast. More advanced systems will be able to recognize and interpret three-dimensional shapes, but both the image processing job and optical sensing technology that must be perfected are significant challenges, particularly to be economical.

Artificial intelligence: Once in the realm of science fiction, but now becoming a buzz word in "high tech" circles, artificial intelligence encompasses a broad field of research and development involving recognition, adaption, and learning systems. This, therefore, also includes the sensor and vision technologies just mentioned, but goes further to more complex data-processing functions that use such senses as inputs. The key effort in this area is to develop high-level programming languages and powerful but economical data processors that can be used to provide "knowledge-based" or "expert" systems. Such systems will have the capability to recognize and interpret data from sensory inputs and adapt their responses to instructions or experience. Such capabilities will make many robotics applications feasible, which are not possible today, such as complex assembly tasks.

Applications

As these advanced features and functions become available to industry, they will be put to practical applications in manufacturing. Remembering that the robot is still just one of the tools in a data-driven manufacturing environment, the principle trends in application will be in the following areas.

Complex functions: By using advanced sensing capabilities, robots in electronics manufacturing will do more than just materials-handling tasks. Assembly jobs, in particular, will receive the greatest attention

Fig. 7 QTAT Line.

and yield the largest benefit. Complex, batch-oriented tasks will no longer be labor intensive, expensive, and difficult to control.

Clean room robots: A specific area of need in the electronics manufacturing industry is to develop high-performance, relatively low-cost robots that can operate in a clean room environment. There are many potential applications that could use such robots, even today, particularly for wafer-handling tasks such as in process, alignment, and test operations. This will require small, high-speed, precise robots with electric drive systems and special grippers that will not contaminate the product they are handling; these robots will need a "class 10" manufacturing environment.

Integrated manufacturing systems: Robots, by themselves, no matter how sophisticated, will not have as major an impact on manufacturing productivity as when they are part of an integrated system that controls the technical, logistical, and administrative data of the production operation. Such systems will tie tools and processes together by automating the data flow as well as the physical movement and fabrication of products.

Factory of the future: Most visions of future factories incorporate all of the trends and technologies mentioned above, but the common ingredient is an integrated or systems approach to manufacturing. The typical image of an "unmanned factory" or a "flexible manufacturing system" is that of a machining operation, since that industry has made the most progress in this direction to date as an outgrowth of its numerical control heritage. However, the same principles and potential apply to electronics manufacturing, and we are beginning to see some examples emerge. Several years ago, there was an automated semiconductor manufacturing facility installed, which integrated the personalization processes of advanced bipolar devices into a computer-controlled, "quick turnaround time" line. As shown in Fig. 7, wafers are handled individually on this QTAT line and routed between tools by computer control on airtracks. Among the many process steps is the individual personalization of each wafer using electrobeam lithography, which requires a large amount of complex data, driven by a central engineering design system. This is a "flexible manufacturing system," which was developed to fabricate multiple, customized device designs in low volume. It is a "factory of the future" in the electronics industry today, which employs automated tools and robots and was only made possible by an integrated approach to computer-aided manufacturing.

Computers and robots have become major products of the electronics industry—but they are also becoming its key tools!

Sights and Sounds of Chaos

Leon O. Chua and Rabinder N. Madan

Abstract

Chaos *is a ubiquitous and robust phenomenon that can occur in almost any man-made or natural system where nonlinearity is present. Any device, circuit, or system can become chaotic due to component aging. In order to stay away from a failure boundary that leads to chaos, it is essential to understand the nonlinear dynamics and the many possible bifurcation routes leading to chaos.*

Chaos is introduced in this paper via actual circuit experiments. Several well-known chaotic circuits are used as vehicles to introduce the many exotic phenomena associated with chaos. The readers are encouraged to build some of these circuits and to be thrilled by the sight of their awesome strange attractors and their sound of chaos.

Introduction

It is ironic that there are presently more laymen who know something about *chaos* than engineers who are responsible for the integrity of their products and who cannot afford to see them fail because of the onset of this ubiquitous nonlinear phenomenon. This unfortunate state of affairs can be explained, on the one hand, by the proliferation of recent expository articles on chaos in newspapers and magazines and, on the other hand, by the sheer lack of interest among tradition-bound engineers to learn about a phenomenon where no sound design would exhibit, at least on paper. Our objective in this article is to demonstrate, through experiments on simple chaotic circuits, that chaos is an exceedingly robust phenomenon and can occur in virtually any physical circuit, device, or system, if driven by a sufficiently strong signal or if one or more parameters are allowed to drift and cross over some boundary in the parameter space, henceforth called the *failure boundary*.

As in motherhood, chaos is an inevitable offspring of *nonlinearity*, which is present in all physical circuits, devices, and systems, to a lesser or greater extent. Just as it would be impossible to have birth control without an understanding of the dynamics of the conceiving process, so, too, would it be impossible to control chaos, let alone prevent it, without an understanding of the various chaos-producing mechanisms.

Or, using the parlance in the chaotic literature, chaos can only be avoided if one has a thorough understanding of the typical bifurcation routes to chaos and is able to identify their telltale manifestations.

On the other hand, there are, in fact, applications when chaos is deliberately sought and designed into a system. For instance, all pseudorandom number generators used in computer centers are among the simplest examples of benign chaotic systems.

What Is Chaos?

For electrical engineers who feel most at home with the traditional sinusoidal steady-state analysis, chaos is best defined in terms of the long-term behavior, or steady-state response, of a circuit, device, or system after the initial transient has decayed to zero, following a change in state, e.g., the opening or closing of a switch, or the application of a signal. For those well-entrenched in classical circuit theory, the steady-state response can assume only one of the following forms:

1. *DC Steady State:* All trajectories in the state space approach an equilibrium point, which need not coincide with the origin.
2. *Periodic Response:* All solutions converge to a periodic waveform having the *same* frequency ω_o as the input signal frequency ω_s; i.e., $\omega_o = \omega_s$.
3. *Subharmonic Response:* All solutions converge to a periodic waveform whose frequency ω_o is a *submultiple* of the input signal frequency ω_s; i.e., $\omega_o = \omega_s/n$, where $n > 1$.
4. *Superharmonic Response:* All solutions converge to a periodic waveform whose frequency ω_o is a *multiple* of the input signal frequency ω_s; i.e., $\omega_o = n\omega_s$, where $n > 1$.
5. *Almost-Periodic (Quasiperiodic) Response:* All solution waveforms are made of periodic components whose fundamental frequencies are *incommensurable* and, hence, are not periodic. For example, $x(t) = \sin \omega_s t + \sin(\sqrt{2}\,\omega_s)t$ is almost periodic since $\sqrt{2}$ is an irrational number.

If we examine the frequency spectrum of the above classes of steady-state waveforms using a spectrum analyzer, we will see that they all exhibit a *discrete* spectrum with one narrow spike centered at each component frequency. Up until recently, few engineers would believe that anything more complicated than the above behavior is possible. In fact, so well-entrenched was this belief that the sheer observation of a steady-state waveform, which gave rise to a *continuous* rather than discrete spectrum, was invariably interpreted by its beholder as a physical noise.

The literature contains many such anecdotes, the most famous of which is contained in a brief sentence, which Van de Pol published in *Nature* in 1927 [1] when he dismissed what we now know is a chaotic phenomenon [2] as an "irregular noise."

We now know that many nonlinear circuits, devices, and systems can exhibit a steady-state response far more complicated than those listed above. In fact, so many exceedingly complex phenomena have been reported in the recent literature that no systematic classification is likely in the foreseeable future. It is for this reason that the following fuzzy, nonscientific but thought-provoking name has been used to embrace all such phenomena: Any physical, economic, or social system whose dynamics evolve in accordance to some *deterministic* law is said to be *chaotic* if its steady-state behavior belongs to *none* of the above-cited traditional categories.

It follows from this definition that any *deterministic* system of equations of motion driven by either a dc or deter-

Reprinted from *IEEE Circuits and Devices Magazine*, Vol. 4, No. 1, pp. 3-13, January 1988.

ministic (not random) input signal is *chaotic* if one or more of its solution waveforms exhibit a *continuous* frequency spectrum.

Numerous chaotic dynamical systems have been reported in the literature ever since Lorenz published his seminal paper in 1963 [3]. Since most of these chaotic systems are hypothetical rather than physical, there were many critics and doubting Thomases who objected adamantly to their validity on the ground that the reported chaotic phenomena could be an artifact of the computation algorithm or round-off errors.

To overcome such criticisms in this paper, only operational *chaotic* electronic circuits will be presented. Moreover, they are rather simple and inexpensive so that the interested reader could easily build them and verify the chaotic phenomena with an oscilloscope and spectrum analyzer. In addition, since several of these circuits operate at the audio spectrum, one could hear the sound of chaos as well. We hope that the sights and sounds of chaos from these simple circuits will be instructive if not entertaining. Those interested in a more detailed mathematical analysis of these phenomena are referred to the original papers listed in the references, as well as to the August 1987 *Special Issue of the Proceedings of the IEEE on Chaotic Systems.*

Simplest Chaotic Autonomous Circuits

A circuit is said to be of *order n* if it contains "n" energy-storage elements and *autonomous* if it contains no ac sources [4]. The most complicated behavior that an autonomous circuit of order less than 3 can exhibit is a *periodic* oscillation. Hence, the simplest autonomous circuit that can become *chaotic* must be at least of order 3. A textbook example of such a circuit is shown in Fig. 1(a): it is made of two linear capacitors, a linear inductor, a linear resistor, and a *nonlinear* resistor described by the v-i curve shown in Fig. 1(b). This nonlinear resistor can be fabricated by using two op amps [5], an op amp and two diodes [6], or two transistors and two diodes [7]. The three energy-storage elements give rise to the following state equations:

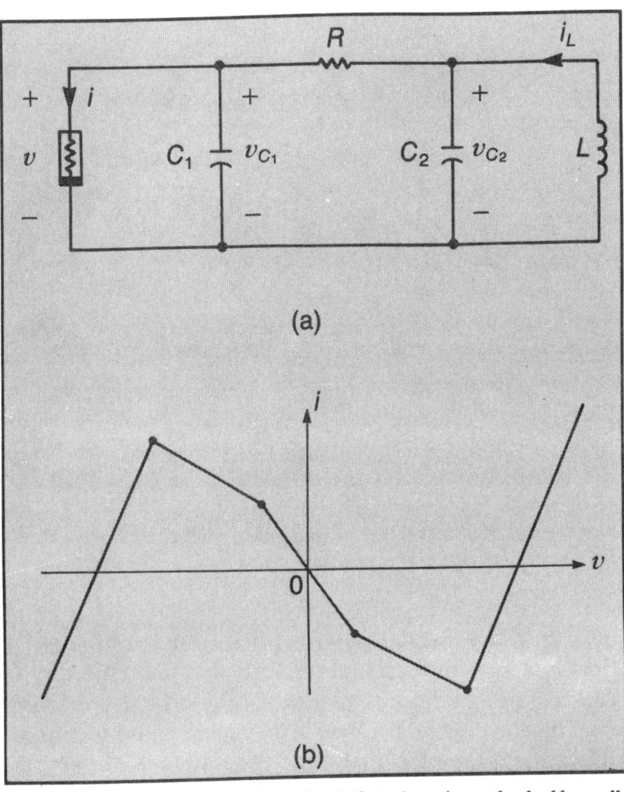

Fig. 1 (a) The autonomous Chua circuit that gives rise to the double scroll. (b) The voltage-current characteristic of the nonlinear element.

$$\dot{v}_{C_1} = \frac{1}{RC_1}(v_{C_2} - v_{C_1}) - \frac{1}{C_1}f(v_{C_1})$$

$$\dot{v}_{C_2} = \frac{1}{RC_2}(v_{C_1} - v_{C_2}) + \frac{1}{C_2}i_L$$

$$i_L = -\frac{1}{L}v_{C_2}$$

The *chaotic* behavior of this equation is now rigorously established, and numerous papers addressing various analytical aspects of the equation have been published [8]–[15]. In fact, the above equation is, to the best of our knowledge, the only chaotic system whose *mathematical analysis* agrees completely with both *experimental observations* and *computer simulations.*

Strange Attractors

Starting from any initial state $[v_{C_1}(0), v_{C_2}(0), i_L(0)]$ near the origin, and after discarding the initial transients, the solution $[v_{C_1}(t), v_{C_2}(t), i_L(t)]$ of the above equation traces out some loci Γ in the $v_{C_1} - v_{C_2} - i_L$ state space, which we call the *steady-state trajectory*. If the steady-state solution is periodic, then Γ would consist of a simple closed loop. Over a wide range of parameters, however, both experiments and computer simulations revealed that Γ seems to keep moving along an endless imaginary track, which criss-crosses but does not touch itself infinitely often, so as to stay confined within some finite preordained region in space. Moreover, while never repeating any point visited earlier, Γ is seen to return *infinitely often* to the vicinity of *every* point it has previously traversed. This strange space-filling trajectory is represented accurately by the colorful fiberglass model shown on the front cover. It is an example of a geometrical object called a *strange attractor* in the literature on *chaos* [16].

A strange attractor is, therefore, just a fancy name to denote a steady-state behavior, which belongs to none of the traditional types alluded to earlier. It is called an "attractor" because any trajectory emanating from an initial point outside of Γ is quickly "sucked" into its interior. It is "strange" because, given any point P belonging to the attractor, *any* trajectory would return to an arbitrarily close vicinity of P, if not actually passing through P, *infinitely* often. Since the trajectory cannot intersect either with itself or with another trajectory, it must weave itself in an incredibly intricate fashion in order to stay confined within a finite volume in space.

If we examine the intersection points between the strange attractor and any horizontal plane that cuts across it, we will find them to cluster along two concentric spirals somewhat reminiscent of that depicted in Van Gogh's "Starry Night." This unique pattern is found at all cross sections, from top to bottom, and therefore gives rise to the name *double scroll*. A three-dimensional perspective of the double scroll is shown in Fig. 2 along with a black shadow representing one of its many projections. To emphasize its "dual" spiral cross sections, an artist's rendi-

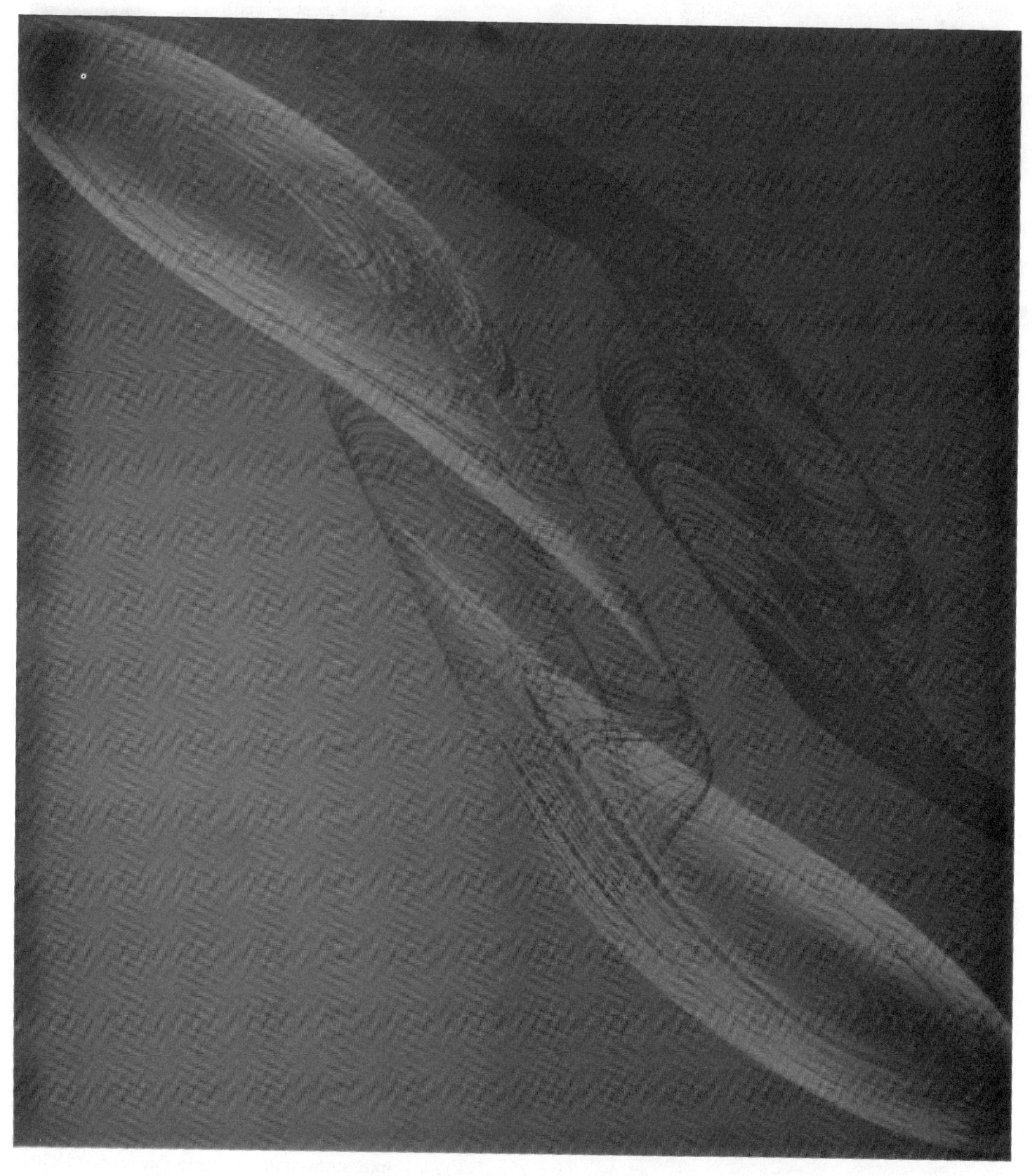

Fig. 2 A three-dimensional perspective of the double scroll and its shadow on the back wall cast by a beam of light in front of it. The double scroll consists of a single trajectory generated by integrating the nonlinear differential equations over a very long period of time. The photo was taken by K. Ayaki and T. Matsumoto.

tion of the double scroll, which highlights only the key geometrical features, is shown in Fig. 3.

Extreme Sensitivity

A typical manifestation of chaos is the *extreme sensitivity* of nearby trajectories to initial conditions. For the double scroll, this means that two trajectories starting near each other at $t = 0$, would, after a sufficiently long period of time, diverge from each other such that one trajectory is coasting near the upper hole while the other is revolving near the lower hole. Because of inevitable round-off errors, this property implies that no long-term chaotic trajectory obtained from computer simulation can represent a true trajectory. Consequently, such trajectories are in fact called *pseudo orbits*.

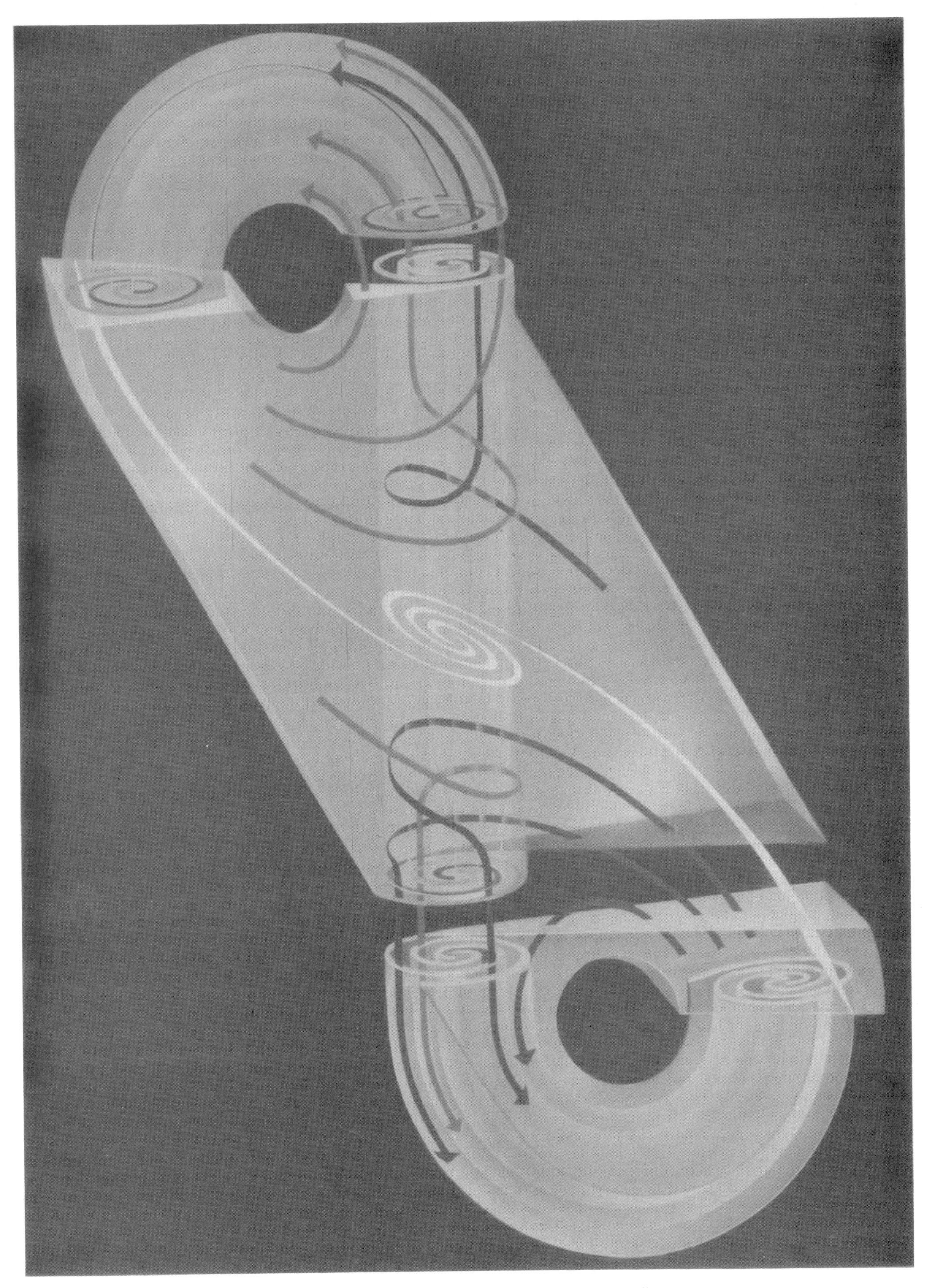

Fig. 3 A caricature depicting the anatomy of the double scroll.

While pseudo orbits may represent a figment of one's imagination, they are all sucked into the same confining volume in space, which we called a *strange attractor*. In other words, strange attractors are not sensitive to initial conditions and represent a truly meaningful picture of the steady-state behavior of chaotic systems.

Several other operational *chaotic* autonomous third-order circuits have been reported in the literature [17]–[21], and no doubt many more will be identified in the future. Though more complicated than the preceding circuit, they can all be built using inexpensive off-the-shelf components. Since they all use relatively low-frequency components, such as op amps, their strange attractors, like the double scroll, are not only visible but also audible. We will briefly describe one such circuit for historical reasons.

Lorenz Circuit

The circuit shown in Fig. 4(a) is called the *Lorenz circuit* [21] because its strange attractor, shown in Fig. 4(b), is qualitatively similar to the classic Lorenz "butterfly" attractor [3]. Details for the construction of this circuit are given in [21]. Although the laboratory circuit version of Fig. 4(a) is quite involved, it is the only known physical system that actually exhibits this famous attractor. The original Lorenz equation [3] associated with this attractor represents only a hypothetical mathematical model of atmospheric turbulence. It was derived from a version of the Navier-Stokes nonlinear partial-differential equation used for weather prediction after making many drastic, if not unrealistic, simplifying assumptions.

Because the Lorenz equation is the most widely referenced equation in the literature on chaos, and because a completely rigorous analysis and characterization of the original Lorenz attractor has not yet been achieved [22], the circuit in Fig. 4(a) could be used as a real-time analog computer to investigate the chaotic behavior of the Lorenz equation. Among other things, our laboratory observations have confirmed that neighboring trajectories associated with the butterfly-shape Lorenz attractor in Fig. 4(b) are extremely sensitive to initial conditions. This observation is, in fact, the basis of the following amusing metaphor usually referred to as the "butterfly effect":

Assume that the Lorenz equation predicted calm and sunshine in Katmandu a week from now, using the present weather data from San Francisco as the initial condition. Suppose this initial condition was changed because of the flapping of the wings of a butterfly in the jet stream above San Francisco bay. Then it follows from the extreme sensitivity property of the Lorenz equation that such an infinitesimal change in the initial condition is sufficient to change the predicted weather in Katmandu from sunshine to storm.

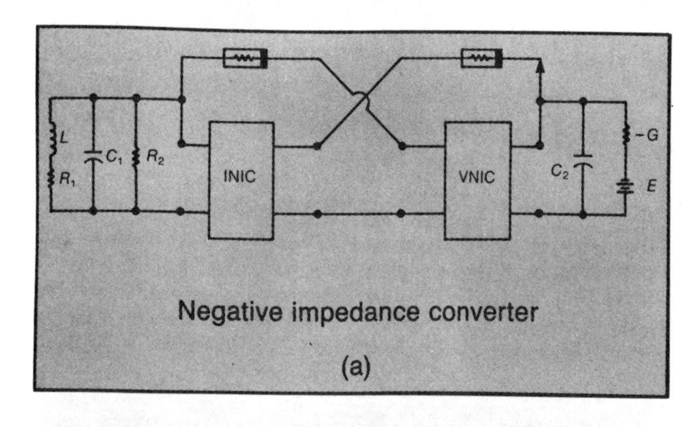

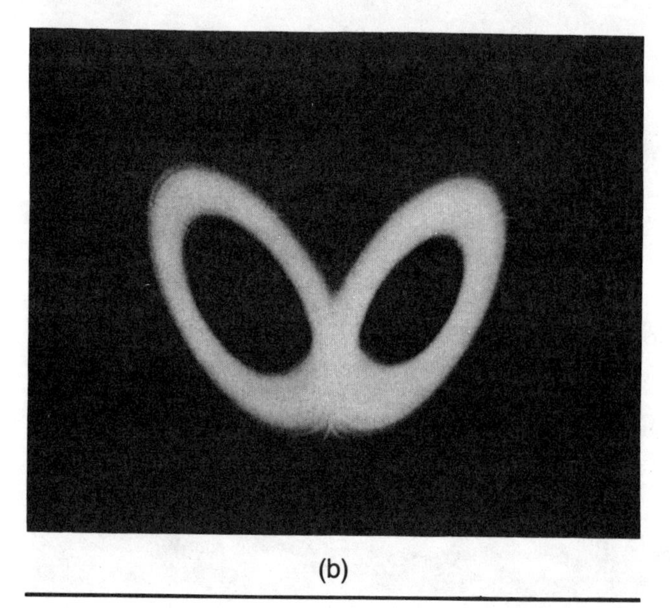

Fig. 4 (a) The Lorenz circuit. (b) The "butterfly" Lorenz attractor.

Lyapunov Exponents

The degree of sensitivity of neighboring trajectories of a third-order autonomous system can be measured by three real numbers called the *Lyapunov exponents*. They can be accurately calculated either from experimental data or from computer simulations. If a system is not chaotic and all trajectories converge to an equilibrium point Q, then the Lyapunov exponents degenerate into the real part of the three eigenvalues associated with the linearized system about Q. In this case, all three Lyapunov exponents are negative numbers.

For chaotic systems, a negative Lyapunov exponent implies that nearby trajectories converge toward each other along the coordinate direction associated with this exponent. Conversely, a positive Lyapunov exponent implies exponential divergence of neighboring trajectories along its associated coordinate direction. A third-order *chaotic* autonomous system has exactly one negative, one zero, and one positive Lyapunov exponent. For the double scroll studied in [6], these three exponents are given by -1.78, 0, and 0.23, respectively.

Fractals

We close this section by pointing out that the three-dimensional fiberglass model of the double scroll on the front cover represents only a myopic view of the strange attractor. In fact, the true double scroll consists of a much more intricate if not bizarre internal structure that resembles a pumicelike object punctuated by infinitely many infinitesimal cavities. Examining the cross sections with a microscope of higher resolving power would reveal a self-similar layered structure, which resembles that found in a French pastry or *Baklava*. Such objects are called *fractals* by Mandelbrot [23]. Since the double scroll is a fractal, its dimension *cannot* be equal to 3 because it is full of tiny holes.

Since it occupies volume in the three-dimensional space, its dimension ought to be greater than 2. One mathematically rigorous way to measure the dimension of a fractal is called the *Lyapunov dimension:* it is a fraction between 2 and 3. For the double scroll analyzed in [6], its Lyapunov dimension was calculated to be equal to 2.13. In contrast with the Lyapunov dimension of 2.06 calculated from a typical Lorenz attractor, one could say intuitively, and with some intellectual satisfaction, that the double scroll is "thicker" than the Lorenz attractor.

Simplest Chaotic Nonautonomous Circuit

A circuit is said to be nonautonomous if it is driven by at least one ac signal. A first-order nonautonomous circuit cannot be chaotic because its state equation can be transformed into an equivalent second-order autonomous system. However, a nonautonomous second-order circuit can become chaotic.

The simplest chaotic nonautonomous second-order circuit is the single-loop *RL*-diode series circuit shown in Fig. 5(a). The chaotic nature of this circuit was first reported by Linsay [24]. Since Linsay's paper contains only *experimental* results, it is not clear which pn-junction model must be chosen in order to duplicate the experimental results. Using the conventional two-element model from SPICE, Azzous et al. were able to reproduce qualitatively the chaotic behavior of this circuit [25], [26].

Subsequently, Matsumoto, Chua, and Tanaka discovered that the circuit of Fig. 5(a) can exhibit chaos when the simplest possible dynamic diode model is chosen; namely, a single *nonlinear* capacitor characterized by a two-segment voltage-versus-charge characteristic, as shown in Fig. 5(b). Consequently, the single-loop *RLC* circuit in Fig. 5(b) represents the *simplest* nonautonomous circuit that can become chaotic [27]. The utter simplicity of this circuit and our embarrassing inability to explain its chaotic behavior rigorously reveals our rather primitive state of the art on chaotic systems. Some recent progress has been made, however, on this circuit, and the interested reader is referred to [28] for details. For now, we will focus our attention on the behavior of this circuit as we tune a single parameter, namely, the *amplitude* E of the sinusoidal input waveform while keeping its frequency $\omega = \omega_s$, and all other parameters, fixed.

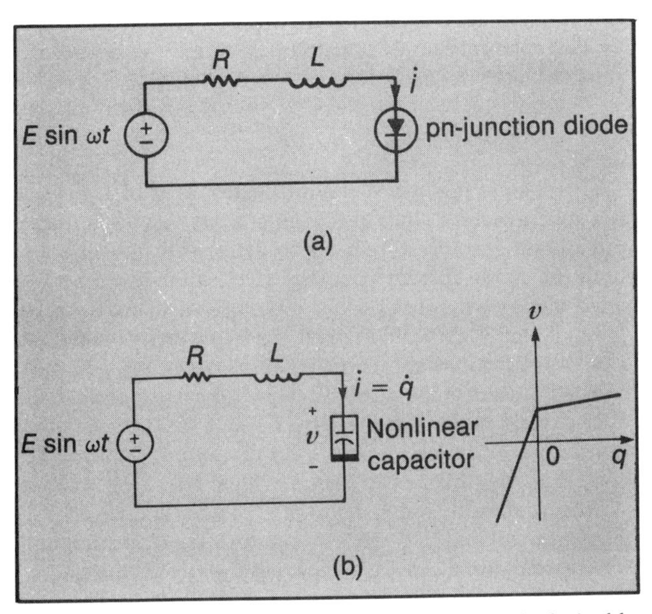

Fig. 5 (a) Series RL-diode circuit. (b) Series RL-C circuit obtained by replacing the pn-junction diode by a nonlinear capacitor characterized by a two-segment piecewise-linear q-v characteristic.

If we sample the diode current $i(t)$ in the *RL*-diode circuit of Fig. 5(a) at a frequency equal to the sinusoidal input frequency ω_s while slowly increasing the amplitude E, we would obtain the *one-parameter bifurcation diagram* shown in Fig. 6. Here, the horizontal axis corresponds to the input *amplitude* E in volts and the vertical axis corresponds to the instantaneous value of the current $i(t)$ measured once every T seconds, where $T = 2\pi/\omega_s$.

If, at some value $E = E_1$, the current response $i(t)$ is periodic at the same frequency as ω_s; then the above sampling process would always yield the same value and only *one* point would appear at $E = E_1$. If, however, at some value $E = E_2$, $i(t)$ is periodic at another frequency ω_s/n and, hence, has a period nT, then the sampled data will not become repetitive until n samples later. Consequently, at $E = E_2$, we would observe "n" distinct points situated along a vertical line centered at E_2. In this case, we say the circuit has an nth-order *subharmonic* response.

Typically, a response corresponding to an nth-order subharmonic, $n = 1, 2, \ldots$, will persist over some nonempty interval $[E_1, E_2]$ so that the bifurcation diagram within this interval would consist of n contiguous branches.

As the order n increases to infinity, the period of the response $i(t)$ tends to infinity and $i(t)$ ceases to be periodic. This limiting situation is manifested in the bifurcation tree of Fig. 6 as a *continuum of points centered at the amplitude E where the samples are taken*. This aperiodic waveform typically gives rise to a *continuous* frequency spectrum and is, therefore, *chaotic*. Again, since such behavior occurs typically over some nonempty interval $[E_1, E_2]$, an entire area above such an interval will be showered with points. Four such chaotic regimes can be identified from Fig. 6.

The bifurcation tree in Fig. 6 resembles an inverted christmas tree with heavy foliage (fuzzy region full of points) interspersed between branches whose number increases consecutively toward the right. For small values of the input amplitude E, we observe a single branch coinciding with the tree stem. This means that over the interval $[0, E]$ where the stem appears, the circuit of Fig. 5(a) has a unique steady-state response having the same frequency as ω_s.

Period-Doubling Route to Chaos

As we increase the amplitude E of the sinusoidal input further, there comes a critical value, called a *bifurcation point*, where the single stem in Fig. 6 abruptly divides into two branches as in a two-prong fork. Further increases in E lead, in even shorter intervals, to $4, 16, \ldots, 2^n, \ldots$ branches before the onset of the first chaotic region. In fact, the bifurcation points converge so rapidly that the higher-order branches preceding the first chaotic regime in Fig. 6 are squeezed beyond the resolution of our monitoring instrument and, hence, are not discernible in Fig. 6. From careful computer simulation analysis, however, we can expand the area in front of each chaotic regime of the bifurcation tree of Fig. 6 and redraw it using a nonlinear scale

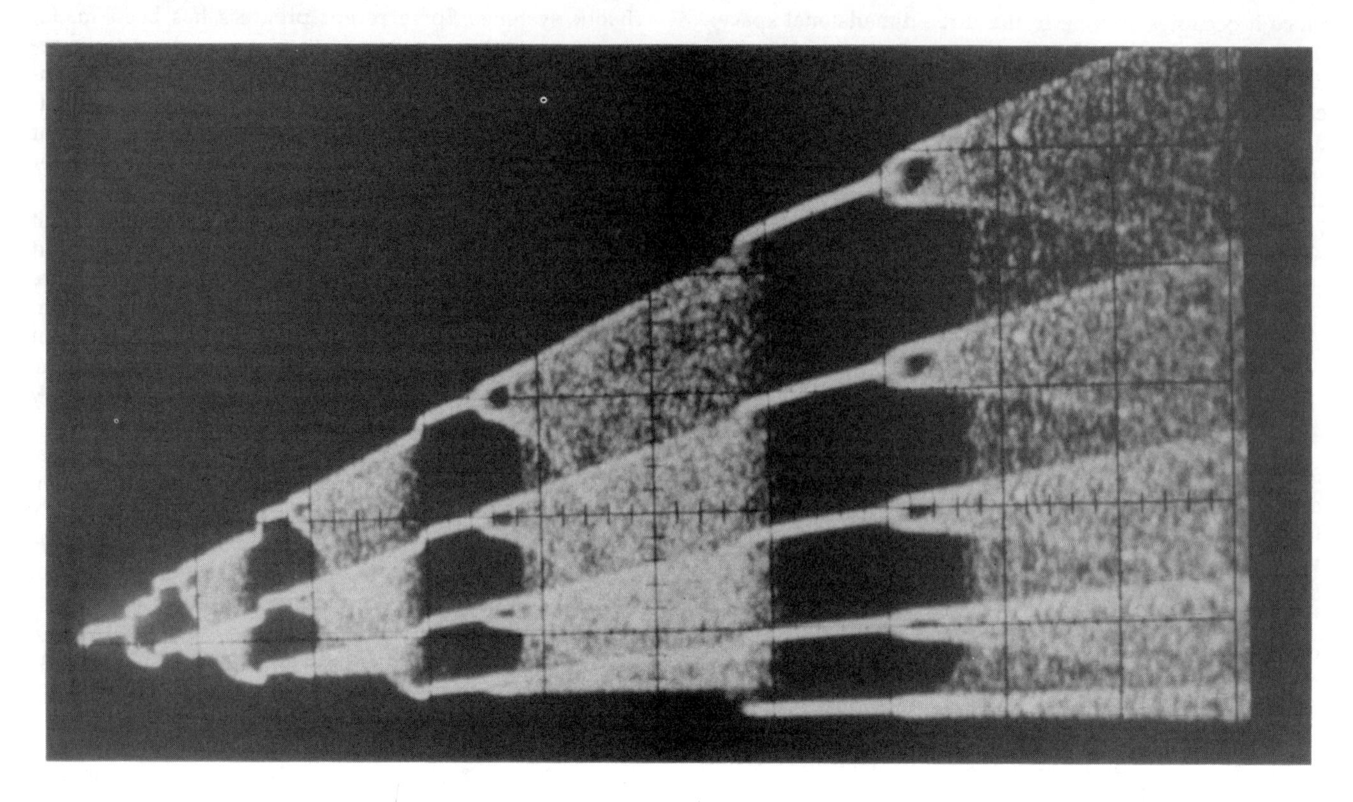

Fig. 6 Bifurcation tree associated with the RL-diode circuit in Fig. 5(a).

as shown in Fig. 7. If we draw a vertical line between each pair of bifurcation points $[\lambda_{n-1}, \lambda_n]$, it would intersect the bifurcation tree in 1, 2, 4, 16, 32, ... , 2^n, ... points, respectively. Such a set of intersection points about $\lambda = \lambda_k$ is called a *one-dimensional Poincare section* at $E = \lambda_k$. It follows from these Poincare sections that the period of the diode current $i(t)$ *doubles* at each bifurcation point. In the limit when $n \to \infty$, $i(t)$ becomes chaotic. This *period-doubling route* to chaos represents one of the most common mechanisms that gives rise to chaos in physical systems.

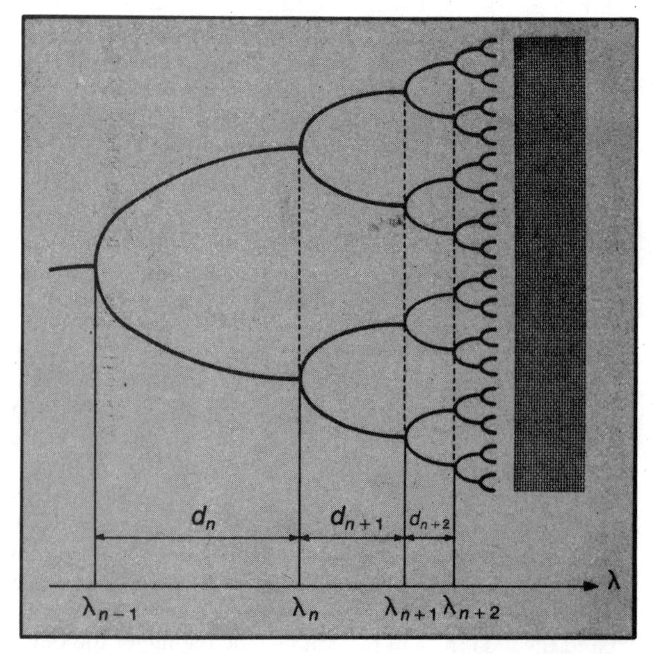

Fig. 7 Expanded bifurcation tree showing in the immediate area preceding each chaotic regime details of the period-doubling process.

While a bifurcation tree of the type shown in Fig. 7 may appear different in shape from one system to another, what is truly amazing is that the respective bifurcation points can be proved to converge toward a chaotic limiting point at a *constant rate* for all systems undergoing a period-doubling process! In particular, if we define $\delta_n \triangleq d_n/d_{n+1} = \lambda_n - \lambda_{n-1}/\lambda_{n+1} - \lambda_n$ where d_n, d_{n-1}, and d_{n+1} denote the distance between consecutive bifurcation points, as indicated in Fig. 7, then $\lim_{n \to \infty} \lambda_n = 4.669201660910299097\ldots$.

This unique convergence rate, called *Feigenbaum's number*, is a *universal* constant in the same sense that Planck's constant is a universal constant. This remarkable discovery by Feigenbaum shows that there is in fact a very strong sense of *order* in chaos.

Two-Dimensional Poincare Section

The chaotic regions in the bifurcation tree of Fig. 6 are not too revealing since the proliferation of points masks out any other interesting property. However, if we sample both the diode current i and the diode voltage v for some fixed value of E chosen arbitrarily within each chaotic regime of Fig. 6, we would obtain the two-dimensional Poincare cross sections shown in Fig. 8.

The Poincare cross section in Fig. 8(a) corresponds to the first chaotic region on the left of Fig. 6 and consists of a continuum of points, which we dub a one-leg crab attractor. Similarly, the cross sections in Figs. 8(b), 8(c), and 8(d) corresponding to the remaining three chaotic regimes in Fig. 6 are dubbed two-, three-, and four-leg crab attractors. These crab attractors are typical examples of *strange attractors* for nonautonomous second-order systems. They are *attractors* because, for any fixed value of E within each chaotic regime, all samples converge to the same continuum of points, regardless of the initial conditions, provided the

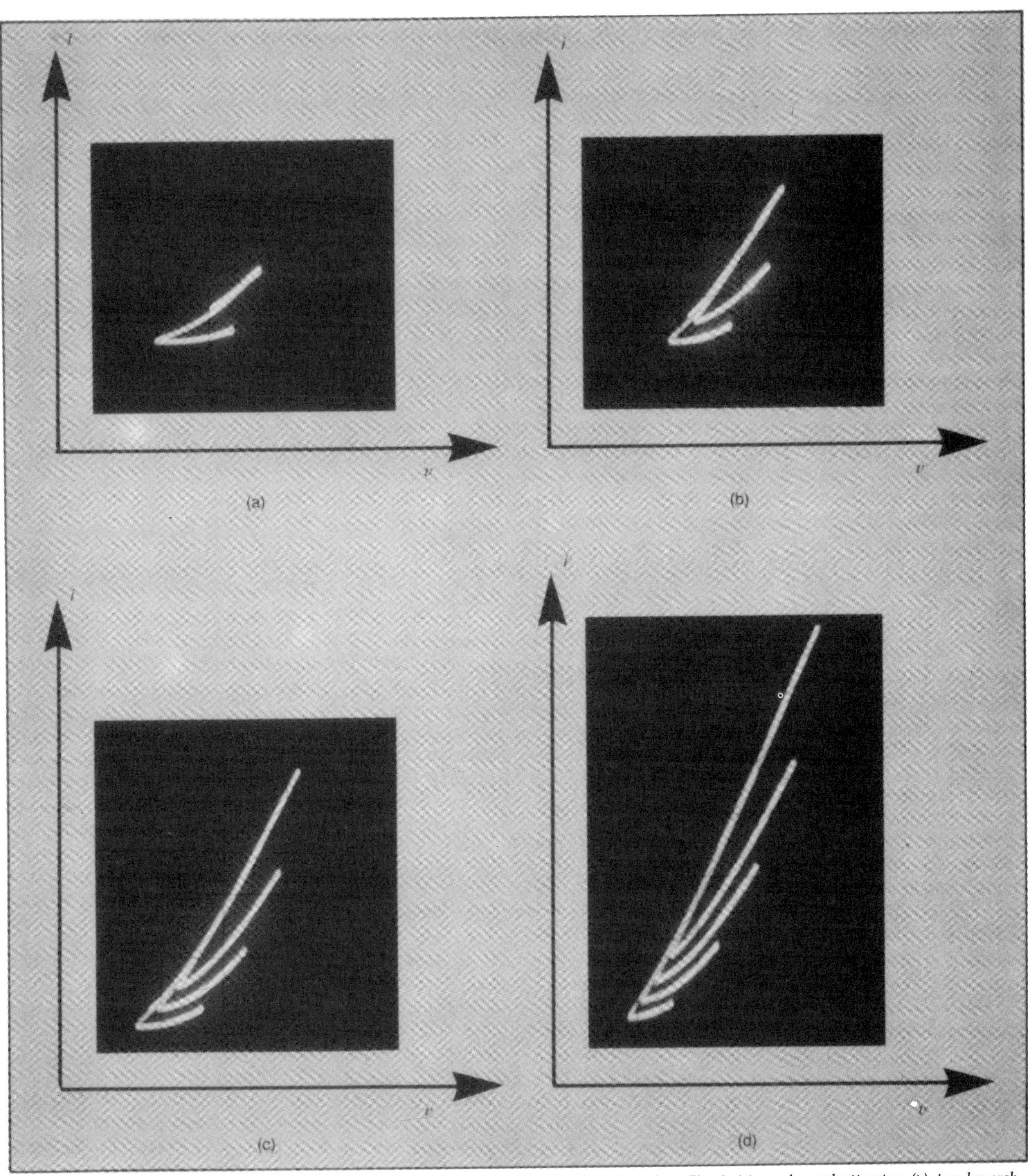

Fig. 8 Two-dimensional Poincare cross sections taken within each of the four chaotic regimes from Fig. 6. (a) one-leg crab attractor. (b) two-leg crab attractor. (c) three-leg crab attractor. (d) four-leg crab attractor.

initial transients are discarded. They are *strange* because, upon closer examination, using microscopes of increasingly higher resolving power, we find each of these legs is, in fact, made up of infinitely many ultrathin layers of points similar to our earlier observations on the double scroll. In other words, the crab attractors in Fig. 8 are, in fact, *fractals*.

Devil's Staircase

Period doubling is not the only route to chaos. There are many other routes, one of which pertains to the neon-bulb circuit shown in Fig. 9. This circuit was first reported by Van der Pol [1] to exhibit an "irregular noise" at various intervals of the capacitance C. We have duplicated Van der Pol's 1927 experiment and identified this noise as *chaos* resulting from a different mechanism called a *devil's staircase* [2].

Unlike period doubling, the one-parameter bifurcation diagram shown in Fig. 10 for the neon-bulb circuit of Fig. 9 consists of output subharmonic responses of *all* consecutive orders (horizontal steps); the responses are separated from each other by narrow gaps (shown shaded)

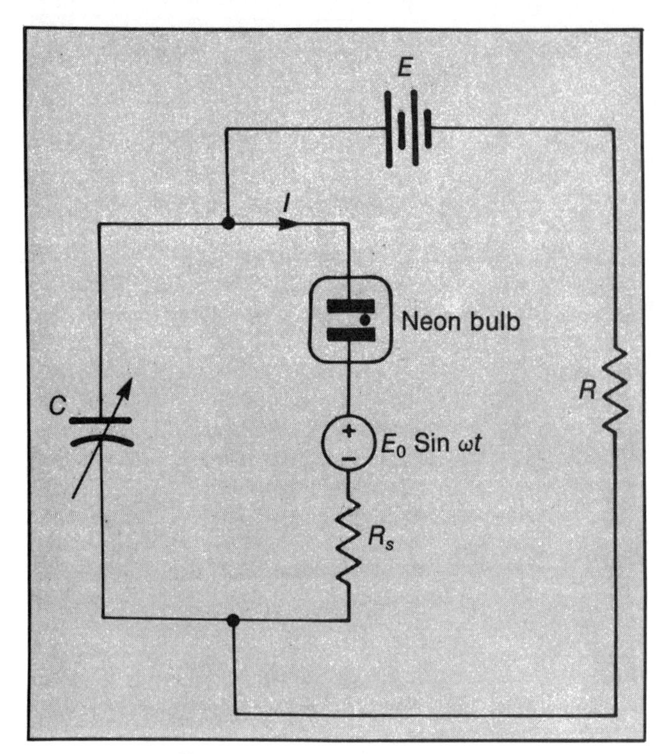

Fig. 9 Van der Pol neon-bulb circuit.

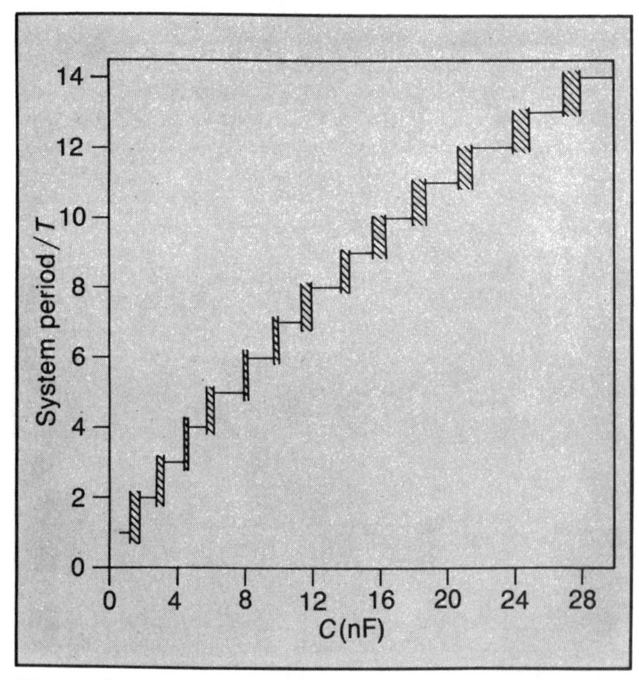

Fig. 10 One-parameter bifurcation diagram showing the order of output subharmonic response as a function of the value of the tuning capacitance C.

where the irregular noise was heard by Van der Pol, using a telephone! Each horizontal step in the ascending staircase in Fig. 10 represents a subharmonic response of order n, where $n = 2, 3, 4, \ldots, 14$. The initial step $n = 1$ corresponds to a unique steady-state response of the same frequency ω as the input sinusoid.

A fine tuning of the capacitance value within each "noisy" shaded interval reveals an extremely intricate, if not bizarre, chain of bifurcation phenomena. For example, a finer bifurcation diagram covering the first chaotic interval ($2 < C < 3.5$) is shown in Fig. 11. Notice that within the gap between the two outermost steps corresponding to $n = 1$ and $n = 2$, respectively, there appear finer ascending staircases, whose steps are in turn separated by still finer gaps. A magnification of these gaps reveals even finer staircases, which look just like their larger parents.

In fact, we can extrapolate our experimental observations to conclude that between every two steps lie finer staircases that are miniaturized clones of the outer ones and that chaos is observed at the outer boundary of each such staircase. Bizarre though they may appear, these weird staircases have been observed from several other circuits [29]–[31] and are all examples of the mathematician's *devil's staircase* [23].

In spite of these mind-boggling nested sequences of devil's staircases, there is, in fact, a beautiful law, which relates the precise locations of these steps. The intrigued reader is referred to [31] for a comprehensive discussion of this law as it applies to a simple two-transistor circuit. The mechanism by which this law leads to chaos is sometimes referred to in the literature as a *period-adding route to chaos* because each new period can be predicted by *adding, rather than doubling, to* the previous period a fixed integer determined by the law.

Phase-Locked Loops Are Potentially Chaotic

We close this section by briefly describing a widely used phase-locked loop (Fig. 12), which can become chaotic. This

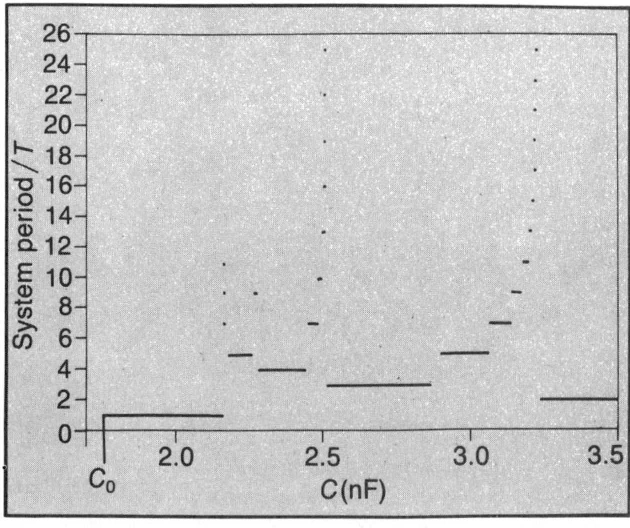

Fig. 11 Devil's staircase between the two steps n = 1 and n = 2.

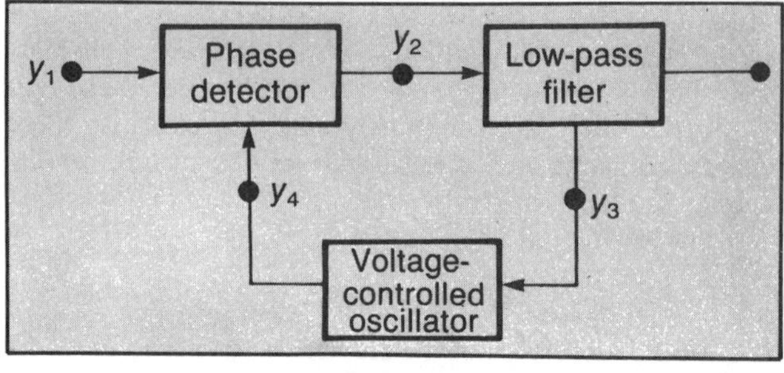

Fig. 12 A standard phase-locked loop.

circuit has been built in the laboratory to function as an *FM demodulator* using an IC module LM565C. The actual circuit used in our experiment is taken directly from a standard IC handbook [32] and is reproduced in Fig. 13. A careful experimental and mathematical analysis of this circuit shows that there exists a certain region in the parameter space where this FM demodulator is chaotic and, hence, will fail to operate as intended [33]. What is even more significant is that the parameter ranges where chaos is found to occur in this circuit are not unrealistic or contrived values but are rather well within the ballpark of typical designs. Consequently, unless one carries out a comprehensive analysis of its associated dynamics, a nominal circuit design, which functions satisfactorily initially, may, in fact, be precariously close to some failure boundary due to chaos.

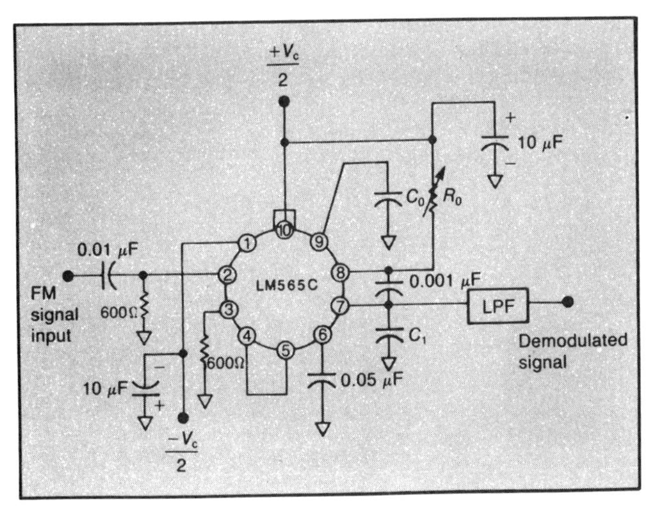

Fig. 13 Handbook version of an FM demodulator.

Concluding Remarks

We have presented various examples showing that *nonlinear* autonomous circuits containing three or more energy-storage elements or nonautonomous circuits containing two or more energy-storage elements can become chaotic. In fact, even circuits containing only *one* capacitor or inductor can become chaotic if their dynamics can be realistically modeled by a *nonlinear difference, rather than differential, equation*—namely:

$$x_{n+1} = f(x_n)$$

A simple example of this type is the first-order synchronized relaxation oscillator circuit described and analyzed in detail in [34]. It turns out that *chaos* sets in when the circuit drifts out of synchronization. Another example is given in [35], which uses a switched-capacitor circuit to realize the *logistic equation* whose chaotic dynamics is well known. In fact, using the techniques described in [36], an infinite variety of chaotic circuits, described by either a first-order or second-order nonlinear difference equation, can be built and investigated.

We close this expository article by remarking that if exceedingly simple electronic circuits can become chaotic over wide parameter ranges, so too can our garden variety of more mundane circuits, which are generally of *higher order* and are *nonlinear*. How far we are actually from the precipice of a *failure boundary* can be answered only by a nonlinear analysis of the underlying dynamics.

Acknowledgment

The authors would like to thank T. Matsumoto, R. Tokunaga, and K. Ayaki for providing the color pictures featured on the front cover and in this article.

References

[1] Van der Pol and J. Van der Mark, "Frequency Demultiplication," *Nature*, vol. 120, no. 3019, pp. 363–364, Sept. 10, 1927.

[2] M. P. Kennedy and L. O. Chua, "Van der Pol and Chaos," *IEEE Trans. Circuits and Systems*, vol. 33, no. 10, pp. 974–980, Oct. 1986.

[3] E. N. Lorenz, "Deterministic Non-Periodic Flow," *J. Atmospheric Science*, vol. 20, pp. 130–141, 1963.

[4] L. O. Chua, C. A. Desoer, and E. S. Kuh, *Linear and Nonlinear Circuits*, McGraw-Hill, 1987.

[5] G. Q. Zhong and F. Ayrom, "Experimental Confirmation of Chaos from Chua's Circuit," *Int. J. Circuit Theory Appl.*, vol. 13, pp. 93–98, Jan. 1985.

[6] T. Matsumoto, L. O. Chua, and M. Komuro, "The Double Scroll," *IEEE Trans. Circuits and Systems*, vol. 32, no. 8, pp. 797–818, Aug. 1985.

[7] T. Matsumoto, L. O. Chua, and K. Tokumasu, "Double Scroll via a Two-Transistor Circuit," *IEEE Trans. Circuits and Systems*, vol. 33, pp. 828–835, Aug. 1986.

[8] L. O. Chua, M. Komuro, and T. Matsumoto, "The Double Scroll Family," *IEEE Trans. Circuits and Systems*, vol. 33, no. 11, pp. 1072–1118, Nov. 1986.

[9] T. Matsumoto, L. O. Chua, and M. Komuro, "The Double Scroll Bifurcations," *Int. J. Circuit Theory Appl.*, vol. 14, pp. 117–146, 1986.

[10] C. Kahlert and L. O. Chua, "Transfer Maps and Return Maps for Piecewise-Linear Three-Region Dynamical Systems," *Int. J. Circuit Theory Appl.*, vol. 15, pp. 23–49, 1987.

[11] M. Broucke, "One Parameter Bifurcation Diagram for Chua's Circuit," *IEEE Trans. Circuits and Systems*, vol. 34, pp. 208–209, Feb. 1987.

[12] L. Yang and Y. Liao, "Self-Similar Bifurcation Structures from Chua's Circuit," *Int. J. Circuit Theory Appl.*, vol. 15, pp. 189–192, Apr. 1987.

[13] C. Kahlert, "The Chaos Producing Mechanism in Chua's Circuit and Related Piecewise-Linear Dynamical Systems," *Proc. of the 1987 European Conference on Circuit Theory and Design*, Paris, Sept. 1–4, 1987.

[14] T. Matsumoto, L. O. Chua, and M. Komuro, "Birth and Death of the Double Scroll," *Physica 24D*, pp. 97–124, 1987.

[15] S. Wu, "Chua's Circuit Family," *Proc. of the IEEE, Special Issue on Chaotic Systems*, pp. 1022–1032, Aug. 1987.

[16] D. Ruelle, "Strange Attractors," *The Mathematical Intelligencer*, vol. 2, pp. 126–137, 1980.

[17] E. Freire, L. G. Franquelo, and J. Aracil, "Periodicity and Chaos in an Autonomous Electronic System," *IEEE Trans. on Circuits and Systems*, vol. 31, pp. 237–247, 1984.

[18] T. Saito, "A Chaos Generator Based on a Quasi-Harmonic Oscillator," *IEEE Trans. Circuits and Systems*, vol. 32, pp. 320–331, 1985.

[19] T. Matsumoto, L. O. Chua, and R. Tokunaga, "Chaos via Torus Breakdown," *IEEE Trans. Circuits and Systems*, vol. 34, pp. 240–253, Mar. 1987.
[20] P. Bartissol and L. O. Chua, "The Double Hook," IEEE CAS, to appear.
[21] R. Tokunaga, M. Komuro, T. Matsumoto, and L. O. Chua, "Lorenz Attractor from an Electrical Circuit with Uncoupled Piecewise-Linear Resistor," *Proc. of ISCAS*, Philadelphia, May 1987.
[22] C. Sparrow, *The Lorenz Equations: Bifurcations, Chaos, and Strange Attractors*, Springer Verlag, New York, 1982.
[23] B. B. Mandelbrot, *The Fractal Geometry of Nature*, W. H. Freeman and Co., San Francisco, 1982.
[24] P. S. Linsay, "Period Doubling and Chaotic Behavior in a Driven Anharmonic Oscillator," *Phys. Rev. Lett.*, vol. 47, pp. 1349–1352, Nov. 1981.
[25] D. Azzouz, R. Duhr, and M. Hasler, "Transition to Chaos in a Simple Nonlinear Circuit Driven by a Sinusoidal Voltage Source," *IEEE Trans. Circuits and Systems*, vol. 30, pp. 913–914, Dec. 1983.
[26] D. Azzouz, R. Duhr, and M. Hasler, "Bifurcation Diagram in a Piecewise-Linear Circuit," *IEEE Trans. Circuits and Systems*, vol. 31, pp. 587–588, June 1984.
[27] T. Matsumoto, L. O. Chua, and S. Tanaka, "Simplest Chaotic Nonautonomous Circuit," *Phys. Rev. A*, vol. 30, pp. 1155–1157, Aug. 1984.
[28] S. Tanaka, T. Matsumoto, and L. O. Chua, "Bifurcation Scenario in a Driven R-L Diode Circuit," *Physica D*, in press.
[29] A. A. Abidi and L. O. Chua, "On the Dynamics of Josephson Junction Circuits," *IEEE J. Elec. Circuits and Systems*, pp. 186–200, July 1979.
[30] L. Q. Qing, F. Guo, S. X. Wu, and L. O. Chua, "Experimental Confirmation of the Period Adding Route to Chaos in a Nonlinear Circuit," *IEEE Trans. Circuits and Systems*, vol. 33, pp. 438–442, Apr. 1986.
[31] L. O. Chua, Y. Yao, and Q. Yang, "Devil's Staircase Route to Chaos in a Nonlinear Circuit," *Int. J. Circuit Theory Appl.*, vol. 14, pp. 315–329, 1986.
[32] National Semiconductor, *Linear Databook*, pp. 9-35—9-39, Oct. 1978.
[33] T. Endo and L. O. Chua, "Chaos from Phase-Locked Loops," *IEEE Trans. Circuits and Systems*, to appear in 1988.
[34] Y. S. Tang, A. I. Mees, and L. O. Chua, "Synchronization and Chaos," *IEEE Trans. Circuits and Systems*, vol. 30, pp. 620–626, Sept. 1983.
[35] A. B. Rodriguez-Vazquez, J. L. Huertas, and L. O. Chua, "Chaos in a Switched-Capacitor Circuit," *IEEE Trans. Circuits and Systems*, vol. 32, pp. 1083–1085, Oct. 1985.
[36] A. Rodriguez-Vazquez, J. Huertas, A. Rueda, B. Perez-Verdu, and L. O. Chua, "Chaos from Switched-Capacitor Circuits: Discrete Maps," *Proc. of the IEEE, Special Issue on Chaotic Systems*, pp. 1090–1106, Aug. 1987.

Introduction to Implantable Biomedical IC Design

Larry J. Stotts

Introduction

Integrated circuits (ICs), particularly complementary metal-oxide semiconductor (CMOS) integrated circuits, are playing an ever-increasing role in implantable biomedical systems. They must be designed to satisfy rigorous reliability, redundancy, and error-checking requirements associated with life-sustaining medical devices. In addition, low-voltage and low-power operation is required for battery-powered systems that may be implanted for as long as 10 years. This article will give a brief overview of some of the major system elements, as well as some design examples.

Physiology of Bioelectrical Events

Electrical potentials exist across the membranes of all living cells, and many cell systems have the ability to propagate a change in these potentials as a means of transferring information or effecting control in the surrounding tissue. Since the human body may be viewed as a volume conductor, these signals can be detected at a distance from their source and, after analysis, can be used as input to control an implantable medical device. In addition, this device may evoke a similar response from most tissue systems through the application of an appropriate electrical stimulus.

The series of reversible changes in membrane potential, which many cells exhibit in response to a stimulus, is known as an "action potential." As shown in Fig. 1, the membrane surrounding the cell serves as a semipermeable barrier to the passage of certain substances and ions. In the resting state, there is a higher concentration of potassium ions inside the cell than in the extracellular fluid. In addition, sodium ions are more abundant outside the cell. This ionic gradient is maintained by virtue of metabolic energy expended by the cell and results in a potential difference across the cell membrane with the inside negative, with respect to the outside. In response to an electrical, mechanical, or chemical stimulus of adequate intensity, there is a rapid reversal in membrane potential accompanied by a sudden influx of sodium ions into the cell and an influx of sodium ions from within the cell. This is known as *depolarization* and is followed by a period of reverse polarization before the membrane is repolarized by reestablishing the original ionic gradient. This wave of depolarization will propagate to adjacent tissue at a speed determined by the cellular structure of the particular conduction system. For example, the propagation speed in the fastest nerves is 150 m/sec, whereas speeds three orders of magnitude slower can be found in portions of the cardiac conduction system.

Batteries for Implantable Medical Devices

Batteries for implantable medical devices should maximize energy density and reliability and should not produce a gas during discharge. Presently, the most popular battery chemistry for this application is lithium iodine. Although these batteries (which can become volatile at elevated temperatures) must be hermetically sealed and must develop a high cell resistance during depletion, they have been used successfully in cardiac pacemakers for over a decade.

A cell is formed when a lithium anode is brought into contact with a cathode composed primarily of iodine. Reaction between these two chemicals produces an electrolyte of solid lithium iodide. As current is drawn from the cell, the lithium is oxidized and the iodine is reduced, forming additional lithium iodide. This lithium iodide, which serves as both electrolyte and separator, accumulates as the cell is discharged, which increases the internal resistance. This results in an initial decline in terminal voltage under load, which is fairly linear (Fig. 2). After sufficient discharge, the cathode will eventually lose most of its iodine and will, itself, increase in resistance; this process produces a "knee" in the discharge curves. It should be pointed out that this resistance can increase substantially at low temperatures, which is usually not a problem since the temperature of the implantable environment is well controlled.

As an example of the effect of current drain on the longevity of an implantable device, consider a device with a current drain of 20 μA and a minimum operating voltage of 2.2 V. A cell resistance of 30 kΩ will drop the initial terminal voltage of 2.8 V by 600 mV; this will occur after 1800 mAh of capacity have been expended. Since 20 μA will discharge the battery at a rate of 175 mAh/year, the minimum operating voltage will be reached in 10.3 years. If the current drain is increased by only 2 μA, the maximum cell resistance decreases to 27 kΩ, which will occur after

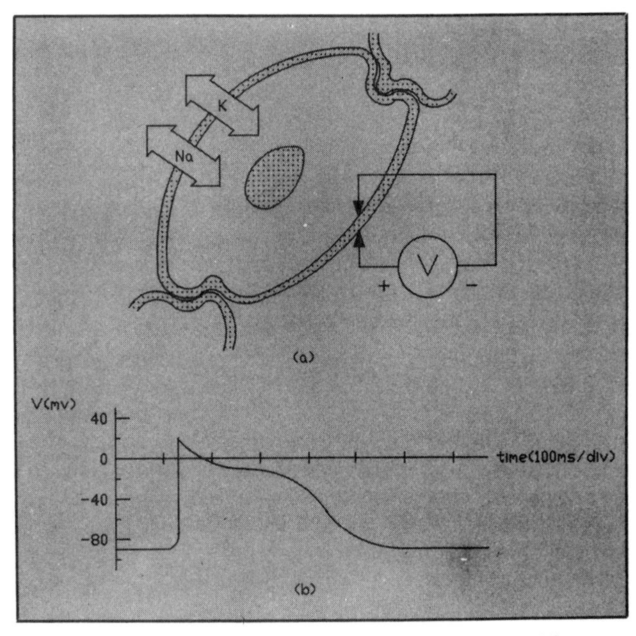

Fig. 1 (a) Cell cross section and (b) cell action potential.

Reprinted from *IEEE Circuits and Devices Magazine*, Vol. 5, No. 1, pp. 12-18, January 1989.

1700 mAh. The increased discharge rate of 193 mAh will, therefore, produce a minimum circuit voltage in 8.8 years.

Low-Voltage Circuit Design

Summary of MOS Device Physics

When the gate voltage of a metal-oxide semiconductor (MOS) transistor is reduced below its threshold voltage, current flow in the device changes from drift to diffusion. This region of operation is known as *subthreshold* or *weak inversion*. While saturation current has a square-law dependence on gate voltage in strong inversion, weak inversion exhibits an exponential dependence on gate voltage, similar to bipolar transistors [1]–[3]. Saturation current for each of these regions of operation may be expressed by the following equations:

$$I_D = \begin{cases} (\mu C_{ox}/2)(W/L)(V_{GS} - V_{TH})^2, \quad V_{GS} > V_{TH} + 2nV_T \\ V_{DS} > V_{GS} - V_{TH} \\ \qquad \text{strong inversion} \quad (1a) \\ I_{DO}(W/L)e^{(V_G - V_{TH})/nV_T}\,[e^{-V_S/V_T} - e^{-V_D/V_T}] \\ V_{GS} < V_{TH} + 2nV_T \\ V_{DS} > 3V_T \\ \qquad \text{weak inversion} \quad (1b) \end{cases}$$

where

μ	= mobility
C_{ox}	= gate oxide capacitance
W/L	= gate width/gate length
I_{DO}	= weak inversion characteristic current
V_T	= kT/q
V_{TH}	= threshold voltage
V_G, V_S, V_D	= gate, source, and drain voltage
n	= weak inversion slope factor

Assuming $V_S = 0$ and $V_D >> nV_T$, Eq. (1) can be simplified to

$$I_D = \begin{cases} K_S(V_{GS} - V_{TH})^2 \quad V_{GS} > V_{TH} + 2nV_T \\ \qquad V_{DS} > V_{GS} - V_{TH} \\ \qquad \text{strong inversion} \quad (2a) \\ K_W\, e^{(V_{GS} - V_{TH})/nV_T} \quad V_{GS} < V_{TH} + 2nV_T \\ \qquad V_{DS} > 3V_T \\ \qquad \text{weak inversion} \quad (2b) \end{cases}$$

where the parameters K_S and K_W are defined by

$$K_S = \left(\frac{\mu C_{ox}}{2}\right)\left(\frac{W}{L}\right) \tag{3}$$

$$K_W = I_{DO}\left(\frac{W}{L}\right) \tag{4}$$

The weak inversion parameter I_{DO} relates to the strong inversion parameters μ and C_{ox} by the expression

$$I_{DO} = \frac{2\mu C_{ox}(nV_T)^2}{e^2} \tag{5}$$

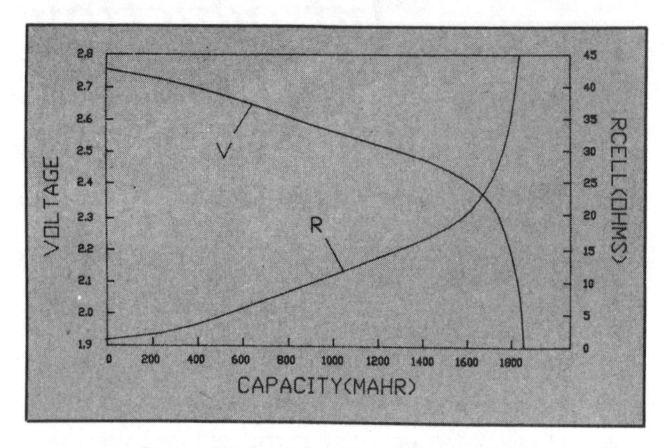

Fig. 2 Lithium-iodine battery characteristics.

The expressions in Eqs. (1) and (2) do not model the finite output resistance. The output resistance due to channel length modulation by the drain-to-source voltage, V_{DS}, in both regions of operation, may be expressed in terms of an extrapolated voltage, V_A, as

$$R_O = V_A / I_{DQ} \tag{6}$$

where I_{DQ} is the quiescent operating current. Equations (1) and (2) could be modified to include output impedance effects by multiplying the given expression for I_D by $(1 + V_{DS}/V_A)$. It should be pointed out that the above equations are greatly simplified but are still very useful for design analysis. The small following signal transconductance defined as the ratio of the small signal drain current to the small signal gate-source voltage, i_D/v_{GS}, can be easily derived by evaluating the partial derivative of I_D with respect to V_{GS} at the Q point.

$$g_m = \begin{cases} I_{DQ}/nV_T; \quad V_{DS} >> nV_T \quad \text{weak inversion} \quad (7a) \\ \sqrt{2I_{DQ}C_{ox}(W/L)} \quad \text{strong inversion} \quad (7b) \end{cases}$$

Figure 3 shows a plot of experimental data for drain current versus V_{GS} for a saturated $250\mu/10\mu$ n-channel device in a typical CMOS process together with the transconductance.

It can be observed from Eqs. (2) and (7) that the model for the drain current as well as the derivative with respect to V_{GS} (i.e., g_m) are both continuous at the transition between weak inversion and strong inversion, which is characterized by the equation

$$V_{GS} = V_{TH} + 2nV_T \tag{8}$$

From (2) and (8), the transition can also be expressed in terms of the transition current as

$$I_D = 2\mu C_{ox}(nV_T)^2(W/L) \tag{9}$$

For an n-channel transistor, the weak inversion–strong inversion transition using $2\mu C_{ox} = 20\ \mu A/V^2$, $V_{TH} = 0.75$ V, $nV_T = 60$ mV, and $W/L = 25$ is characterized by $I_D = 1.8\ \mu A$ and $V_{GS} = 0.87$ V.

The maximum voltage gain possible for a single common source transistor amplifier, obtained when driving an ideal current source load, is $A_V = g_m R_O$. It can be seen from (6) and (7a) that this reaches and maintains the maximum in weak inversion and is equal to $A_{V(\max)} = V_A/nV_T$. Although

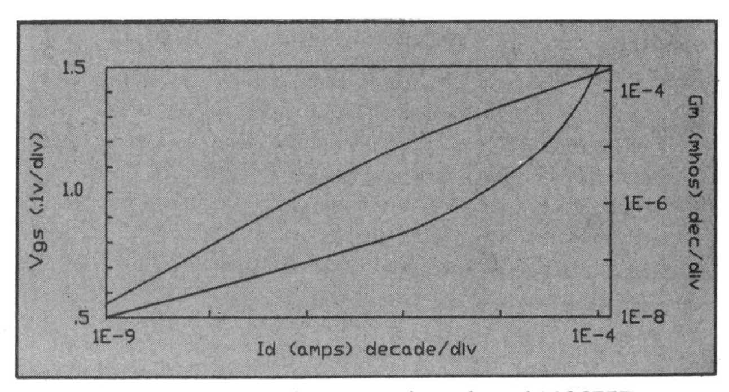

Fig. 3 Current and g_m curves for n-channel MOSFET.

large gains are attainable, weak inversion operation is not advantageous in all circuit applications, as demonstrated by the following example.

Consider the current mirror in Fig. 4a, which is assumed to have mismatches in both size and threshold voltages. If the threshold offset voltage is defined as $V_{OS} = V_{TH1} - V_{TH2}$, it follows from (2a) and (2b) that the output current of the current mirror may be approximated by

$$I_{out} = \begin{cases} I_{IN}(K_{W2}/K_{W1})(1 + V_{OS}/nV_T) & \text{weak inversion (10a)} \\ I_{IN}(K_{S2}/K_{S1})[1 + 2V_{OS}/(V_{GS} - V_{TH})] & \text{strong inversion} \end{cases} \quad (10b)$$

Since (8) requires $(V_{GS} - V_{TH}) > 2nV_T$ in strong inversion, the mismatch error produced in the current mirror for a given offset voltage is reduced in strong inversion. This effect can be seen with the data plotted in Fig. 4b. The wide variation at very low currents is due to a mismatch in leakage currents, which were not included in the model. The current mirror accuracy can be improved, at the expense of higher saturation voltages, by reducing the *W/L* ratio so that it operates either in the transition region or in strong inversion.

DC-DC Converters

One of the difficulties in designing an implantable system that requires large currents over short intervals around the lithium-iodine battery is that the high cell resistance severely limits the amount of current that can be supplied to a low-impedance load. The traditional solution to this is to first charge a capacitor relatively slowly with a small current from the battery, then discharge this capacitor into the external load. In addition, many applications require an output voltage greater than the battery terminal voltage. In these cases, a capacitive dc-to-dc converter can be used to charge the output capacitor to a multiple of the battery voltage.

In order to illustrate the basic principles behind capacitive voltage multipliers, consider the circuit of Fig. 5, which is a voltage doubler that has a load current, I_L, being drawn from the output capacitor.

If the battery resistance is small, then the pump capacitor, C_p, is charged to the battery voltage, V_{bat}, during Phase I. During this phase, charge is supplied by the output capacitor C_O to maintain the load current I_L. During Phase II, capacitor C_p is reconnected, as shown via the switches. Since this is the only time during the cycle that the output capacitor is supplied with charge, the average current from C_p can be represented by $I = C_p(V/T) = C_pVf$, where T is the period of the switching waveform, f is the frequency, and V is the peak-to-peak ripple voltage on C_p. The average capacitor current is equal to the average load current, I_L. Therefore,

$$C_pVf = I_L \quad (11)$$

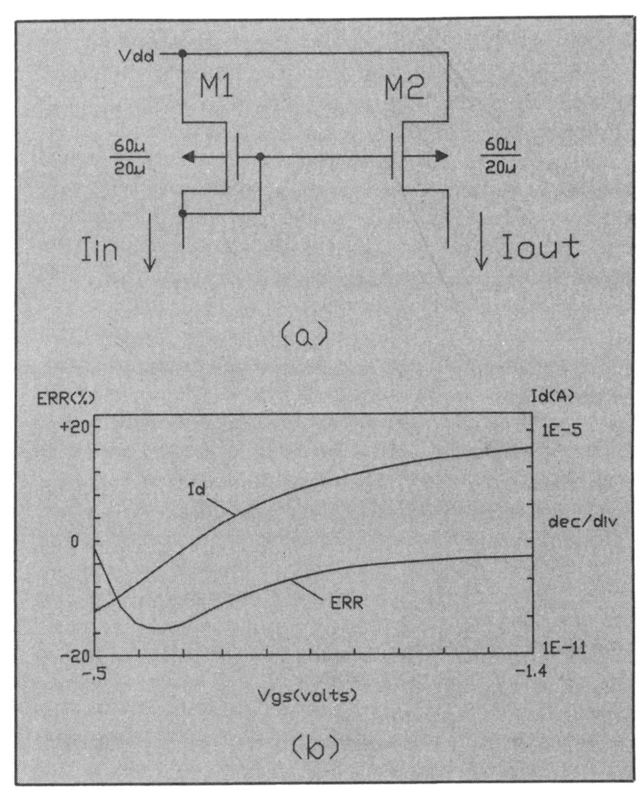

Fig. 4 Current source ratio error versus current.

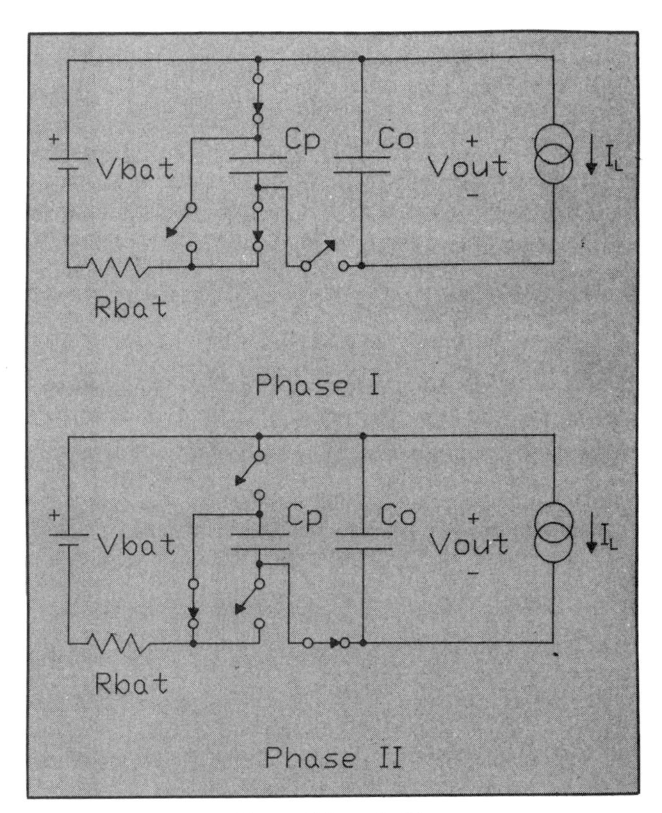

Fig. 5 Voltage doubler.

If we define R_m to be the average output resistance of the voltage multiplier, V/I_L, then

$$R_m = \frac{1}{C_p f} \tag{12}$$

The average battery current may be calculated by summing the charge supplied by the battery during each phase. For the doubler shown, this is C_pV for both cycles. The average battery current, I_{bat}, is, therefore,

$$I_{bat} = 2(C_pV)f = 2I_L \tag{13}$$

In other words, for each charge Q supplied to the load, $2Q$ is removed from the battery. This multiplier configuration has an energy transfer efficiency, defined as $\eta = (I_L V_{out})/(IV_{bat})$, where V_{out} is the average output voltage. It is easy to show that the efficiency approaches 100 percent for low values of ripple.

A problem arises when the battery resistance of this circuit (or switch resistances for that matter) becomes large enough so that the pump capacitor does not fully charge or discharge to the steady-state period value during each cycle. In the case where the battery is heavily filtered, it follows from (13) that the average voltage that the pump capacitor will charge to is $V_{bat} - 2I_LR_{bat}$. The maximum output voltage will thus be

$$V_{out} = 2V_{bat} - 4R_{bat}I_L - I_L/(fC_p) \tag{14}$$

The total equivalent charging resistance, therefore, is $R_m = 4R_{bat} + 1/(fC_p)$. In general, for this type of voltage multiplier with n multiplication stages and, thus, with an output voltage equal to n times the input battery voltage, the equivalent circuit will be a voltage source, V_{OC}, and a series resistance, R_m, which can be approximated by

$$V_{OC} = nV_{bat} \tag{15a}$$

$$R_m = \begin{cases} (n-1)/(fC_p) & f << 1/(R_{bat}C_p) \\ n^2R_{bat} + (n-1)/(fC_p) & f >> 1/(R_{bat}C_p) \end{cases} \tag{15b}$$

Although (15) is only an approximation, it predicts experimental data quite closely, as shown in Fig. 6. Finally, it should be pointed out that a practical implementation of the voltage multiplier of Fig. 5 in CMOS requires the use of well switching and level translating circuitry.

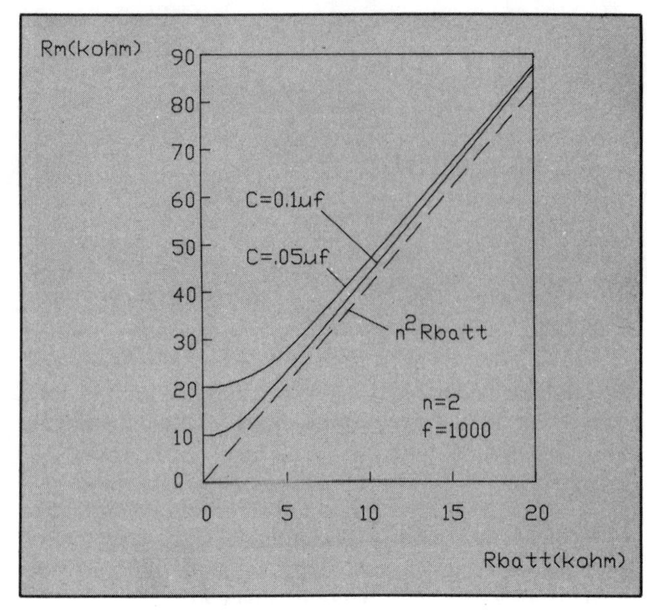

Fig. 6 Charging resistance versus battery resistance.

Very Low Frequency Filters

The slow propagation times of electrical activity within the body mean that the signals of interest are low in frequency, typically ranging from 1 Hz to 1 kHz. Realization of the resulting large time constants in a small area is required for integrated active filters. For example, consider the integrator blocks shown in Fig. 7, which could be used in a variety of filter topologies. Filters in this frequency range will require equivalent integrator "$R_{eq}C$" time constants of 160 ms to 160 μs. Assuming a practical upper limit on capacitance per pole of 20 pF, this would require equivalent resistances in the range of 8000 MΩ to 8 MΩ.

The availability of processes with polysilicon or thin-film resistors with sheet resistances as high as 75 kΩ² [4] make integrated filters above a few hundred hertz possible with the integrator of Fig. 7a. Tuning can be accomplished by switching the input resistor into the circuit with a variable duty cycle [5] or by using a single-ended Gilbert current multiplier in series with the input resistor to provide voltage tuning [6]. MOS transistors may also be used in place of the resistor, with a differential configuration providing increased linearity [7]. Filter frequencies will, in most cases, still be restricted to 100 Hz and above, due to the large transistor areas.

Another type of integrator is shown in Fig. 7b and utilizes the transconductance of a CMOS transistor [8]. The integrator time constant is equal to C/g_m. In weak inversion, the transconductance is independent of the size of the transistor, but very low bias currents are required for filter frequencies below 100 Hz. For example, a 10-Hz filter would require the transconductance state to be biased at 75 pA. Larger bias currents are possible if the transconductance stage is operated in strong inversion, but large transistor size will again limit the practical lower frequency that can be integrated.

Finally, the integrator shown in Fig. 7c uses a switched capacitor to simulate a resistor. The integrator time constant is equal to (C/fC_u), where f is the switching frequency. Very low time constants can effectively be integrated with this configuration. For example, a filter frequency of 10 Hz can be achieved with a unit capacitor of 1 pF switched at a clock rate of 1,257 Hz. It is important to use "parasitic insensitive" configurations [9] to compensate for leakage currents in very low frequency filters. This technique has been used to successfully integrate biomedical filters in the 1 to 100 Hz range.

The lower frequency range that nonswitched capacitor filters are based on (Fig. 7b) can be practically integrated and can be reduced by using *transconductance reduction* and *capacitance multiplication* techniques. A method that reduces the transconductance of a differential stage by the factor $(n + 1)/(n - 1)$ is shown in Fig. 8 [10].

Capacitance multiplier circuits can be classified into two basic categories, as shown in Fig. 9. In Fig. 9a, a voltage controlled–voltage source (VCVS) is used in series with the capacitor to effect an increase in capacitance by a factor of $(A + 1)$. Figure 9b uses a current controlled–current source (CCCS) in parallel with the capacitor to increase the capac-

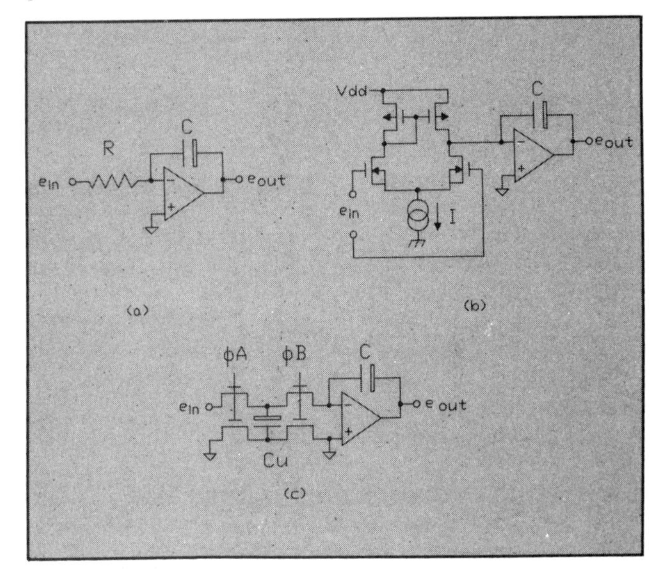

Fig. 7 Integrator stages.

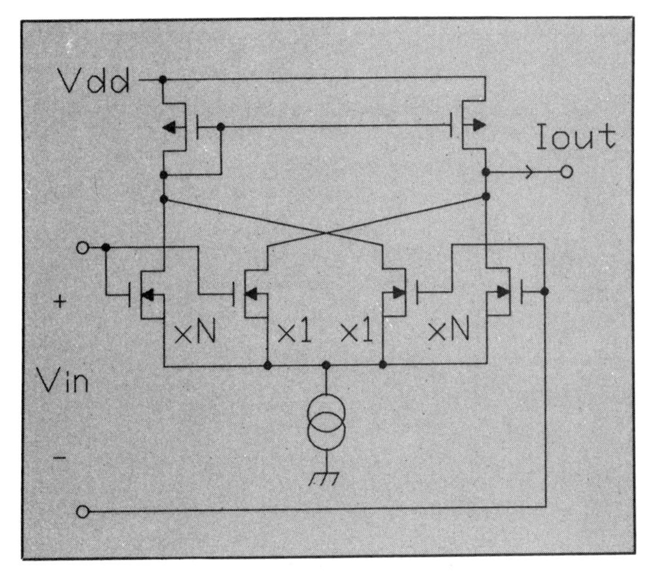

Fig. 8 Transconductance reduction.

itance by $(B + 1)$. Circuits based on Fig. 9a are required to generate a voltage A times larger than the input voltage, which decreases the dynamic range in low-voltage applications. Circuits based on Fig. 9b have increased dynamic range at low supply voltages, but this comes at the expense of increased bias current. The most obvious application of Fig. 9b uses a "current mirror" as the CCCS (as shown in Fig. 9c), without the dc bias circuits. The drawback of this configuration is that the dynamic resistance of the diode-connected FET adds an unwanted pole in the transfer function. This effect can be reduced with the circuit of Fig. 9d, which uses a differential input op-amp. The effective input conductance is

$$G_{in} = 1/R_f + sC[1 + (R_f/R_s)] \tag{16}$$

A similar circuit is shown in Fig. 9e. The effective input resistance for this circuit is

$$Z_{in} = (R_sR_f)/(R_s + R_f) + 1/[sC(1 + R_s/R_f)] \tag{17}$$

Finally, a switched capacitor technique similar to this circuit is shown in Fig. 9f [11]. Other switched capacitor realizations of this technique have also been reported [12]. Both capacitance multiplication and transconductance reduction circuits have been used in biomedical applications in the 10 to 500 Hz range and have been particularly useful in integrating antialiasing filters.

Low-Voltage Logic Design

The average current drain of a clocked CMOS inverter may be approximated by the following equation:

$$I = C_OV_{DD}f + I_{max}[1 - (V_{THn} + V_{THp})/V_{DD}](T_R f) \tag{18}$$

where

C_O = load capacitance
f = clock frequency
V_{THn}, V_{THp} = threshold voltages
I_{max} = peak device through current
T_R = clock rise time

The first term is the current required to drive the load capacitance, and the second term is the current that flows during the brief time that both CMOS transistors are conducting. It is clear that the current drain may be reduced by operating at low supply voltages and clock frequencies. In addition, the conduction current may be reduced dramatically by operating the circuit at supply voltages near the sum of the thresholds. Another method to reduce this current is to clock the p-channel and n-channel devices with nonoverlapping clocks, so that there is never a direct current path between the supply rails. This is particularly useful in low-voltage design where rise and fall times can become quite slow. These techniques have been successfully used in the design of a low power-level translator [13].

In general, total current drain can be reduced by using architectures that power down unused circuitry when not in use. The block diagram shown in Fig. 10 represents such a system. It is an 8-bit microcomputer with read-only memory (ROM) and random access memory (RAM), which has been tailored to be used in a wide variety of implantable biomedical applications [14]. It features a bidirectional, pulse position modulated communications channel, parity checking of each instruction with a hardware response to a detected error, and a unique memory structure that allows multiple input/output (I/O) ports to share and exchange program information. Other features include a low-current crystal oscillator, six addressable timers, multiple I/O ports, a unique memory patching function that allows external patching of ROM programs with patches in RAM, and test modes to facilitate production testing of the devices.

In order to conserve current, the processor can be put into a low current (SLEEP) mode via software where only the oscillator and timers are running. Timer outputs or external inputs at certain I/O ports can reactivate (WAKEUP) the processor. When this happens, the processor resumes execution at the next instruction after the sleep request. Tailoring the processor architecture to the overall system in this manner reduces the continuous current drain of 10 μA to less than 3 μA running system software. The minimum operating voltage for this circuit is 1.7 V.

The microcomputer was fabricated in a low-voltage sili-

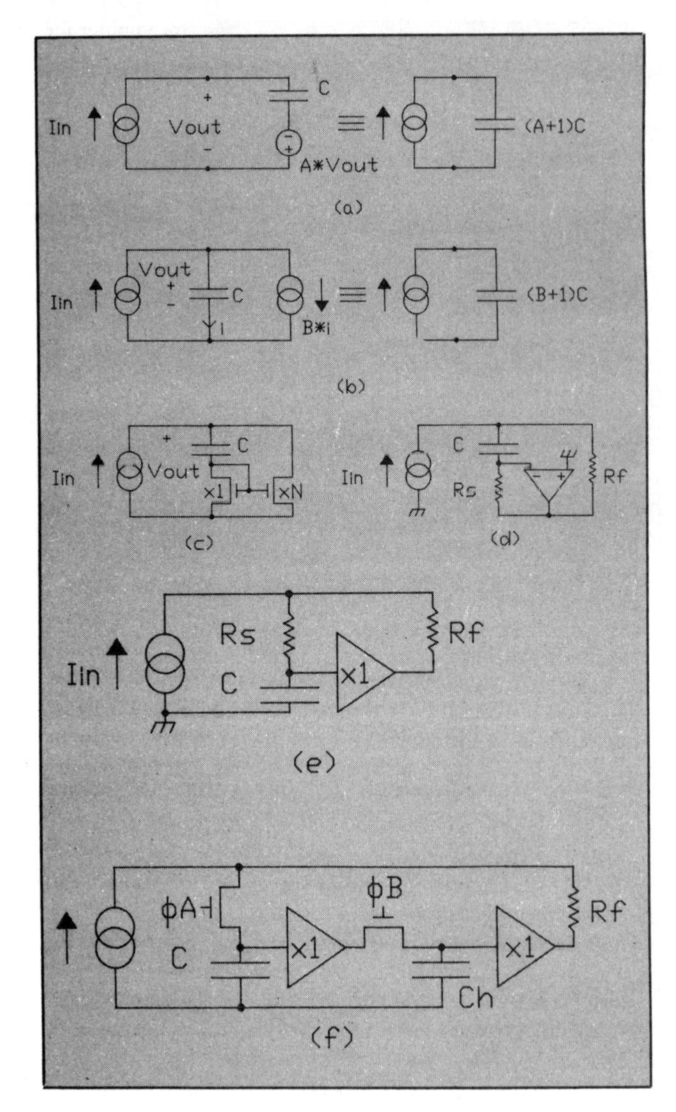

Fig. 9 Capacitance multiplication.

con gate CMOS process, which incorporated a high level of reliability screens. A photograph of the device showing major circuit areas is shown as Fig. 11. The die size is approximately 5.1 mm × 5.1 mm square and contains approximately 20,000 transistors. This design has also been converted to a "standard cell" and has been used in more complex, mixed analog and digital system chips [15]. In particular, one of the chips has been used to implement an advanced single-chip dual-chamber pacemaker, which is in commercial production.

References

[1] Y. Tsividis and P. Antongnetti, *Design of MOS VLSI Circuits for Telecommunications*, Prentice-Hall, 1985.

[2] R. Swanson and J. Meindl, "Ion-Implanted Complementary MOS Transistors in Low-Voltage Circuits," *IEEE J. Solid-State Circ.*, vol. 7, no. 2, pp. 146–153, Apr. 1972.

[3] E. Vittoz and J. Fellrath, "CMOS Analog Integrated Circuits Based on Weak Inversion Operation," *IEEE J. Solid-State Circ.*, vol. 12, no. 3, pp. 224–231, June 1977.

[4] D. Stone, J. Schroeder, R. Kaplan, and A. Smith, "Analog CMOS Building Blocks for Custom and Semicustom Applications," *IEEE J. Solid-State Circ.*, vol. 19, no. 1, pp. 55–61, Feb. 1984.

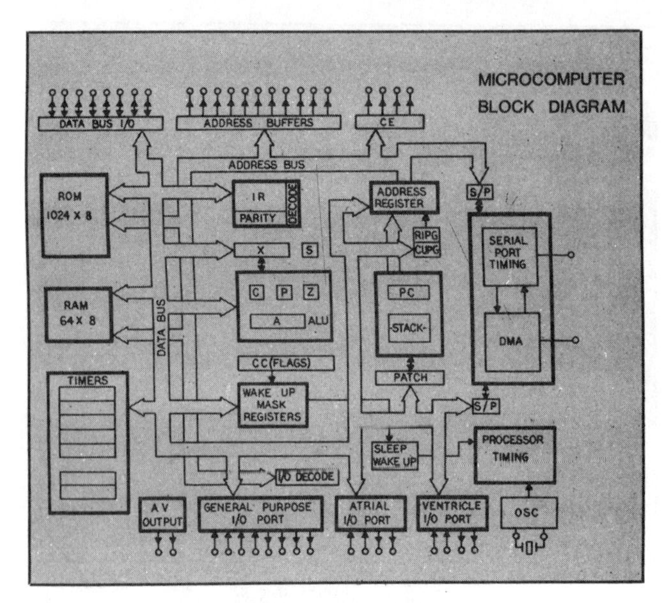

Fig. 10 Microcomputer block diagram.

[5] R. Geiger, P. Allen, and D. Ngo, "Switched-Resistor Filters—A Continuous Time Approach to Monolithic MOS Filter Design," *IEEE Trans. Circ. Syst.*, vol. 29, pp. 306–315, May 1982.

[6] E. Scratchley, "Single-Ended-Input/Single-Ended-Output Four-Quadrant Analog Multiplier," *IEEE J. Solid-State Circ.*, pp. 394–395, Dec. 1972.

[7] M. Banu and Y. Tsividis, "Fully Integrated Active RC Filters in MOS Technology," *IEEE J. Solid-State Circ.*, vol. 18, no. 6, pp. 644–651, Dec. 1983.

[8] R. Geiger and E. Sánchez-Sinencio, "Active Filter Design Using Operational Transconductance Amplifiers: A Tutorial," *IEEE Circ. & Dev. M.*, vol. 1, no. 2, pp. 20–32, Mar. 1985.

[9] R. Gregorian and W. Nicholson, "MOS Switched Capacitor Filters for a PCM Voice Codec," *IEEE J. Solid-State Circ.*, vol. 14, no. 6, pp. 970–980, Dec. 1979.

[10] P. Garde, "Transconductance Cancellation for Operational Amplifiers," *IEEE J. Solid-State Circ.*, pp. 310–311, June 1977.

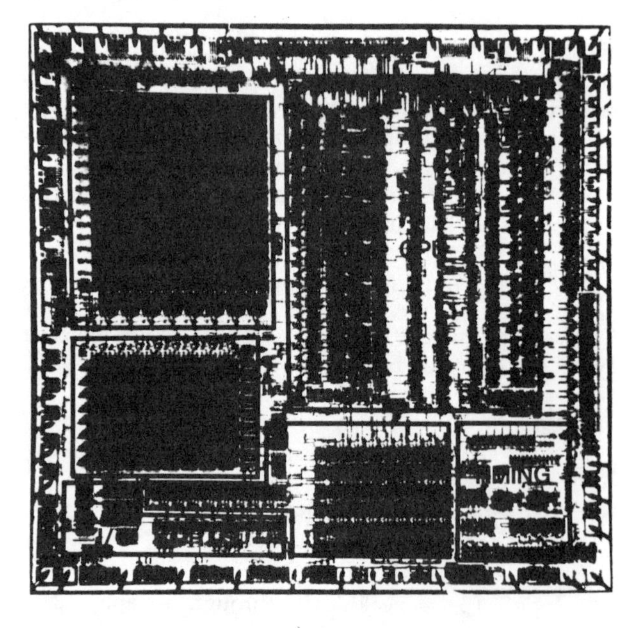

Fig. 11 Die photograph.

[11] R. Cuppens, H. DeMann, and W. Sansen, "Simulation of Large On-Chip Capacitors and Inductors," *IEEE J. Solid-State Circ.*, vol. 14, no. 3, pp. 543–547, June 1979.
[12] K. Nagaraj, "A Novel Parasitic Insensitive Switched Capacitor Technique for Realizing Very Large Tune Constants," *IEEE Custom Integrated Circuits Conf. Digest of Technical Papers*, pp. 12.4.1–12.4.3, May 1988.
[13] U.S. Patent 4,663,701, *Low Voltage Level Shifter*, granted to L. Stotts, May 5, 1987.
[14] L. Stotts, J. Miner, and R. Baker, "An 8b Microcomputer for Implantable Biomedical Applications," *ISSCC Digest of Technical Papers*, pp. 14–15, Feb. 1985.
[15] L. Stotts and K. R. Infinger, "An 8-Bit Microcomputer with Analog Subsystems for Implantable Biomedical Applications," *IEEE Custom Integrated Circuits Conf. Digest of Technical Papers*, pp. 4.6.1–4.6.3, May 1988.

Oversampled Data Conversion Techniques

The mix of sigma-delta modulation and DSP produces an explosion in the variety and complexity of a-d and d-a converters implemented in MOS technology.

Vladimir Friedman

Digital signal processing techniques combined with the forty-year-old principle of sigma-delta modulation have been highly fruitful when applied to a-d and d-a converters [1]. These converters are well-suited for implementation in MOS technology. The area of the analog section can be minimized, and the digital-signal-processing (DSP) algorithms implemented economically in MOS. As a result, the number of a-d and d-a devices using the combination of sigma-delta modulation and DSP has increased significantly over the last few years. And the devices have evolved to more complex and sophisticated architectures. A knowledge of the most commonly used sigma-delta modulators, their performance, and their range of applications will aid the system engineer in designing such a system.

Basic Modulator Structures

In a single-loop, asynchronous, sigma-delta modulator, an impulse of height A is generated at the output Y each time the analog integrator output becomes equal to or greater than 0 (Fig. 1a). This pulse is fed back to the integrator input, where it is subtracted from the input signal. For a dc input signal equal to x ($x \leq A$), the output of the integrator is a sawtooth waveform of slope 1/x (Fig. 1b). The output pulses are generated every A/x seconds, and the average number of pulses during a fixed interval (the interval corresponding to the Nyquist rate, for instance) is proportional to the value of the dc input signal.

In synchronous modulators (which can be implemented easily by switched-capacitor networks), a triggering mechanism allows the generation of output pulses only when the external clock C is active (Fig.2). This causes overshoot in the domain of positive voltages. Immediately after each clock pulse the integrator output Z is the same for the synchronous modulator as for the asynchronous one, but the output pulses are no longer equally spaced in time. During a fixed interval, however, their average number remains the same as for the asynchronous modulator, and that number is proportional to the dc input signal. The signal z (Fig. 2b) contains a large number of high frequency components, which will be present in the spectrum of the output signal as well.

The value of the a-d converter output Y_0 could be generated by averaging. For a clock rate 128 times higher than the Nyquist rate the output value of the converter is:

$$Y_0(N) = Y(n) + Y(n-l) + ... Y(n-127).$$

This output sample is generated every 128 clock periods from the bits coming from the modulator during the interval. (Decimation, i.e., a reduction of the sampling rate, is performed at the same time as the averaging.) The range of the output values is 0 to 128. Though the output of the modulator is a one-bit stream, the a-d converter has a little over seven bits of resolution (which is actually attainable if the noise is sufficiently low). This is the essence of oversampling systems: data samples with several bits of resolution can be obtained from a one-bit oversampled data stream.

If the input is a random signal, it can still be considered a constant over short intervals of time because the sampling rate is much

Reprinted from *IEEE Circuits and Devices Magazine,* Vol. 6, No. 6, pp. 39-45, November 1990.

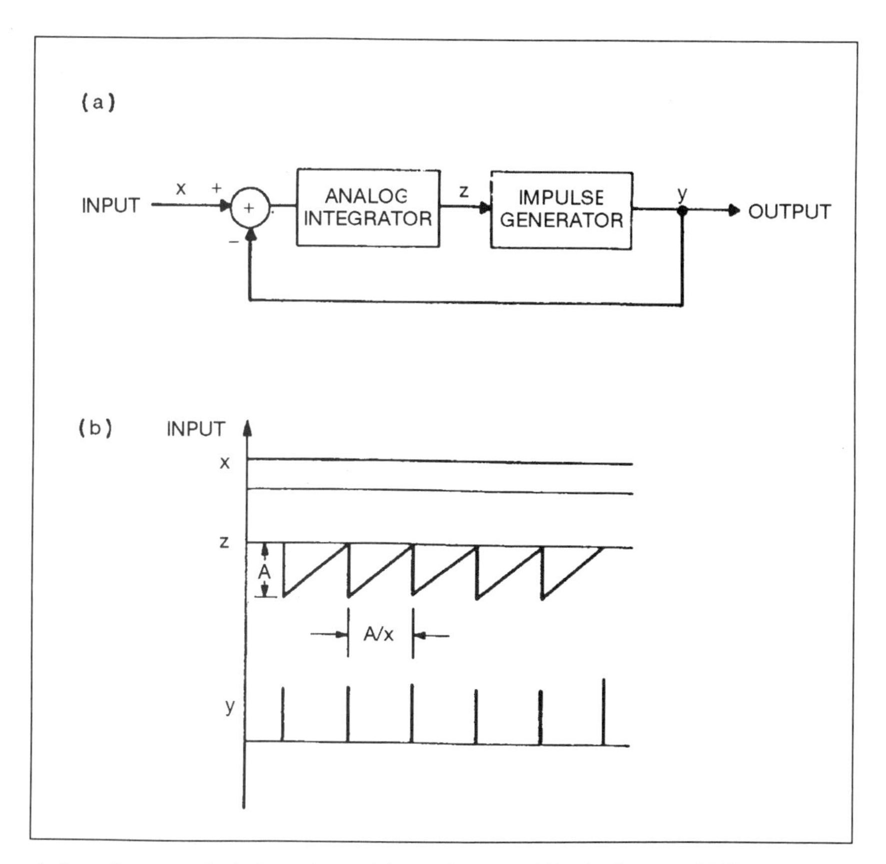

1. Asynchronous single-loop sigma-delta modulator. (a) Block diagram. (b) Waveforms.

higher than the Nyquist rate. The output of the analog integrator will look like the waveform z in Fig. 2b and will change slowly in time with the voltage of the input signal. The output of the digital low-pass filter is a replica of the input signal.

A typical implementation of a single-loop sigma-delta modulator performs the analog integration with an adder having a delay element in its feedback loop (Fig. 3a). This corresponds to a switched-capacitor-network implementation. The output pulses are generated by a one-bit quantizer, which is usually a clocked comparator circuit. The same structure can be used for d-a conversion. When the input x(n) is sampled data, the integrator is implemented with digital add-and-dump structures. The high-frequency noise contained in the one-bit data stream is subsequently removed by analog filters, which generate the continuous output signal.

A double-loop sigma-delta modulator does a double integration of the difference between the input signal and the output bit stream, which provides better performance (fig. 3b). A second feedback loop (represented by a dashed line in the figure) improves the modulator's stability. The waveform of the quantizer input q(n) is no longer a simple sawtooth waveform; it contains parabolic elements. The average number of pulses in a fixed interval still follows the input signal, but the high- frequency noise spectrum has a different shape.

Performance Analysis

A good way to evaluate the performance of $\Sigma\Delta$ modulators is to replace the one-bit quantizer with an additive white noise source. (The input signal has a sufficiently large random component to make the noise source a good representation of even a one-bit quantizer [9].)

If q is the quantizer step, the power of the noise source

$$\bar{e}^2 = \frac{q^2}{12} \tag{1}$$

is uniformly distributed in the band $(0, f_s)$ with the noise spectral density:

$$E(f) = \frac{q^2}{12 f_s} \tag{2}$$

Doubling the sampling frequency will reduce the power of the noise source in the signal band $(0, f_0)$ by 3 dB due to a smaller noise spectral density (Eq. 2). This is true for any quantizer.

If X(z),E(z) are the respective z-transforms of the input signal and the noise source, for the single-loop modulator we have

$$Q(z) = z^{-1}\frac{X(z) - Y_1(z)}{1 - z^{-1}}$$

from which we obtain the output as equal to

$$Y_1 = Q(z) + E(z)$$

$$Y_1(z) = z^{-1}X(z) + E(z)(1 - z^{-1}) \tag{3a}$$

Similarly, the output of the double-loop modulator is

$$Y_2(z) = z^{-1}X(z) + E(z)(1 - z^{-1})^2 \tag{3b}$$

The first term in each of Eq. (3a) and Eq. (3b) is the delayed input; the second term is the output noise. Its spectral density is obtained by substituting $z = \exp(j2\pi f/f_s)$

$$E_1(f) = 2E(f)\sin\left(\pi\frac{f}{f_s}\right) \tag{4}$$

The noise spectral density of double-loop modulators is lower than that of single-loop modulators for frequencies near dc, where the signal band is (Fig.4). The noise becomes larger than that of the single-loop modulator for high frequencies, so filtering requirements for the noise outside the signal band is greater, as well. If the sampling rate is doubled, the maximum signal frequency f_0 will scale down by a factor of two with respect to f_s (to f'_0 in Fig.4). The total noise power in the signal band will decrease, and the decrease will be more pronounced for the double-loop modulator because of the parabolic shape of its noise density curve (compared to the single-loop modulator's near-linear curve). Higher resolution can be obtained after filtering the noise outside the signal band, which is a benefit that is larger for double-loop modulators. The total noise power for single-loop (N_1) and double-loop (N_2) modulators, respectively, is obtained by summing the appropriate noise density from Eq. 4 in the band $(0, f_0)$

$$N_1 \approx \frac{\overline{e}^2 \pi^2}{3} \left(\frac{2f_o}{f_s}\right)^3$$

$$N_2 \approx \frac{\overline{e}^2 \pi^4}{5} \left(\frac{2f_o}{f_s}\right)^5 \quad (5)$$

Doubling the sampling frequency leads to a reduction in the noise power of 9 dB for single-loop modulators versus 15 dB for double-loop modulators. For both types, 3 dB of the reduction is due to the lower noise power density; the remainder to the noise shaping by the modulators that pushes most of the noise outside the signal band. As a result, a double-loop modulator can run at a lower sampling rate than a single-loop modulator and still attain the same signal-to-noise performance. This is an important advantage because oversampling rates are in the megahertz range and the speed of the analog CMOS circuits is one of the factors limiting the signal-to-noise performance of these converters. It accounts for the current popularity of the double-loop configuration in spite of its more complex structure and the early popularity of the single loop [8, 11, 13].

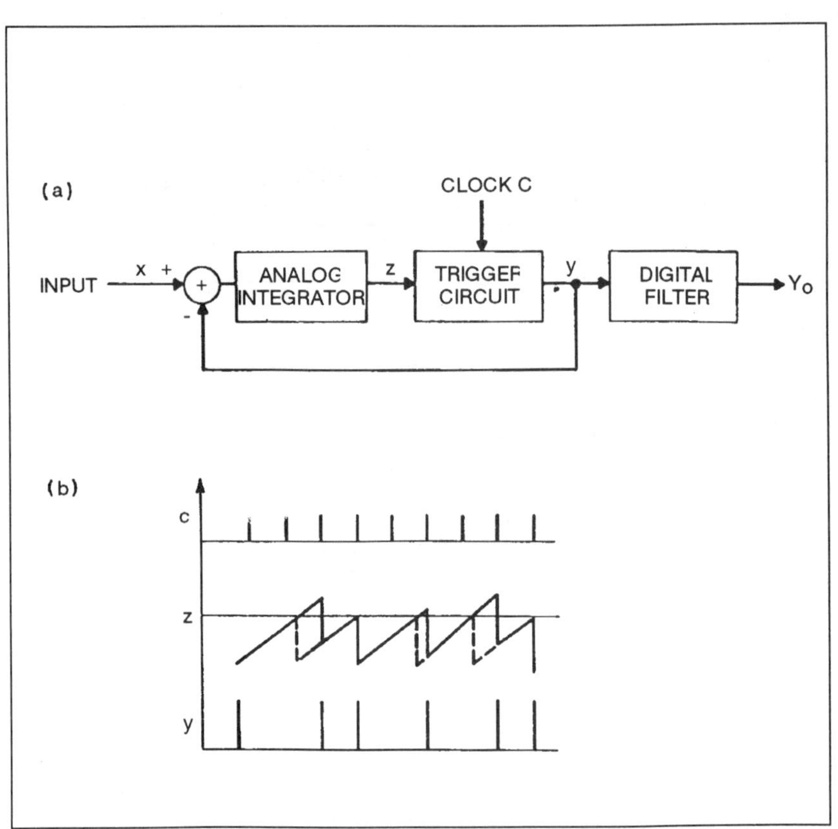

2. Synchronous single-loop sigma-delta modulator. (a) Block diagram. (b) Waveforms.

Using a one-step quantizer and the feedback loop makes $\Sigma\Delta$ modulators very robust. Computer simulations, based on a mathematical model, show that a variation of 10 to 15 percent in the gains of the integrators for double-loop modulators will not cause a noticeable degradation in converter performance. The limits are even larger for single-loop configuration.

More elaborate theories based on an equivalent circuit [4] or on a pure mathematical model [9, 10] have characterized single-loop modulators precisely. For double-loop configurations, there are no good theoretical models besides the approach used here, so computer simulation plays an important role in design [5].

The most important difference between the model described here and a computer simulation using the ideal model is the predicted behavior of sigma-delta modulators when the input is a dc value. The one-bit output stream of the single-loop modulator is periodic or quasiperiodic under these conditions; its spectrum is no longer continuous, but discrete. The double-loop modulators display a similar behavior for certain DC input levels [6]. On idle voice channels, these limit cycles are perceived as tones and are noticeable at levels much lower than the white noise or the noise characteristic of the $\Sigma\Delta$ modulators.

Several techniques are used to combat this effect. Among them are dithering, which randomizes the discrete spectrum of these cycles, and the application of a dc offset, which pushes the fundamental frequency out of the signal band so that it will be attenuated by the subsequent filtering [11].

For double-loop modulators the signal-to-noise performance becomes worse for large signal levels. The simulation shows that the amplitude of the quantizer input signal becomes higher as the dc input level approaches the value of the quantizer step. For dc inputs equal to the quantizer step, the quantizer input signal may become infinite. We've measured a peak signal-to noise ration of 83 dB, which is typical for double-loop modulators with an oversampling ratio of 128.

Most advanced work is currently focused on higher-order modulator structures capable of achieving higher resolution. Triple-loop modulators are not stable, so two other architectures have been proposed. The first approach is multi-stage noise shaping (MASH) [16, 17]. One way of implementing this technique is to obtain third-order noise shaping by using three single-loop modulators (Fig.6). The quantizer input $\Delta_1(z)$ of the first stage is used as input for the second modulator, and $\Delta_2(z)$ is used as input for the third modulator.

The output Y(z) is a sum of the three modulators outputs multiplied by the z-transform weight functions $w_1(z)$, $w_2(z)$, and $w_3(z)$. Y(z) is no longer a one-bit data stream.

The noise from the second quantizer $E_2(z)$ appears twice in the output signal, first as quantization noise at the output of the second modulator $Y_2(z)$, then as a noise component of the signal $\Delta_2(z)$ at the output of the third modulator $Y_3(z)$. The z-transform weighting functions $w_2(z)$, $w_3(z)$ were chosen such that the two noise components coming from the second modulator cancel each other. Weighting function $w_1(z)$ was selected to cancel the noise of the first modulator. Therefore, the output contains a replica of the input and noise coming from the third modulator $E_3(z)$ multiplied by the z-transform weight

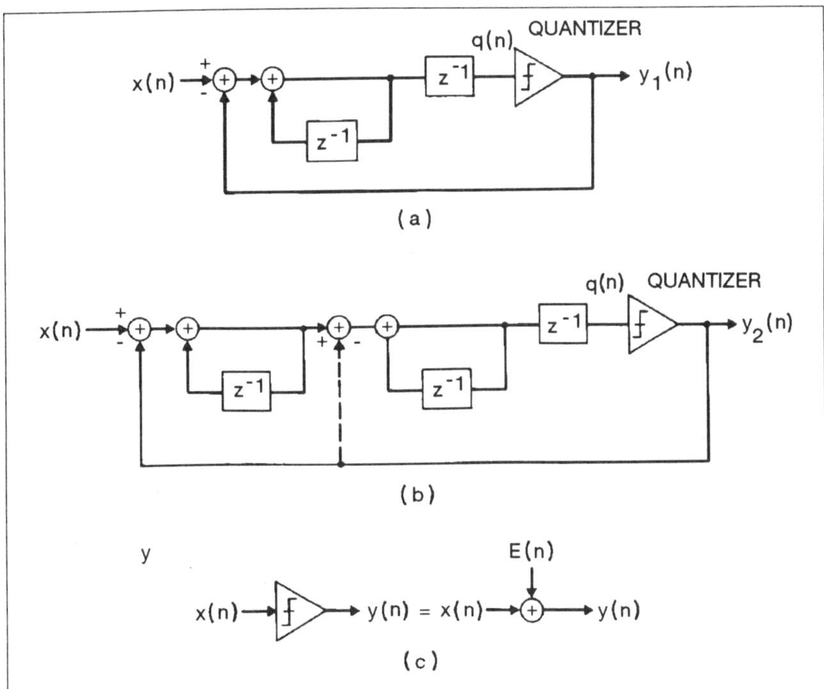

3. (a) Single-loop and (b) double-loop sigma-delta modulator architectures, and their (c) quantizer equivalent.

factor $w_3(z)$:

$$Y(z) = z^{-3}X(z) + (1 - z^{-1})^3 E_3(z) \tag{6a}$$

Similar results are obtained with a double-loop modulator followed by a single-loop modulator. (This procedure can be generalized for higher order noise shaping.)

Because this method is based on the cancellation of the noise coming from the first two modulators through different paths, any variation in the modulator gains will lead to incomplete noise cancellation. The resulting residual noise will appear at the output as a first- or second-order quantization noise. To avoid this, the gains of the different paths must be matched to less than 1 percent, which is much less than the 10 percent required for double-loop modulators.

The second approach for an N-order oversampled modulator consists of a feed-forward structure with the coefficients A_i (i =1, 2...N) and a feedback structure given by the coefficients B_i (Fig. 7) [15 ,18]. The transfer functions $H_A(Z)$, $H_B(Z)$ of the two structures are polynomials in $1-z^{-1}$ with the coefficients A_i and B_i, respectively. The adc and dac are usually reduced to a single bit. Replacing the quantizer with an additive noise source E(z) produces an output equal to

$$Y(z) = \frac{H_A(z)}{H_A(z)+1-H_B(z)}X(z) + \frac{1-H_B(z)}{H_A(z)+1-H_B(z)}E(z) \tag{6b}$$

The idea is to design $1-H_B(z)$ as a stop-band filter having the transmission zeros in the signal band (Fig. 8) to minimize the noise contribution in Eq. (7a). The rest of the noise is removed by subsequent digital filters. $H_a(z)$ is chosen to ensure stability.

The signal band is usually a hundredth of the sampling frequency, so small variations in the gain values of the coefficients B_i may modify the positions of the zeros of transfer function $1-H_B(z)$ significantly, and with them the noise performance of the modulator. The gains must be matched very precisely, one of the main limitations of all architectures used for higher-order noise shaping.

Signal Processing for A-D Converters

In a typical oversampled a-d converter, an anti-aliasing filter at the input attenuates frequencies larger than $f_s/2$ (Fig. 9a). Its cutoff frequency is much higher than the top of the signal band, so the frequency band is unaffected by changes in the cutoff frequency resulting from temperature and process variations. This is an important advantage of the oversampling technique because the anti-aliasing filter is a continuous filter implemented with RC networks, and control of resistor values in MOS technology is not very good.

PCM samples can be obtained from the output of the modulator by filtering the noise and undersampling the signal down to the Nyquist sampling rate (Fig. 10a). Using FIR filters for this decimation function is attractive because they reduce the multiplication operation to an addition (since the input to the filter is only one bit wide in the case of single-loop and double-loop modulators). For large oversampling rates practical considerations, such as the lengths of the filter coefficients and the number of taps, make it uneconomical to implement a low-pass-filter FIR with a cutoff frequency much smaller than the sampling frequency. This is why decimation is usually performed in two stages (Fig. 9a).

The first stage performs a decimation of the modulator output to an intermediate frequency f_I. A simple FIR-filter implementation has zeros at multiples of the frequency f_I because the noise around these frequencies is aliased into the signal band (Fig. 10b). This class of filters has the z-transform

$$H(z) = \frac{1}{N_I^k}\left(\frac{1-z^{-N_I}}{1-z^{-1}}\right)^k \tag{7a}$$

and an amplitude characteristic

$$H(\omega) = \frac{1}{N_I^k}\left(\frac{\sin N_I \omega T}{\sin \omega T}\right)^k \tag{7b}$$

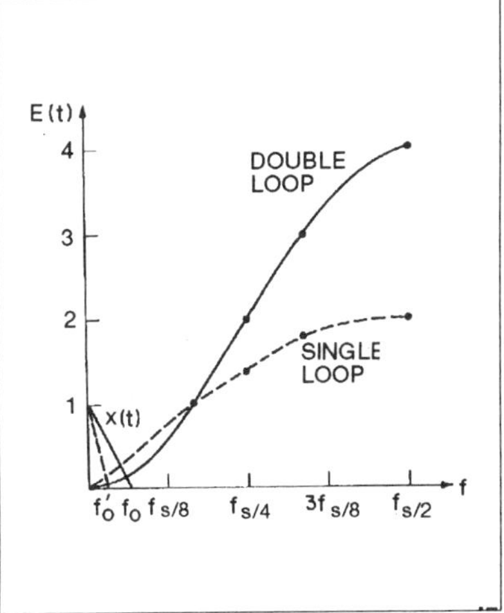

4. Noise spectra of the ΣΔmodulators.

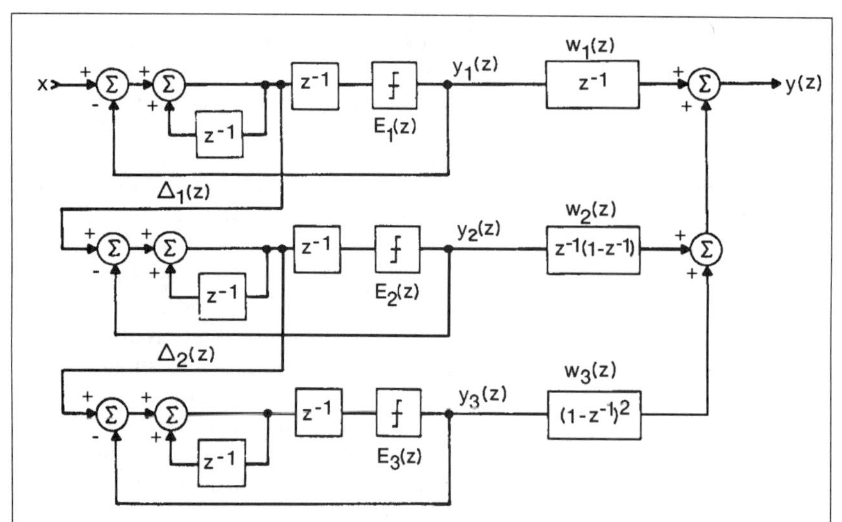

5. *MASH architecture for third-order noise shaping.*

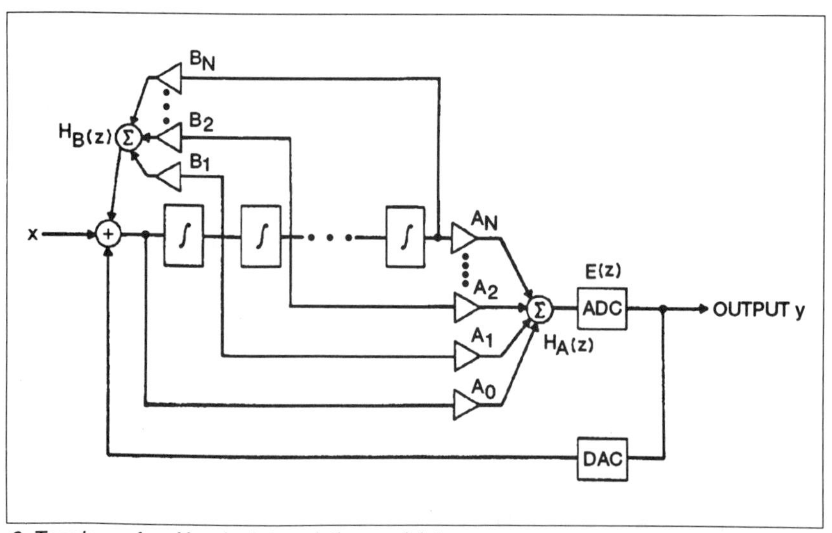

6. *Topology of an N-order interpolative modulator.*

where $N_1 = f_s/f_1$ and $T = f/f_s$.

The transmission zeros will alias over dc, so most of the noise aliased in the signal band is concentrated toward the high frequency end f_N. For k = 1 this filter achieves the averaging operation described early in this article. The envelope of this filter's frequency characteristic (the dashed line in fig. 10b) decreases by 6 dB per octave in the low-frequency domain. This compensates for the increase of the noise density function of the single-loop $\Sigma\Delta$ modulator (Fig. 10a). The envelope of the noise spectrum at the output of the filter before the decimation is almost flat (Fig. 11). The noise that is aliased in the high-frequency region of the signal band from the multitude of aliasing frequencies is larger than the in-band noise, and it degrades the performance of the A/D converter.

For single-loop modulators, the noise aliased in the signal band can be made negligible by using a sinc square filter (k = 2), whose envelope has a slope of 12 dB per octave because of the filter's multiple zeros. For the contribution of the high-frequency noise components aliased into the signal band to be negligible, the order of multiplicity of these transmission zeros (k) must be greater than the shaping order of the modulator [2, 3]. This corresponds to a sinc square function (k = 2) with its a triangular impulse response (fig. 12a) for a single-loop modulator and a sinc cube function (k = 3) with its parabolic impulse response (fig. 12b) for a double-loop modulator. These functions can be implemented using accumulate and dump structures or by computing the filter coefficients [7].

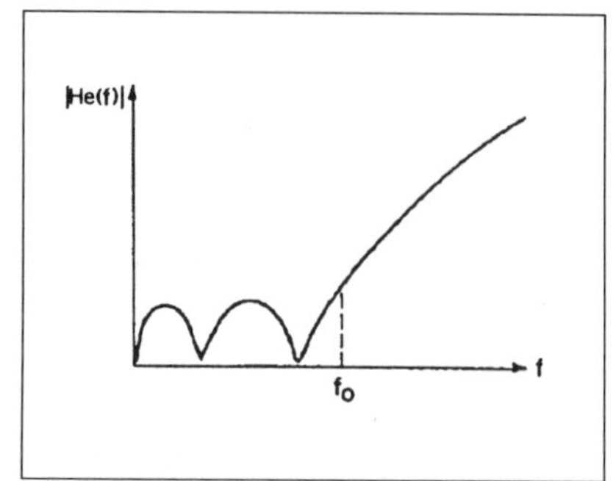

7. *Quantization noise spectrum at low frequency of the N-order interpolative modulator.*

The second decimation stage has a decimation ratio $N_F = f_I/2f_N$ (f_N is the Nyquist frequency). If linear phase is not required, this stage can be implemented efficiently with an IIR filter, which will remove the remaining noise (Fig. 10c) and compensate for the attenuation at high frequencies introduced by the first decimation stage.

The intermediate frequency f_I is a design parameter. Because the power of the noise aliased in the signal band is negligible, f_I should be made as small as possible. This will decrease the computational complexity of the low-pass IIR section, which is proportional to f_I [7].

Signal Processing for A-D Converters

The D/A converter uses the $\Sigma\Delta$ modulation technique with the role of the digital and analog sections interchanged (Fig. 9b). The one-bit stream is generated by a digital $\Sigma\Delta$ modulator, while the filtering of the noise is done by the analog section. This change of roles produces some significant differences. Notably, the process of raising the sampling frequency generates images of the input spectrum. First the sampling frequency is raised to an intermediate frequency f_R by repeating each input sample N_F times, where $N_F = f_R/2f_N$. This is equivalent to attenuating the images by a sinc function (Fig. 13a) given by the equation

$$S(f) = \frac{1}{N_f} \frac{\sin(N_F \pi \frac{f}{f_R})}{\sin(\pi \frac{f}{f_R})} \quad (8a)$$

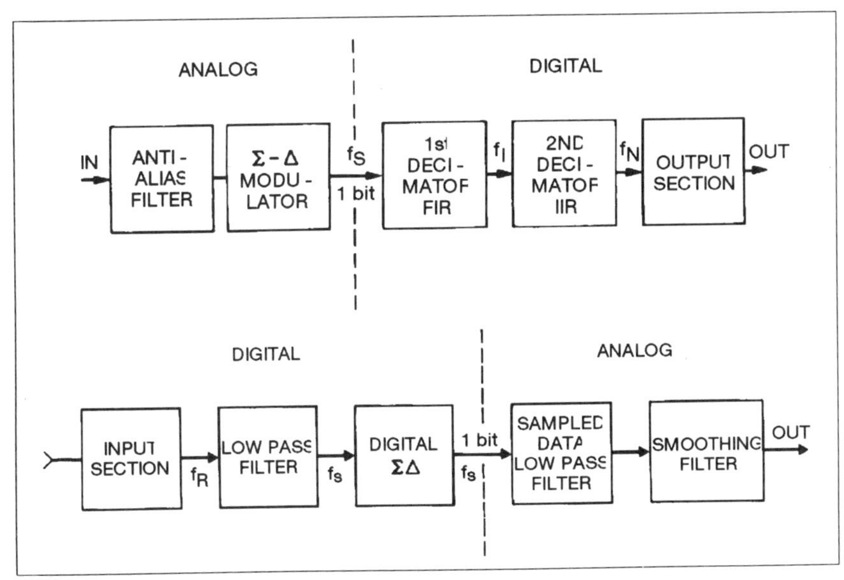

8. Block diagram of a typical oversampled (a) a-d converter and (b) d-a converter.

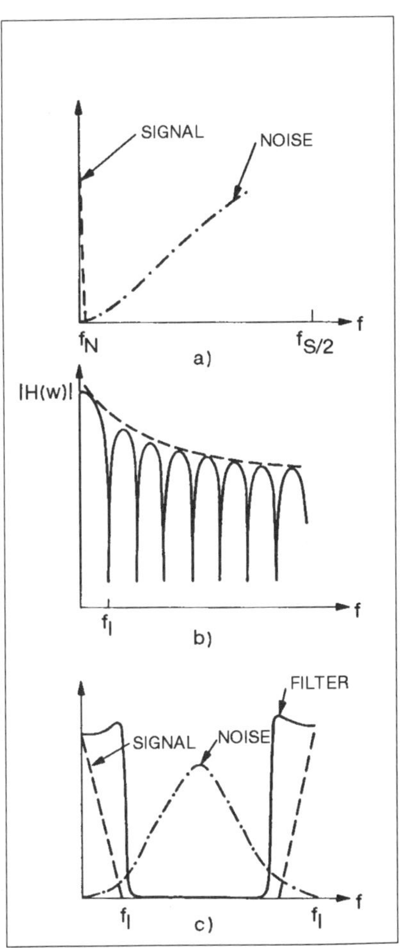

9. Encoder signal processing. (a) Signal and noise at the output of the modulator. (b) Frequency characteristic of the sinc filter. (c) Signal and noise at the output of the FIR decimator.

The residues of the images of the input signal spectrum in the (0, f_R) band are removed by a low pass IIR filter. The output signal is applied to a digital sigma-delta modulator, with each input sample being repeated $N_R = f_s/f_R$ times. Images of the input signal spectrum are created at multiples of f_R frequencies. They are attenuated by a sinc function (Fig. 13b):

$$S(f)=\frac{1}{N_R}\frac{\sin(N_F\pi\frac{f}{f_S})}{\sin(\pi\frac{f}{f_S})} \qquad (8b)$$

It is the task of the analog section to provide enough filtering to attenuate the residues of these images and the noise generated by the digital ΣΔ modulator outside the signal band (Fig. 13c).

Performance and Application

Converters based on the oversampling technique are ideally suited for MOS technology. The size and complexity of the analog section is reduced to the minimum. Because of the high sampling rate, sample and hold circuits are not necessary, and designing the anti-aliasing filter is easy. There are no differential nonlinearity errors, which distort low-level signals. Also, rather large temperature and process variations in the analog-circuit parameters will not affect converter performance. Most of the signal processing is done in the digital domain as this allows very good control of the frequency characteristic. In addition, digital circuits scale with technology much better than analog circuits. For technologies less than 1.5 μ, converters that use oversampling techniques require less silicon area than comparable charge-redistribution devices.

The power dissipation of ΣΔ devices is higher than that of charge-distribution devices because of the large oversampling rate. For the same reason, these devices are limited to low-frequency (0-100 kHz) applications such as voiceband and ISDN.

One of the first ΣΔ devices implemented in MOS technology was a voiceband codec using a single-loop modulator with a 4- MHz sampling rate [11]. Its resolution was close to 13 bits for low input levels. A similar device using a double-loop modulator at a 1-MHz sampling rate exhibited 14-bit resolution for the peak of the curve and close to 15 bits for low input levels. A modulator of this type designed for ISDN applications achieved 12-bit resolution in the 120-KHz band using a 15-MHz sampling frequency [13]. Higher-order noise-shaping architectures can result in 16-bit resolution or better over the whole audio band [14, 16], and recent work has raised the top of the frequency band to 100 KHz [17]. Such performance cannot be attained as efficiently in MOS technology using other conversion techniques.

Because of the reduced functionality of the analog section, the crosstalk between channels is very small. This makes it possible to implement devices that contain several converters [8, 12]. Another area of application is that of the DSP containing such converters in which the digital filter operations necessary for conversion are programmed together with more complex tasks as echo cancellation or voice compression. For submicron CMOS technologies the area necessary to implement an IIR or FIR filter in the digital domain is less than that of the corresponding switched-capacitor network. This makes it possible to replace some low-frequency analog signal processors with similar devices containing an a-d converter, a DSP, and a d-a converter. The inherent programmability of these devices should make possible a faster turnaround than is presently achievable with analog design. ***CD***

References

1. D.J. Goodman, "The Application of Delta

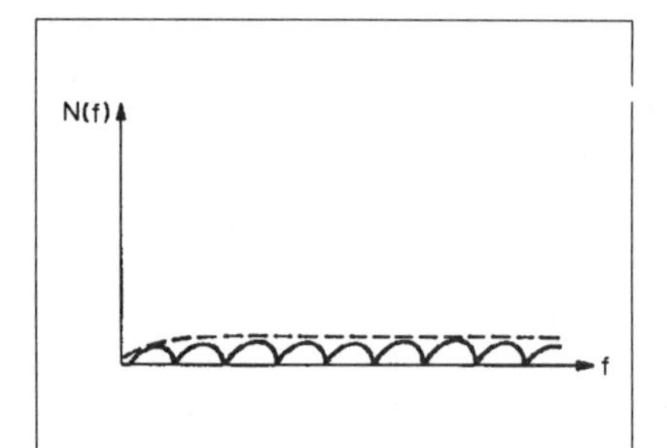

10. The noise density at the output of an averaging FIR filter for single-loop modulators has a nearly flat envelope.

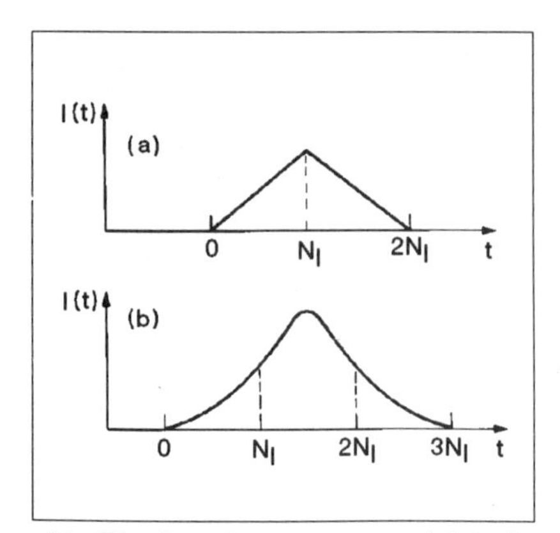

11. The impulse response of (a) sinc square filter and a (b) sinc cube filter.

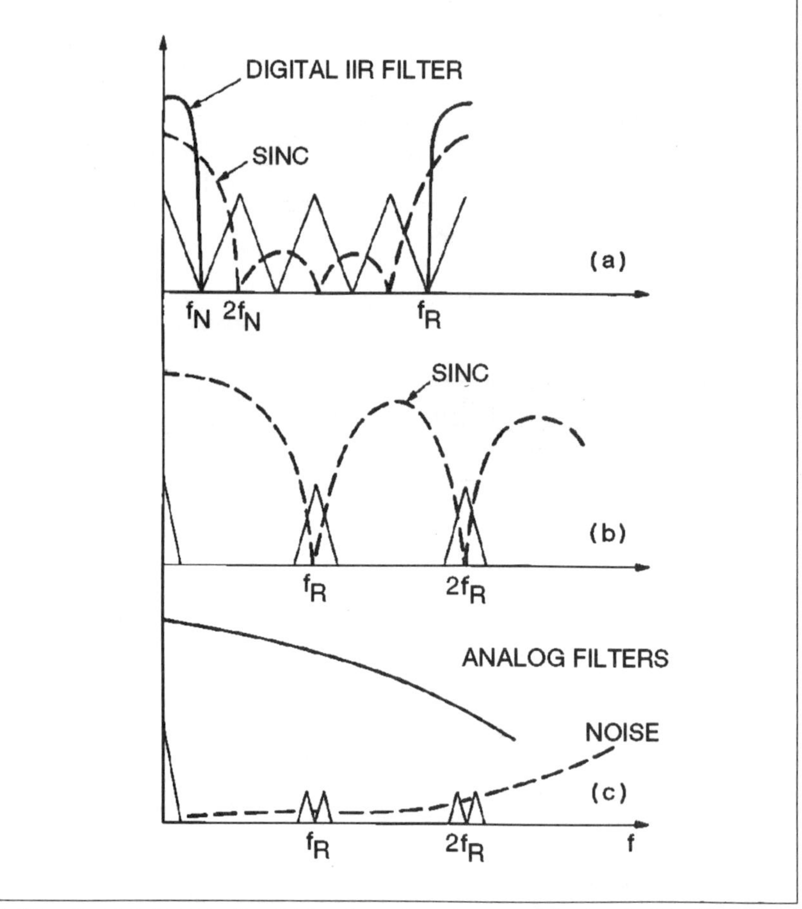

12. Decoder signal processing. (a) Input spectrum images at the input of the IIR filter. (b) Input spectrum images at the input of the ΣΔ modulator. (c) Signal and noise spectrum at the output of the modulator.

Modulation to Analog-to-Digital PCM Encoding," Bell Syst. Tech. J., 48, p. 321-343, 1969.

2. J.C. Candy, "Decimation for Sigma Delta Modulation," IEEE Trans. Commun., COM-34, p. 72-76, 1986.

3. J.C. Candy, "A Use of Double Integration in Sigma Delta Modulation," IEEE Trans. Commun., COM-33, p. 249-258, 1985.

4. J.C. Candy, B.A. Wooley, O.J. Benjamin, "A Voiceband Codec with Digital Filtering," IEEE Trans. Commun., COM-29, p. 815-830, 1981.

5. B.P. Agrawal, K. Shenoi, "Design Methodology for Sigma-Delta Modulation," IEEE Trans. Commun., COM-31, p. 360-369, 1983.

6. V. Friedman, "The Structure of the Limit Cycles in Sigma Delta Modulation," IEEE Trans. Commun., COM-36, p. 972-979, 1988.

7. V. Friedman, D.M. Brinthaupt, D-P. Chen, T.W. Deppa, J. Elward, Jr., E.M. Fields, H. Meleis, "A Bit-Slice Architecture for Sigma-Delta Analog-to-Digital Converters," IEEE J. on Selected Areas in Commun., 6, no.3, p. 520-526, 1988.

8. V. Friedman, D.M. Brinthaupt, D-P. Chen, T.W. Deppa, J. Elward, Jr., E.M. Fields, J.W. Scott, T.R. Viswanathan, "A Dual-Channel Voice-Band PCM Codec Using ΣΔ Modulation Technique," IEEE J. Solid State Circuits, SC-24, p. 274-280, 1989.

9. R.M. Gray, "Oversampled Sigma-Delta Modulation," IEEE Trans. Commun., COM-35, p. 481-489, 1987.

10. R.M. Gray, "Spectral Analysis of Quantization Noise in a Single-Loop SigmaDelta Modulator with DC Input," IEEE Trans. Commun., COM-37, p. 588-600, 1989.

11. T. Misawa, J.E. Iversen, L.J. Loporcaro, J.G. Ruch, "Single-chip per Channel Codec with Filters", IEEE J. of Solid State Circuits, SC-16, p. 333-341, 1981.

12. B.H. Leung, R. Neff, P.R. Gray, R.W. Brodersen, "A Four-Channel CMOS Oversampled PCM Voiceband Coder," Proc. Int. Solid State Circuits Conf., p. 106-107, Feb. 1988.

13. R. Koch, B. Heise, "A 120-KHz Sigma/Delta A/D Converter," Proc. Int. Solid-State Circuits Conf., p. 138-139, Feb. 1986.

14. K. Matsumoto, E. Ishii, K. Yoshitate, K. Amano, R.W. Adams "An 18-bit Oversampling A/D Converter for Digital Audio," Proc. Int. Solid State Circuits Conf., p. 200-201, Feb. 1988.

15. W.L. Lee, C.G. Sodini, "A Topology for Higher Order Interpolative Coders," Int. Symposium on Circuits and Systems, p. 459-462, 1987.

16. Y. Matsuya, K. Uchimura, A. Iwata, T. Kobayashi and M. Ishikawa, "A 16b Oversampling A/D Conversion Technology using Triple Integration Noise," Proc. Int. Solid-State Circuits Conf., p. 48-49, Feb. 1987.

17. M.Rebeschini et al., "A 16-bit 160-KHz CMOS A/D Converter using Sigma-Delta Modulation," Proc. Custom Integrated Circuits Conf., p. 6.1.16.1.5, May 1989.

18. D.R. Welland, B.P. Del Signore, E.J. Swanson, T. Tanaka, K. Hamashita, S. Hara, K. Takasuka, "A Stereo 16-Bit Delta-Sigma A/D Converter for Dogotal Audio," J. Audio Eng. Soc., 37, No.6, p. 476-485, 1989.

CHAPTER 3

OPTICAL TECHNOLOGY

Photochemical Processing of Semiconductors: New Applications for Visible and Ultraviolet Lasers

J. Gary Eden

Abstract

A wide variety of semiconductor and metal films have recently been grown, doped, or etched by laser photochemical processes. While still early in its development, laser-driven processing of semiconductor devices appears promising as a supplement to existing processing techniques and particularly in the fabrication of III-V compound and custom devices.

Introduction

While lasers have been used for over two decades to drill, weld, and cut various materials, it has only been in the last six years that the potential for laser processing of semiconductors has been demonstrated. The development of this field has been spurred by the growing demands being placed on semiconductor processing technology as increased packing densities continue to shrink the size of individual devices toward 1 μm and beyond. One barometer of its rapid growth is the number of metal or semiconducting films that have been deposited by laser-assisted processes. First demonstrated by a research group headed by Dan Ehrlich, Tom Deutsch, and Richard Osgood at MIT Lincoln Laboratory in 1979, laser photochemical deposition has come to encompass at least 17 elements of the periodic chart, which have been grown as thin films in elemental or compound form (for example, As in GaAs) [1].

Laser-assisted deposition, etching, or doping of films all draw upon one or both of two of the unique characteristics of laser radiation: its spatial coherence and spectral purity. The former is related to the ease with which laser beams can be focused down to spot diameters that are roughly equal to the wavelength of the laser light. The "monochromaticity" or narrow spectral output of the laser, on the other hand, guarantees photons having well-defined energies that allow one to initiate specific chemical reactions in the desired medium. Figure 1 illustrates the basic experimental configuration for laser processing. The beam from a laser enters a reactor through a window and either strikes or passes just above (and parallel to) the substrate. One or more gases (or a liquid) are introduced to the reactor and are flowed over the substrate. The fundamental notion underlying laser processing is that, in the absence of laser radiation, the gas (or liquid)/substrate interface is chemically inert. However, by careful choice of the laser wavelength, gas, and substrate, the laser will initiate a chemical reaction that culminates in the deposition of a film or in etching or doping of the substrate itself.

Laser chemical materials processes can be separated into one of two categories, depending on which of the three media shown in Fig. 2 (gas, molecular layer adsorbed on surface, or substrate) absorb strongly at the laser wavelength. *Pyrolytic* (or thermally driven) *processes* are those in which the gas is transparent to the incoming radiation, but the substrate is strongly absorbing. Therefore, the laser provides a localized source of heat and polyatomic molecules in contact with the surface and can be broken down (pyrolyzed) if the temperature exceeds a certain value that varies from molecule to molecule. One advantage of this approach over conventional resistance (or RF) heating of the substrate is that laser heating can be confined close to the surface by simply adjusting the laser wavelength. Also, the highest temperature region is confined laterally to an area determined by the thermal conductivity of the substrate. If a laser pyrolytic process is used to grow a film, then the technique is known as *laser chemical vapor deposition* (LCVD) and, except for the source of heat, is otherwise similar to conventional chemical

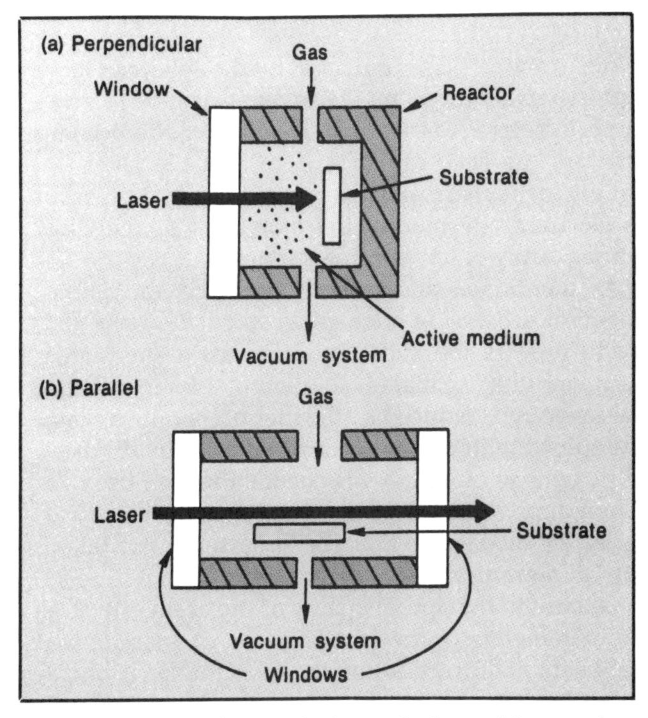

Fig. 1 Generalized diagrams for laser-assisted materials processing: (a) for the laser beam at normal incidence to the substrate and (b) for the beam passing parallel to the surface. The "active medium" can be a gas or liquid.

Reprinted from *IEEE Circuits and Devices Magazine*, Vol. 2, No. 1, pp. 18-24, January 1986.

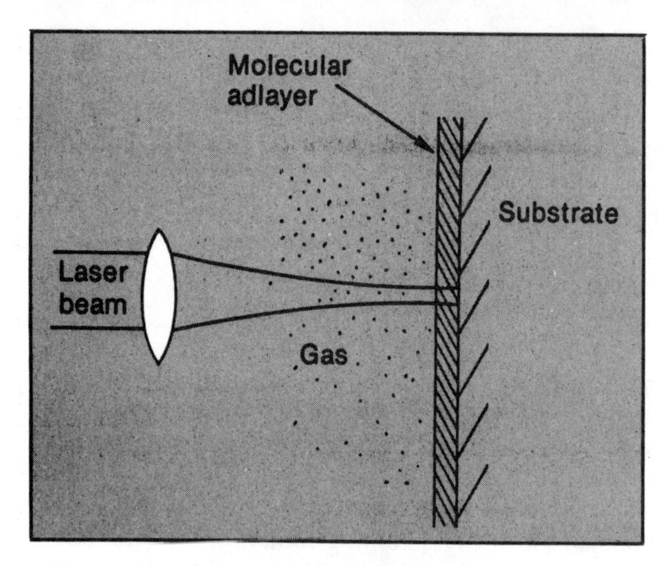

Fig. 2 Drawing of the interaction of a focused laser beam with a substrate showing the molecular adlayer on the surface (after Ref. [5]).

vapor deposition (CVD). A wide variety of metal (W, Ni, Ti, etc.) and semiconductor films have been grown in this manner. [2]

If the substrate is transparent and the gas or molecular adlayer absorbs strongly at the laser wavelength, the process is said to be *photolytic* or *photochemical*. Photochemical processes in the gas phase involve photons having sufficient energy to break chemical bonds within the gaseous molecule, thereby producing molecular or atomic fragments that react with the surface. The energy of the chemical bond to be broken determines the maximum allowable wavelength (i.e., minimum photon energy) of the laser.

Absorption of laser radiation by the adsorbed layer that inevitably forms on the substrate can be *pyrolytic* or *photochemical*, depending on whether the primary effect of the laser radiation is to heat the layer (resulting in the thermal dissociation of the molecules) or to electronically modify the molecules so that they subsequently react with the surface.

As mentioned above, both pyrolytic and photochemical processing offer *spatial selectivity* — the ability to process the substrate over just a small area. Although the "spatial" resolution of a laser chemical process can be reduced by the thermal conductivity of the substrate (for pyrolytic processes) or by diffusion of gaseous photofragments outside the laser beam, to a first approximation, the region of the substrate affected by the laser is that portion actually irradiated. The minimum-laser focal spot size (diameter) is related both to the optical quality of the beam as well as the parameters of the focusing lens. Assuming that the beam is diffraction-limited (i.e., that as it propagates, its diameter increases no faster than that due to diffraction), then the focal spot size is given by $1.2\lambda f \cdot 1/d$ where λ is the wavelength of the laser radiation and f and d are the lens focal length and diameter, respectively. For visible and ultraviolet lasers, submicron spot sizes are readily attainable. Consequently, devices can be patterned directly onto a substrate without the need for conventional photolithographic masking techniques. This process has come to be known as *laser direct writing*.

A further attractive aspect of photochemical processing is its ability to exploit the unique absorption spectra of molecules to selectively influence only a *particular* species. For example, Fig. 3 shows the absorption spectra for three of the Column III A alkyls [trimethyaluminum (TMA), trimethylgallium (TMGa), and trimethylindium (TMIn)] in the ultraviolet. Since peak absorption for each occurs at a different wavelength, one can discriminate between the three by careful choice of the laser wavelength.

The remainder of this article will emphasize laser photochemical processes; another article in this issue by Dr. Susan D. Allen deals with LCVD in more detail.

Film Growth

Elemental Films by Photodissociation or Photoionization

A wide variety of photochemical processes have been shown to be suitable for depositing metal and semiconductor films. Far and away, the most prevalent choice is photodissociation in which the absorption of one or more photons by the molecule causes it to unravel or photodissociate. That is

$$MX + n\hbar\omega \rightarrow M + X \quad (1)$$

where $\hbar\omega$ is the photon energy, MX is generally a polyatomic molecule, M and X are either atoms or molecular radicals, and n is an integer. As an example, thin ($\lesssim$1-μm-thick) cadmium films have been grown at room temperature on various substrates by photodissociating dimethylcadmium ($Cd(CH_3)_2$) with

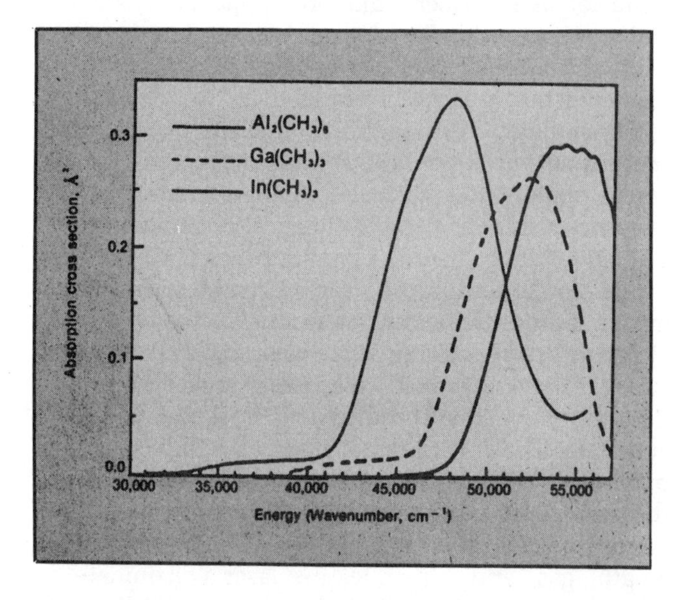

Fig. 3 Absorption spectra for the Column III A alkyls trimethylaluminum (in dimer form), trimethylgallium, and trimethylindium in the ultraviolet (after Ref. [7]).

an ultraviolet laser. Since the energy of one ultraviolet photon is sufficient to break both Cd-CH_3 bonds, free cadmium atoms are produced in the laser beam and, subsequently, migrate to the surface to produce a film. The residual methyl radicals are simply pumped out of the system. One attractive aspect of this approach is that the vapor pressure of $Cd(CH_3)_2$ at room temperature is several Torr, while that for metallic cadmium is 10^{-10} Torr!

Often, it is not necessary for the laser to completely free the atom of interest from the host or "parent" molecule. Once one chemical bond is broken, the remaining molecular fragment is either unstable and eventually breaks down into its atomic constituents, or it may react with the surface to form the desired film. Germane (GeH_4), for example, decomposes into $GeH_2 + H_2$ upon simultaneously absorbing two 248 nm ($\hbar\omega = 5.0$ eV) photons. The direct production of Ge atoms from GeH_2 is unlikely since the absorption of two more photons is required. Rather, GeH_2 and GeH_3 radicals (the latter produced from GeH_2 by collisions) impinge on the substrate, free hydrogen is liberated, and a metallic germanium film is formed. Depending on the parent molecule and the wavelength of laser chosen, the absorption of several photons may be required.

One of the most fascinating characteristics of laser chemical processing is that one can frequently control the fragments produced by varying the laser wavelength. Consider the Column III A alkyl trimethylaluminum ($Al_2(CH_3)_6$: TMA), for example. For low intensity, ultraviolet lamps or lasers, the molecule is prone to absorb one photon, and its absorption spectrum is shown in Fig. 3. In this case, the molecule photodissociates and the products of the photoprocess (Al atoms and CH_3 radicals) are electrically neutral. However, at higher laser intensities, an additional process known as *multiphoton ionization* (MPI) occurs if the laser wavelength is tuned to a resonance of the aluminum atom. As illustrated in Fig. 4, the MPI resonances of TMA are sharp (and lie in the blue portion of the visible spectrum); however, at these wavelengths, not only is TMA photodissociated, but the free Al atom is simultaneously photoionized. For this process to occur requires that the molecule simultaneously absorb 5 photons. That is

$$Al_2(CH_3)_6 + 5\hbar\omega \rightarrow Al^+ + e^- + 6CH_3 \qquad (2)$$

The advantage of producing charged products is that the aluminum ions can be attracted to the substrate by an electric field, allowing one to discriminate against the neutral fragments. In summary, the degree of selectivity required for a given process will largely determine the choice of photochemical process.

Compound Films

Compound semiconductor and insulator films can be deposited by utilizing photodissociation of a molecule to initiate a gas phase reaction. Low temperature deposition of Si_3N_4 was demonstrated at the Colorado State University by irradiating mixtures of silane (SiH_4) and ammonia (NH_3) with a 193 nm excimer laser. Silane absorbs weakly at 193 nm, but ammonia rapidly decomposes, and the resulting gas phase reaction produces Si_3N_4 on a nearby substrate. The stoichiometry of the silicon nitride is sensitive to the relative partial pressures of SiH_4 and NH_3. Similarly, SiO_2 films have been grown by the laser-initiated reaction of SiH_4 with N_2O [3].

$$SiH_4 + 2N_2O + \hbar\omega \rightarrow SiO_2 + \text{Products} \qquad (3)$$

Recently, a group at Bell Labs has also succeeded in growing InP by photolyzing a mixture of In- and P-alkyls with a laser whose wavelength was also 193 nm.

Adsorbed Phase Reactions

Under certain conditions, the primary photochemical process occurring in film deposition is the interaction of laser radiation with the layers of molecules that are adsorbed onto the substrate. Though quite thin, such adsorbed layers often absorb ultraviolet (UV) radiation much more strongly than an equivalent number of molecules in the gas phase [4]. Photodissociation of adsorbed molecules occurs in much the same way as it does in the gas phase. For dimethylcadmium

$$Cd(CH_3)_2 \text{ (Adsorbed)} + \hbar\omega \rightarrow Cd\text{(Ads.)} + 2CH_3\uparrow \qquad (4)$$

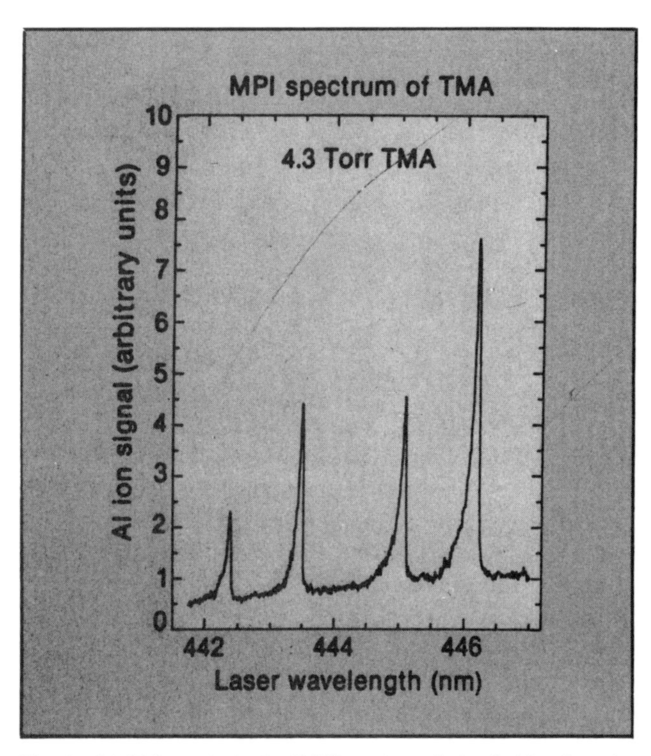

Fig. 4 Multiphoton ionization (MPI) spectrum of trimethylaluminum in the visible (blue). This spectrum was obtained by measuring the number of electrons (and Al ions) produced as the laser wavelength was varied. Note the sharp resonances as opposed to the continua of Fig. 3.

While (4) is similar to the gas phase reaction, the production of a film from an adsorbed layer offers higher spatial resolution, since the mobility of the metal atom is restricted because it is on the surface and not in the gas phase [5].

Table 1 lists many of the elemental and compound films that have been grown to date and the laser wavelength and "parent" molecule that are involved. All offer the advantage of low-temperature operation as compared with conventional processing techniques such as CVD. This feature is of growing importance as devices increasingly depend on faster, yet more temperature, sensitive materials such as GaAs. Also, the desire to minimize the migration or redistribution of dopants into adjoining devices and warpage of wafers dictates that lower temperature processing techniques be sought [6].

Laser-Triggered CVD

One recent development that demonstrates the promise of photochemical processing to permit lower temperature operation is the observation of laser-initiated or "triggered" CVD. Normally, pyrolysis does not occur below a specific temperature, which depends on the gas pressure and flow rate. For 5 percent GeH_4 in He at a total pressure of 225 Torr, this temperature is ~440°C. It has been shown, however, that if the substrate is illuminated with a weak fluence of 193 nm radiation for 10–20 sec, chemical vapor deposition will occur for substrate temperatures at least 100°C lower than is normally possible. Figure 5 shows the rate at which a Ge film grows on GaAs at 350°C when the substrate is irradiated for 20 sec with a UV laser beam. Each sinusoidal cycle represents the growth of ~0.6 μm of film. Note that growth does not start until after the laser is turned *off*. The effect of substrate temperature on the film growth rate is given in Fig. 6. Temperature *decreases* to the right, and the data points to the *left* of the dashed line indicate Ge films that were grown by conventional CVD (no laser). To the right of the dashed line, however, is a region where films will not normally grow, but illuminating the surface briefly with 193 nm laser radiation allows for the CVD process to proceed.

Etching

Laser-assisted etching of films occurs along similar lines. It generally involves photodissociating a polyatomic molecule containing a halogen atom such as fluorine or bromine, which then chemically attacks the surface. The result is an etched region that closely matches the laser beam cross section. For example, chlorine gas (Cl_2) can be photodissociated into Cl

Parent Molecule(s)	Material Deposited	Laser Wavelength (nm)
SiH_4	Si	9,000–11,000
SiH_4/NH_3	Si_3N_4	9,000–11,000
SiH_4/C_2H_4	SiC	9,000–11,000
$Cd(CH_3)_2$	Cd	337–676
$Mn_2(CO)_{10}$	Mn	337.4–356.4
$Mo(CO)_6$	Mo	260–270
$Cr(CO)_5$	Cr	257, 260–270
$Fe(CO)_5$	Fe	257
$W(CO)_5$	W	257
$Cr(CO)_6$	Cr	257
$Cd(CH_3)_2$	Cd	193, 257
$Zn(CH_3)_2$	Zn	193, 257
$Al(CH_3)_2$	Al	193, 257
$Ga(CH_3)_3$	Ga	257
$TiCl_4$	Ti	257
$Pt(PF_3)_4$	Pt	248
SiH_4	Si	193, 248
GeH_4	Ge	193, 248
WF_6	W	193
SiH_4/N_2O	SiO_2	193
SiH_4/NH_3	Si_3N_4	193
$Al_2(CH_3)_6/N_2O$	Al_2O_3	193
$Zn(CH_3)_2/NO_2$ (or N_2O)	ZnO	193 (248)
$In(CH_3)_3/P(CH_3)_3$	InP	193

Table 1 Elemental and compound films deposited by laser photochemical process.

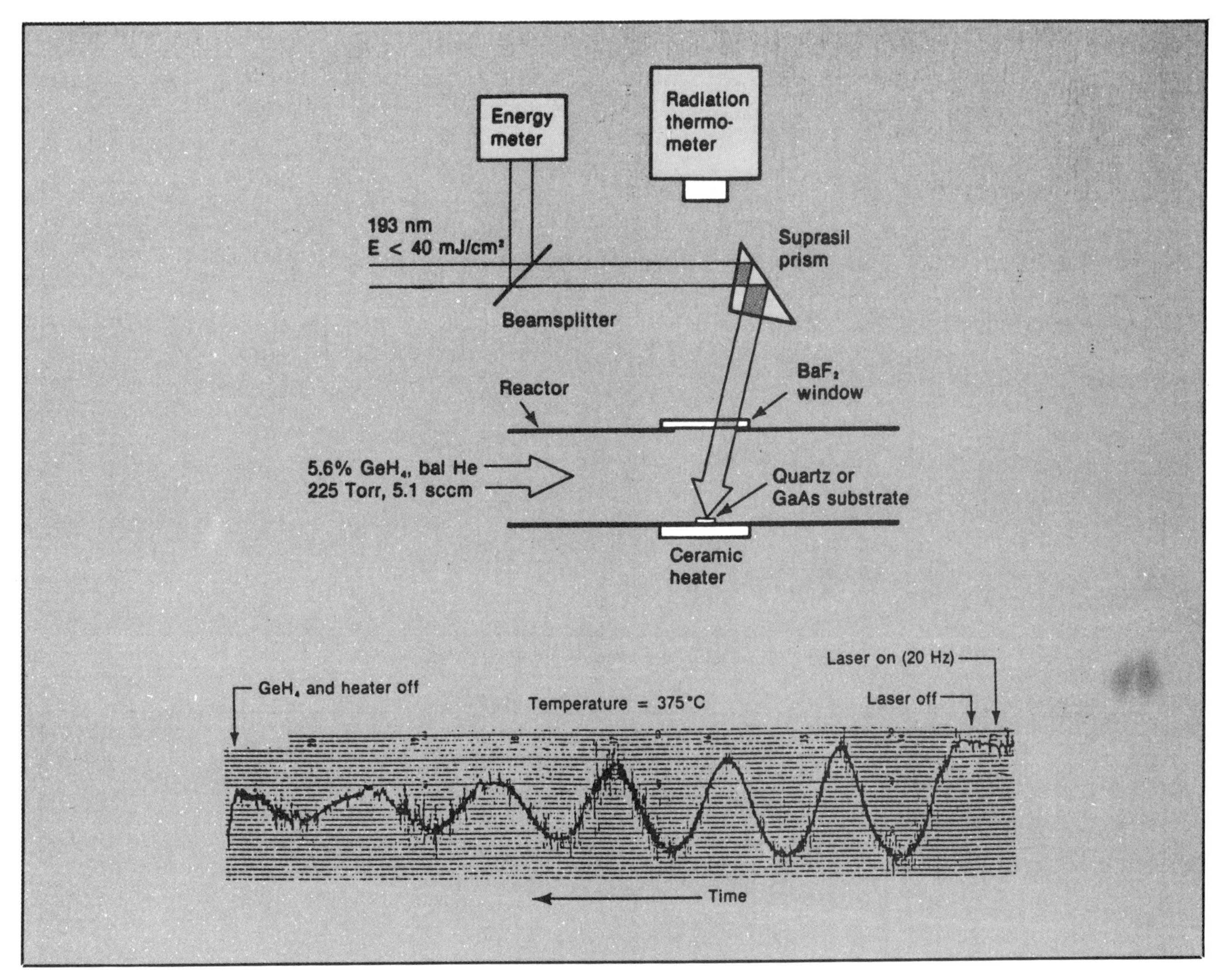

Fig. 5 *Growth of a germanium film on GaAs at 350°C by laser-triggered CVD. A schematic diagram of the apparatus used in this experiment is shown at the top of the figure. Growth does not begin until the laser is turned off, but film growth will not occur at this temperature without the laser radiation. Each 360° oscillation in the curve (bottom of the figure) represents an additional 0.62 μm in the thickness of the film. At the end of the run, the total film thickness is ~3.6 μm. This curve was obtained by monitoring the infrared radiation from the GaAs substrate as the Ge film grew.*

atoms with a visible or UV laser. In the presence of a silicon or SiO_2 surface, silicon tetrachloride, a volatile molecule, is formed and leaves the surface to be pumped out of the reactor. Laser etching is highly anisotropic (i.e., yields sharp vertical walls) and etching rates of 1000 Å-sec^{-1} are easily obtained. Table 2 lists several of the etching systems that have been demonstrated thus far. Note that the etchant can also be a liquid such as KOH.

Doping

Two different approaches to laser-assisted doping of semiconductors have been followed. One—in situ doping—involves photodissociating the parent molecule for the dopant at the same time that the crystal is being grown. Germanium and silicon films have been doped, for example, by photodissociating GeH_4 (or SiH_4) and TMA simultaneously [8]. The more commonly used technique, however, is to accomplish two tasks with the laser. The dopant atom is

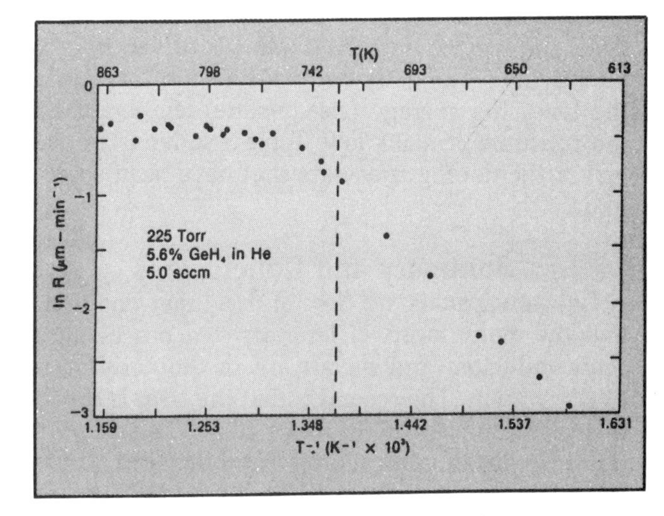

Fig. 6 *Variation of the Ge film growth rate with substrate temperature. The region to the left of the vertical dashed line is one in which conventional CVD growth occurs. To the right of this threshold, irradiation of the surface is necessary to initiate (but not sustain) growth.*

Substrate	Etchant(s)	Laser Wavelength (nm)
Si	Br_2, Cl_2, HCl, XeF_2, KOH	488
Si	Cl_2	308, 337
Si	COF_2	193
Ge	Br_2	488, 514
GaAs	Cl_2, CCl_4, H_2SO_4:H_2O_2(Aqueous), KOH, CH_3Br, CH_3Cl, HNO_3	488
GaAs	CH_3Br, CH_3Cl	193
GaP	KOH	488
InP	CF_3I, H_3PO_4, HCl:HNO_3	488
CdS	H_2SO_4:H_2O_2, HNO_3	488
SiO_2	Cl_2	488
SiO_2	CCl_2F_2	248
W	COF_2	193
Al	H_3PO_4:HNO_3:$K_2Cr_2O_7$	488
Ag	Cl_2	337, 355

Table 2 Summary of several substrates etched by laser process.

Substrate	Parent Molecule (Dopant)	Laser Wavelength (nm)
Si	BCl_3 (boron)	257–514
	PCl_3 (phosphorous)	257–514
	BCl_3	193, 351
	PCl_3	193, 351
	$B(C_2H_5)_3$ (boron)	193
GaAs	BCl_3, PCl_3	193, 351
InP	$Cd(CH_3)_2$ (cadmium)	257–514

Table 3 Laser-assisted doping of semiconductors.

first released by photodissociation of the donor molecule. At the same time, the impinging laser beam heats the surface and the dopant is driven into the underlying material by normal diffusion. By pulsing the laser, the average (steady-state) temperature of the substrate remains low. Table 3 summarizes several of the doping processes that have been studied to date.

Summary and Conclusions

Two conclusions are clear at this time. The first is that the many facets of laser-assisted processing of semiconductors and metals are in the early stages of exploration. The second is that the laser is capable of performing all of the basic processing steps required in device fabrication without the need for photolithographic techniques. The boundaries of the field are set only by the number of semiconductor and metal-containing compounds having reasonable vapor pressure (~1 Torr) at several hundred degrees centigrade or less. One of many areas demanding attention in the next few years lies in determining the photofragments produced when a molecule is irradiated at different wavelengths. However, it is certain that, as substrate temperature looms as a larger obstacle because of decreasing device geometries, the laser will certainly play a larger role in semiconductor processing.

References

[1] D. J. Ehrlich and J. Y. Tsao, "A Review of Laser-Microchemical Processing," *J. Vac. Sci. Technol.*, pp. 969–984, 1983.

[2] S. D. Allen, A. D. Trigubo, and Y. C. Liu, "Laser Chemical Vapor Deposition Using Continuous Wave and Pulsed Lasers," *Opt. Eng.*, vol. 23, pp. 470–474, 1984.

[3] P. K. Boyer, G. A. Roche, W. H. Ritchie, and G. J. Collins, "Laser-Induced Chemical Vapor Deposition of SiO_2," *Appl. Phys. Lett.*, vol. 40, pp. 716–719, 1982.

[4] C. J. Chen and R. M. Osgood, Jr., "Spectroscopy and Photoreactions of Organometallic Molecules on Surfaces," *Proc. Mat. Research Soc.*, vol. 17, pp. 169–175, 1983.

[5] D. J. Ehrlich, R. M. Osgood, Jr., and T. F. Deutsch, "Photodeposition of Metal Films with Ultraviolet Light," *J. Vac. Sci. Technol.*, pp. 23–32, 1982; R. M. Osgood and T. F. Deutsch, "Laser-Induced Chemistry for Microelectronics," *Science*, vol. 227, pp. 709–714, 1985.

[6] S. C. Su, "Low Temperature Silicon Processing Techniques for VLSI Fabrication," *Solid State Technology,* vol. 24, pp. 72–82, Mar. 1981.
[7] H. H. Gilgen, C. J. Chen, R. Krchnavek, and R. M. Osgood, Jr., "The Physics of Ultraviolet Photodeposition," in *Laser Processing and Diagnostics,* ed. by D. Bäuerle, New York: Springer, pp. 225–233, 1984.
[8] R. W. Andreatta, C. C. Abele, J. F. Osmundsen, J. G. Eden, D. Lubben, and J. E. Greene, "Low-Temperature Growth of Polycrystalline Si and Ge Films by Ultraviolet Laser Photodissociation of Silane and Germane," *Appl. Phys. Lett.*, vol. 40, pp. 183–185, 1982.

Laser-Enhanced Plating and Etching for Microelectronic Applications

Robert J. von Gutfeld

Abstract

Lasers can be used in several ways to bring about a large enhancement in the rate of both deposition and material removal on a very localized scale. The local rate enhancements using a focused beam make it possible to produce patterns as small as 1 μm or less without the use of masks. This paper describes the use of lasers to obtain patterns using plating and etching techniques. The enhanced rates result from accelerated electrochemistry and mass transport of the electrolytic solutions resulting from local heating produced by laser absorption at the workpiece-electrolyte interface. Examples of laser electroplating include results for nickel, copper, and gold with rates as high as ~1–5 μm/s. This is ~3 orders of magnitude higher than conventional electroplating rates. Similarly, high rates of electro-etching have been observed, particularly for such metals as stainless steel. Several applications of this technique will be described, particularly as they pertain to VLSI circuit boards and connectors. A recent modification of the laser plating technique combines a laser with a high-velocity jet of electrolyte to obtain still higher rates of deposition. This scheme will be described in some detail for gold and copper plating with observed rates as high as 20 and 50 μm/s, respectively, for the two metals.

Laser-enhanced deposition and etching are techniques of increasing interest in the area of microelectronics as a means for achieving maskless patterning for circuit design and repair. These techniques allow localized areas to be accessed for either deposition or removal of material without the costly and time-consuming process of any number of lithographic steps normally required to obtain patterning. In the present discussion, we will limit ourselves to material processing in a liquid environment. The local rate enhancement of the process is due to thermal effects produced by focused laser radiation absorbed at the liquid-solid interface [1]–[6]. Localization is controlled by the size of the focused laser spot, the speed with which the laser moves relative to the workpiece undergoing plating or etching, and thermal parameters of the workpiece. The plating and etching processes to be described are generally nonlinear with temperature so that very large enhancements can occur locally. Simultaneously, some background plating or etching occurs in those regions not subject to the laser radiation but at rates that are orders of magnitude smaller. Patterns can be traced by means of relative motion between the laser and the workpiece in one of several possible ways. For example, a scanning mirror can move the laser relative to a fixed substrate, or a stationary beam can be made to scan along a substrate attached to a computerized, movable XYZ arm. The laser wavelength is chosen so that minimal optical absorption occurs in the liquid (electrolyte) with substantial absorption in the substrate.

Experiments to be reported here have been undertaken using a continuous wave (cw) argon ion laser (multiline) directed into a cell consisting of glass or quartz walls open at the top (Fig. 1). An electrode containing a hole for the light to pass through is biased positively with respect to the workpiece (cathode) for electroplating. A third electrode, the reference electrode, is used to control the potential close to the cathode and to minimize potential drops due to current-resistance losses in the electrolyte. Early experiments on the laser-enhanced electrodeposition of nickel resulted in plating rates up to 10^3 times greater than nonirradiated regions using focused laser spots several microns in diameter. It was also shown in these experiments that substrates such as tungsten and molybdenum, which are normally difficult to plate by conventional techniques, can be laser plated quite readily with excellent adhesion between film and substrate. It is believed that the thin oxide layers common to those metals on exposure to air are rapidly dissolved by laser heating just prior to deposition, thereby providing a clean surface for the strong adhesion. These experiments also provided data on very high speed etching of stainless steel with the stainless steel anode irradiated by the laser. Rates up to 10 μm/s were achieved in the electroetching of holes 25–100 μm in diameter, a rate over three orders of magnitude faster than the background etch rate.

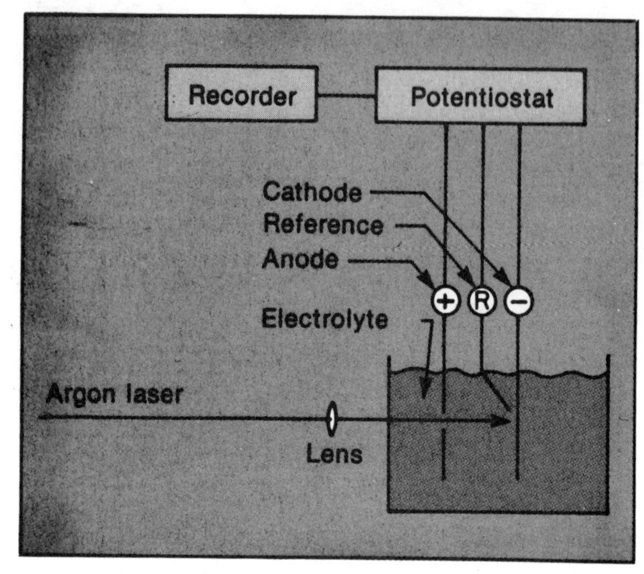

Fig. 1 Schematic of quartz cell showing the three electrodes. The laser is shown passing through the anode, focused onto the cathode.

A detailed investigation of laser-enhanced electrodepositions was subsequently undertaken to elucidate more fully the mechanisms giving rise to the large observed plating enhancements [3]. For this

Reprinted from *IEEE Circuits and Devices Magazine*, Vol. 2, No. 1, pp. 57-60, January 1986.

study, the system Cu/Cu^{++} was used together with special microelectrodes designed to obtain plating current-voltage relationships, both with and without laser irradiation (polarization curves). The active cathode area was approximately equal in size to the incident focused laser spot, from 200 to 500 μm in diameter. As shown in Fig. 1, the three-electrode system was used to obtain the values of the cathode-reference electrode potential difference. This represents the potential drop or overpotential occurring across a narrow region of electrolyte in which the electrochemical dynamics occur. These studies revealed that the laser has an important effect over a wide range of overpotentials. At low values, where the plating rate is limited by electrochemical kinetics, a local increase in temperature leads to more rapid kinetics expressed by Arrhenius-type terms in the equations relating the current to the overpotential. At higher overpotentials, plating becomes mass-transport controlled; that is, the plating rate is limited by how fast the locally depleted ions can be replenished. In conventional electroplating (etching), the ion replenishment occurs via diffusion driven by the concentration gradient. With the addition of laser power (on the order of 10^3 W/cm^2), the electrolyte-cathode interface becomes heated to temperatures on the order of 50–100°C above ambient, depending of course on the substrate's thermal and optical parameters. This gives rise to large temperature gradients and strong fluid convection currents, which are equivalent to a local microstirring of the electrolyte. This provides ion replenishment, hence the high plating rates.

There is an additional effect that the laser heating causes to enhance plating rates, namely a local shift in the value of the rest potential. Although the thermodynamics of this effect will not be discussed in detail here, it can be described in terms of a local effective potential at the liquid-solid interface, which is temperature-dependent. Absorbed laser radiation increases the temperature locally and causes a local potential change. This gives rise to a potential difference between the heated and unheated regions. This potential difference represents a local thermobattery and can cause local plating and simultaneous etching without the application of an external EMF. For certain electrolytes such as copper sulfate, the laser-heated region undergoes local plating [3], [4]. However, overall charge conservation is maintained by simultaneous etching from the relatively large unheated (cooler) regions, but at a much slower rate. This simultaneous plating-etching phenomenon is a special form of exchange plating, a type of plating commonly observed when a metal is submerged into a solution containing ions more noble than that of the metal. For a metallized glass substrate undergoing thermobattery deposition, plating continues in the laser-irradiated portion until a narrow peripheral region is fully etched away, producing an electrically isolated central spot. For certain metal-electrolyte interfaces, the thermobattery has the opposite polarity and thus produces plating in the periphery with etching in the centrally heated region. This effect has been experimentally observed for a nickel-plated substrate immersed in a $NiSO_4$ solution [3].

Our recent interests have centered on obtaining high-speed gold patterning using laser-enhanced techniques. Gold is especially important in the microelectronics industry since this metal is used to provide electrical contact surfaces on both circuit boards and pin connectors [7], [8]. With conventional plating techniques, masking of contact parts is particularly problematic and expensive. Consequently, masking is often not used. This results in gold deposition in areas where it is not needed, a highly wasteful use of precious metals. High-speed localized plating, such as that afforded by laser-enhanced techniques, can be utilized to bring about considerable cost reductions by (1) reducing the amount of precious metal deposited by limiting the plating to needed areas, (2) reducing the amount of time needed to plate the part, and (3) reducing the amount of space needed to accommodate the plating line. The space saving comes about through the use of much smaller plating tanks, which in turn results from the much smaller time interval the part needs to be in the bath to reach the desired plating thickness. This is particularly true for parts moving along a continuous belt-driven line. The smaller tanks also bring about an additional saving through the need for a much smaller gold inventory.

Work on laser-enhanced gold plating on predeposited 2-μm-thick nickel-coated beryllium copper substrates, a material commonly used for connector pins and pads, has led to plating rates as high as 1 μm/s. Patterns in the form of localized spots on the order of 0.5–1.0 mm in diameter serve as excellent electrical contact areas and can readily be deposited by the laser-enhanced plating technique. Depositions have been made under a variety of electrochemical and laser parameters. Examination of both cross-sectioned deposits and top surfaces of the same deposits using scanning electron and optical microscopy reveals excellent morphologies so long as the current density is kept below a value that provides a continuous resupply of ions to the substrate (cathode). Operation above this value, known as the *limiting current density*, results in porous films with poor adhesion; both effects due in part to evolved hydrogen entrapped within the film.

A method to promote faster ion replenishment (mass transport) in order to further increase the plating rates for gold deposition has recently been devised using forced convection [9] called *laser-jet plating*. Here the electrolyte reaches the cathode in the form of a free-standing jet stream emanating from a small nozzle via a pressurized chamber with the jet directed onto the cathode. The anode is contained within the chamber and is electrically in contact with

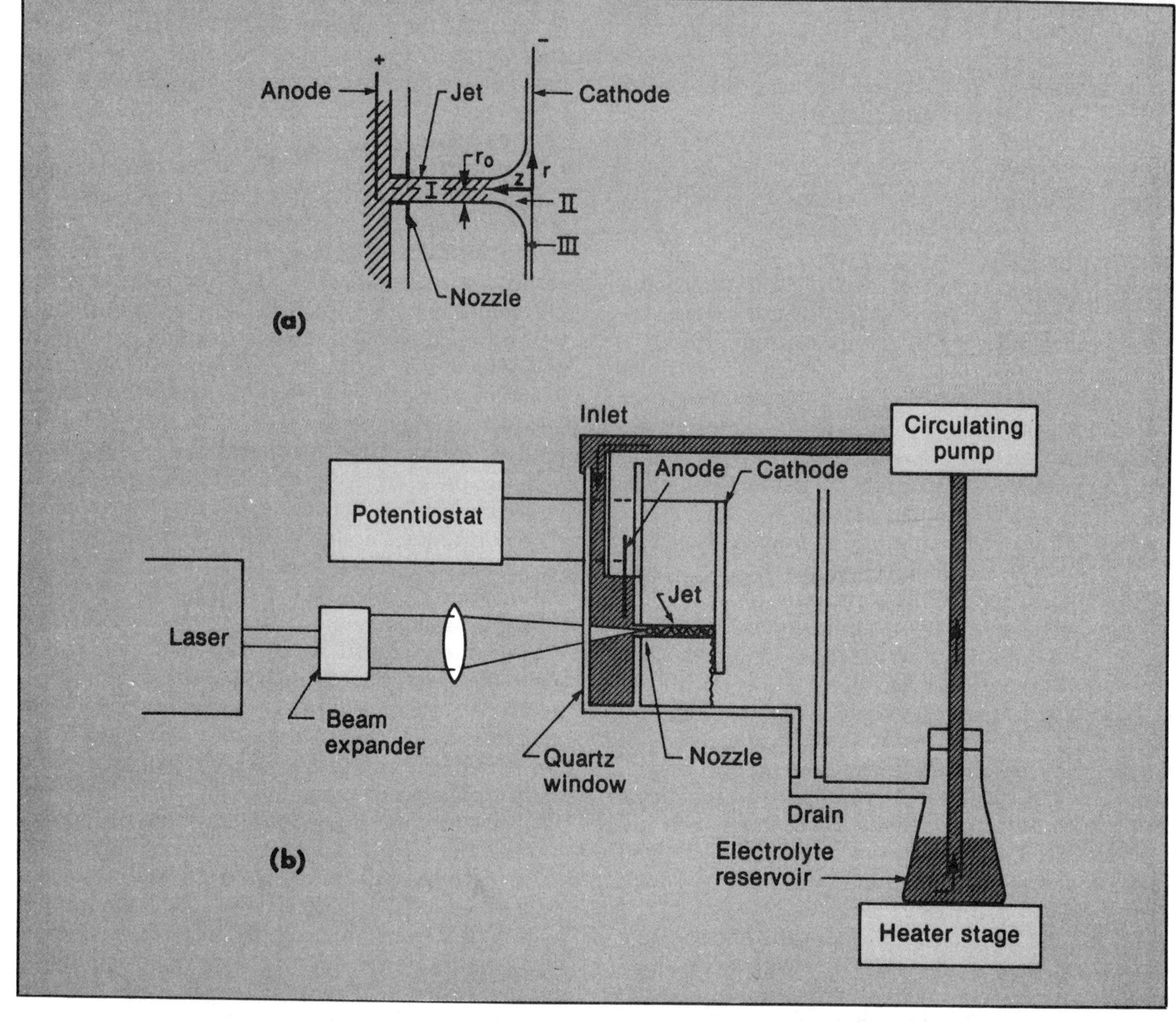

Fig. 2 General scheme for laser-enhanced jet plating. Nozzle with emerging jet is shown in (a). Region II is the stagnation region, that portion of the stream that resupplies the cathode with ions. III is the boundary-layer region, a thin fluid layer that carries very little current. The overall scheme is shown in (b) with a description given in the text.

the cathode via the conducting jet stream. As shown in Fig. 2, the laser is focused into the cell and directed into the jet stream, the latter acting not only as a source of ions but also as a light guide for the laser beam. The cathode is attached to a computer-controlled movable arm to provide relative motion between the laser-jet and the workpiece undergoing patterning. It has been found that under optimal operating conditions, using a commercially available gold-plating solution containing 4 Troy ounces of gold/gallon, plating rates of up to 20 μm/s can be achieved with the use of 20 W of laser power aimed into the jet. Even at these high plating rates, the deposits are found to be free of voids and well adhered to the substrate. In contrast, samples prepared under identical conditions, but without the laser, exhibit extensive void formation irrespective of film thickness with extremely poor adhesion to the substrate. We believe that laser irradiation produces sufficient local heating to partially anneal samples during film growth while also providing an enhanced number of nucleation sites to produce a denser deposit. Increased sites make void formation less likely, since growth can proceed more uniformly. As previously mentioned, the laser also helps to dissolve thin oxide layers on the substrate, which results in improved adhesion of the films undergoing deposition.

Laser-jet studies on copper deposition have also been undertaken. These data differ somewhat from those observed for gold in that the laser does not increase the rate of deposition but does improve metallurgical quality of the deposits [10]. This improvement is based again on a decrease in the number of voids observed both in cross-sectioned samples as well as in the smoother surface morphology. A lower electrical resistance was found for the copper using the laser-jet. The reason for the lack of a rate enhancement with laser power is that the plating efficiency for Cu/Cu^{++} is already close to 100 percent below the limiting current density. This means that virtually all of the plating current is utilized to reduce the copper ions from Cu^{++} to Cu°. Thus, the plating rate cannot be increased further. Jet-plated gold without laser irradiation is only approximately 20 percent efficient. For that system, the laser acts to (1) increase the efficiency of the plating by reducing the available current for side reactions such as hydrogen evolution, and (2) improve the film quality [11].

In summary, experiments on laser-enhanced plating and etching have demonstrated the ability to obtain local deposition/removal rates on the order of from 10^2 to 10^4 times faster than the conventional techniques standard in the industry. The laser absorption also gives rise to improved metallurgical properties and increased adhesion for laser-deposited films. High-speed maskless patterning using both laser and laser-jet plating techniques offers substantial technical and cost advantages. So far, the results discussed in this paper have been demonstrated in a laboratory environment. It remains to be seen to what extent these laser techniques find their way into future materials processing technologies.

References

[1] R. J. von Gutfeld, E. E. Tynan, R. L. Melcher, and S. E. Blum, "Laser-Enhanced Electroplating and Maskless Pattern Generation," *Appl. Phys. Lett.*, vol. 35, p. 651, 1979.

[2] R. J. von Gutfeld, E. E. Tynan, and L. T. Romankiw, "Laser-Enhanced Electroplating and Etching for Maskless Pattern Generation," Extended Abstract No. 472, p. 1185, *Electrochem. Soc.*, 1979.

[3] J.-Cl. Puippe, R. E. Acosta, and R. J. von Gutfeld, "Investigations of Laser-Enhanced Electroplating Mechanisms," *J. of Electrochem. Soc.*, vol. 128, p. 2539, 1981.

[4] R. J. von Gutfeld, R. E. Acosta, and L. T. Romankiw, "Laser-Enhanced Plating and Etching: Mechanisms and Applications," *IBM J. of Res. & Develop.*, vol. 26, p. 136, 1982.

[5] R. J. von Gutfeld, "Laser Processing of Integrated Circuits and Microelectronic Materials," in *Laser Applications*, vol. 5, pp. 1–67, Academic Press, J. F. Ready, Ed., 1984.

[6] R. J. von Gutfeld, "A Review of Laser-Enhanced Plating and Etching for Electronic Materials," *Denki Kagaku*, vol. 52, p. 452, 1984.

[7] M. H. Gelchinski, L. T. Romankiw, and R. J. von Gutfeld, External Abstract 82-2, p. 206, *Electrochem. Soc.*, Detroit, MI, 1982.

[8] L. T. Romankiw, M. H. Gelchinski, R. E. Acosta, and R. J. von Gutfeld, "Maskless, Selective, Laser-Enhanced Plating and Etching," in *Proc. of the Symposium on Electroplating Engineering and Waste Recycle, New Developments and Trends*, vol. 83, no. 12, p. 66, 1983.

[9] R. J. von Gutfeld, M. H. Gelchinski, L. T. Romankiw, and D. R. Vigliotti, "Laser-Enhanced Jet Plating: A Method of High-Speed Maskless Patterning," *Appl. Phys. Lett.*, vol. 43, p. 876, 1983.

[10] R. J. von Gutfeld and D. R. Vigliotti, "High-Speed Electroplating of Copper Using the Laser-Jet Technique," *Appl. Phys. Lett.*, vol. 46, p. 1003, 1985.

[11] M. H. Gelchinski, L. T. Romankiw, D. R. Vigliotti, and R. J. von Gutfeld, "Electrochemical and Metallurgical Aspects of Laser-Enhanced Jet Plating of Gold," to be published, *J. of Electrochem. Soc.*, Nov. 1985.

James Brannon

EXCIMER LASER ABLATION AND ETCHING

Making a Mark in Many Applications

From the earliest days of the laser, scientists and engineers have curiously explored what happens when intense pulses of light interact with solid surfaces. The results—often brilliant, sometimes catastrophic—were sufficiently interesting to generate research efforts that continue unabated to this day.

The technological potential of pulsed-laser/solid interaction was soon realized, and efforts were centered on ways to produce "controlled" laser damage of surfaces that would prove beneficial. These efforts were clearly successful, and pulsed lasers, mainly ruby, YAG, and CO_2, have been used for more than two decades to drill, weld, and cut a variety of materials. Although industrially useful, these long-wavelength lasers are incapable of producing high-resolution features (on the order of 1 μm) without resorting to expensive or exotic optics, and they are also incapable of initiating efficient photochemical reactions—two desirable features for material processing. The discovery of the rare-gas halide (RGH) excimer laser in 1975 quickly changed this situation. The realization of the technical importance of this laser resulted in the first commercial product only a year later.

Today, excimer laser ablation and etching techniques have found use in applications ranging from semiconductor processing to correcting human vision problems. This article discusses these techniques, as well as some relevant engineering issues. The progress of excimer lasers in patterning technology is briefly presented, and the article concludes by discussing several applications of ablation and etching that have high potential industrial use and significance.

Reprinted from *IEEE Circuits and Devices Magazine*, Vol. 13, No. 2, pp. 11-18, March 1997.

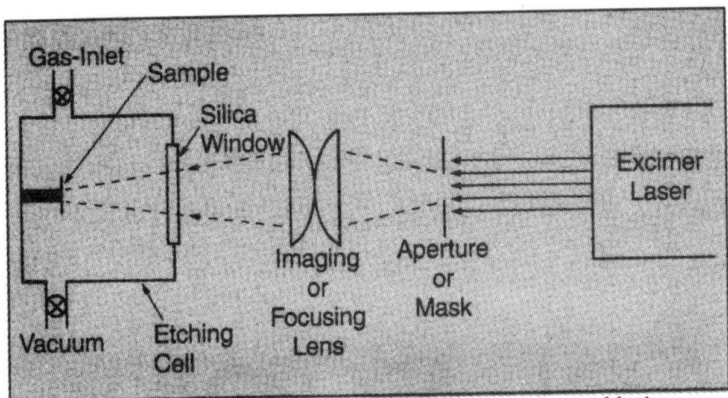

1. Schematic of apparatus for perfoming excimer laser ablation or etching. Ablation is often performed in vacuum, while laser etching requires the presence of an etchant (usually a gas) in the cell. For film deposition, a substrate is placed inside the cell near the sample target.

Excimer Lasers

Excimer lasers are based on the electronic excitation of a molecular gas. This results in the emission of high-intensity, short-duration pulses of ultraviolet light [1]. Commercial devices can emit up to 1 joule in a singe pulse of 15-30 nsec duration. Light is emitted at one of five discrete wavelengths ranging from 193 to 351 nm, depending upon the gas mixture used. This deep UV output is readily absorbed by materials and can initiate photochemistry in a broad range of gases and liquids—desirable characteristics in laser etching applications. Light at these wavelengths can also be patterned with speckle-free, sub-micron accuracy, which explains the semiconductor industry's interest in excimer laser lithography. The high intensity UV pulses permit large and very rapid surface-energy depositions of 10^3-10^6 J/cm^3, conditions under which many materials break down and ablate. Because the RGH laser's active material is gas, the device can be fired at high repetition rates. Several hundred millijoules per pulse at 1000 Hz may be possible in future devices. The high intensity and high repetition rate result in high average power output, with 100 watt lasers currently available. A final advantage is large beam area. Beams of up to 10 cm^2 emitted directly from the laser are available, facilitating large-area processing.

As a result of the excimer laser's strengths, many materials-processing capabilities have been demonstrated over the last decade [2]. In addition to precision ablation and etching, excimer lasers have been used for sub-micron lithography, patterned material deposition from both gas-phase and solid-state precursors, selective doping of semiconductors, texturing of surfaces for adhesion promotion, surface planarization for topographic control in microelectronic fabrication, and deposition of thin films by ablation of bulk precursors. Several of these applications will be discussed later in the article.

Ablation Versus Etching

In excimer laser processing the terms *ablation* and *etching* are often used interchangeably, but there are discernible physical differences even though a similar experimental apparatus is used for both techniques (see Fig. 1). Ablation refers to material being ejected by the sole interaction of the high-intensity laser pulse with the material. Ablation is usually described in terms of physical mechanisms such as vaporization, spallation, exfoliation, or shock effects. Ablation is often performed in vacuum or air. Examples include polymer ablation, removal of thin metal films from an insulating substrate, and deposition of high-T_c superconducting thin films by ablation of bulk targets. Table 1 lists several materials that undergo ablation by excimer laser radiation [2].

Etching, by contrast, is a chemically assisted laser removal process that results from laser irradiation in conjunction with a gas or liquid (or even a solid) that reacts with the material being

Table 1: Examples of Materials Ablated by Excimer Laser Radiation

Material	nm)	Ablation Rate (Å/pulse)	Fluence (mJ/cm^2)
Aluminum	308	1200	200
Gold	308	500	30
Copper	308	1000	80
Polyimide	248	1200	150
PMMA	193	3000	200
Polycarbonate	248	400	100
Polystyrene	248	150	100
c-silicon	248	2000	2000
Gallium arsenide	248	5	200
Pyrex	193	1000	2000
Silicon Monoxide	248	6000	600
Alumina	248	200	1000
Lithium Niobate	308	1000	1500
$YBaCu_2O_3$	248	1000	1000

Material	λ (nm)	Chemical	Etch Rate (Å/pulse)	Fluence (mJ/cm^2)
aluminum	308	Cl_2	1000	500
Copper	308	Br_2 (liq.)	500 (Å/min)	—
Copper	248	Cl_2	100	200
Molybdenum	193	NF_3	0.2	60
Tungsten	193	COF_2	0.5	80
c-silicon	308	Cl_2	0.1	300
Poly-silicon	193	NF_3	0.5	250
Gallium Arsenide	193	HB_r	20	20
Gallium Arsenide	248	HNO_3	1	20
Indium Phosphide	248	CH_3I/H_2	5	600
Glass	248	CF_2Br_2	1	1000
Pyrex	193	H_2	1500	500

Table 2. Examples of Materials Etched by Excimer Laser Radiation and the Listed Chemical

processed. Etching can be a more controlled technique than ablation, offering greater material selectivity and lower laser intensities. Representative applications include 248 nm etching of silicon in the presence of gaseous chlorine, etching of silica by excimer laser excitation of gaseous CF_2Br_2, and etching of gallium arsenide using liquid etchants such as KOH, H_2SO_4, and HNO_3 [3]. Table 2 lists many substances that have been etched by excimer laser radiation [2].

Ablation

Generally, there are two classes of laser ablation mechanisms: thermal and electronic (or non-thermal). Thermal processes rely on an intense laser pulse to very rapidly heat a surface at rates on the order of 10^{10} K/sec. These processes can produce expansion and vaporization from solid and melted regions. Hydrodynamic expansion is particularly apt in ablation of biological tissue that contains water. Laser heating of the water creates very high pressures that cause the tissue to break [4]. If melting occurs, liquid droplets may be accelerated from the surface by hydrodynamic sputtering [5]. Similarly, "backward" momentum transfer to the molten surface from rapidly escaping particles can cause material splatter in a phenomenon known as the "piston" effect [6]. Alternatively, the transiently melted surface may become cooler than the melted sub-surface regions because of evaporation. This non-equilibrium situation may cause material to explode outward as the hotter sub-surface region rapidly expands against the cooler surface [7].

Rapid heating may also cause ablation due to exfoliation or spallation. In a solid comprised of layers of materials with different thermal properties, exfoliation may result from the thermal shock occurring when heat attempts to rapidly flow from one layer to another. Cracking or breaking at this interface may occur. If a mismatch in mechanical properties also exists (at an air/solid interface for example), then spallation may occur. Rapid heating of the solid will cause internal tension due to reflection of a thermal pressure wave at the interface. If the tension stress exceeds the mechanical strength of the material, breakage and spallation occurs [8].

Electronic mechanisms do not principally rely on heating. Two of these quantum-type processes have been discussed widely. In the first, laser photons excite (by either single or multiphoton absorption) and break the bonds of the solid, causing ejection of material [9]. In the second, photo-excitation (again, single or multiphoton) creates electron-hole pairs. Once created, the potential energy of the pair may be coupled directly into kinetic energy of the atoms via a radiationless pathway. The energized atoms are able to overcome the surface binding energy, and material is ejected [10].

Laser ablation of a given material can involve both thermal and electronic mechanisms. For example, during the interaction of the leading edge of the laser pulse with the solid, non-thermal processes may predominate because significant heating has not yet occurred. As the rest of the pulse is absorbed, the temperature rises rapidly and thermally activated mechanisms may commence. It should be noted that laser excitation of a strongly absorbing solid will always result in substantial heating regardless of the ablation mechanism. Although non-thermal processes may occur, there is no way to "turn-off "the thermal processes. Which type of process becomes dominant is ultimately determined by the detailed chemistry and physics of the solid.

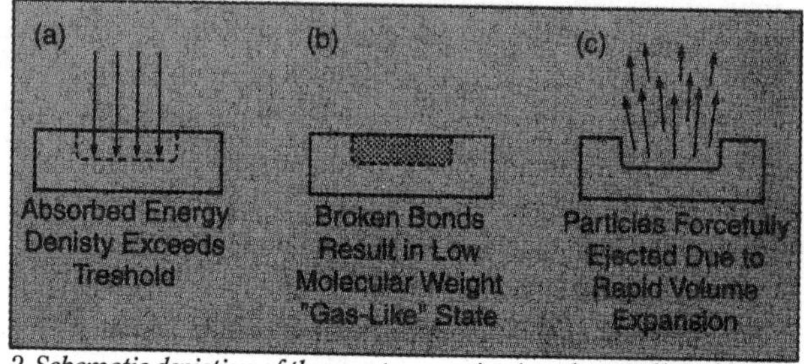

2. *Schematic depiction of the events occurring in polymer photoablation.*

Table 3. Correlation of Relative Ablation Efficiency with 248 nm Absorption Coefficient (α) and Thermal Diffusivity (κ)

Material	α(cm^{-1})	κ(cm^2/sec)	Rel. Efficiency
Silicon	1.8×10^6	0.78	Poor
Copper	9.0×10^5	1.11	Poor
CuCl	1.6×10^5	0.003	Good
SiO_2	<10	0.01	Poor
Polyimide	2.0×10^5	0.001	Good
Polyurethane	7.0×10^4	0.001	Good
Plexiglass	5×10^3	0.001	Poor
Teflon	200	0.001	Poor

Excimer-laser ablation of organic polymers is particularly significant [9]. UV photoablation of polymers was discovered in the early 1980s by researchers in both the U.S. and Japan. Since then, the process has spurred intense scientific and technological interest because: 1) little or no thermal damage is observed in the area surrounding the ablated region, 2) large ablation depths per pulse are produced, and 3) both spatial patterning and the ablated depth can be controlled precisely.

Polymer ablation depends on two key conditions, as shown in Fig. 2. First the polymer must absorb light strongly at the laser's wavelength. Clean, precise ablation usually requires linear absorption coefficients of at least 10^4 cm^{-1}. Second, ablation occurs only after the polymer has absorbed a minimum energy per unit volume, i.e., the laser intensity must exceed a threshold value [11]. Ablation is easily produced in polymers. DuPont's Kapton, for instance, will ablate at a rate of approximately 1500 Å/pulse at a 248 nm fluence of 0.5 J/cm^2 using a 20 nsec pulse duration.

In general, the ease with which ablation can be initiated depends upon the thermal, optical, and quantum-mechanical properties of the solid. Two key factors are the absorptivity and thermal diffusivity. Copper, for example, is highly absorbing, but thick samples still require an incident fluence of 1-2 J/cm^2 for ablation to occur. This is because the laser-supplied thermal energy is quickly removed from the irradiated region due to copper's high thermal diffusivity. In contrast, thin copper films (on the order of 10^3 Å) on insulating substrates will easily undergo ablation at fluences of approximately 0.1 J/cm^2. Here, the thermal energy is confined in the thin film by the dielectric and is thus available to produce ablation.

Silicon is also a good example of the importance of absorptivity and thermal diffusivity. Both crystalline and amorphous silicon possess very high absorption coefficients at excimer laser wavelengths. Crystalline silicon, with thermal properties similar to that of copper, has a 248 nm ablation threshold near 1 J/cm^2 for thicknesses very much greater than silicon's characteristic thermal diffusion length (about 3 µm for a 20 nsec pulse). Amorphous silicon has a thermal diffusivity approximately two orders of magnitude less than crystalline silicon's and will undergo ablation at a 248 nm fluence of less than 0.1 J/cm^2.

Thermal insulating materials, which have very low thermal diffusivities by definition, may or may not ablate easily, depending upon their absorptivity. SiO_2, for example, is very transparent and requires a substantial fluence prior to ablation (which makes it a useful UV optical material). CuCl (cuprous chloride), in contrast, is highly absorbing and requires only a mild fluence to undergo ablation at 248 nm ($\geq$25 mJ/cm^2) [12].

These ideas on absorptivity and thermal diffusivity are qualitatively summarized in Table 3, which displays the relative 248 nm ablation efficiency (depth removed per incident fluence) for several materials. Those substances that have a large absorption coefficient and small thermal diffusivity generally possess a high ablation efficiency.

Etching

Pulsed laser etching has many of the same physical interaction mechanisms as laser ablation. But laser etching requires an active medium to be in contact with the solid since chemical reac-

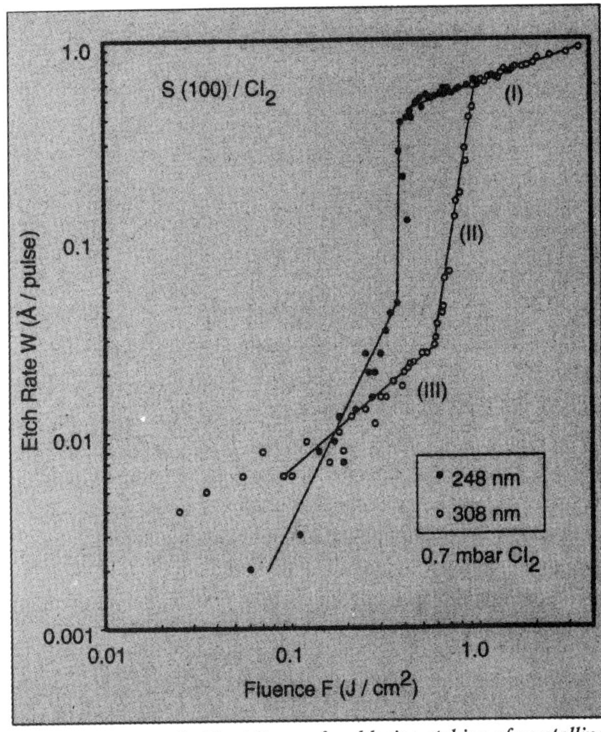

3. Etch rate versus incident fluence for chlorine etching of crystalline silicon at 248 and 308 nm. In the high fluence regime (I), the etch mechanism is dominated by surface melting. The low fluence regime (III) is dominated by low temperature photochemical processes. Regime II is transitional between thermal and non-thermal effects. Reprinted with permission from: W. Sesselmann, E. Hudeczek & F. Bachmann, Journal of Vacuum Science & Technology, Vol. 7, No. 5, p1284–1294 1989. American Vacuum Society.

tions serve to initiate the laser material removal process. The input energy required to initiate these reactions can be relatively small, so etching is often chosen over ablation whenever it is desirable to minimize the thermal loading on the system—assuming an appropriate chemical system can be found.

Laser etching occurs in the following steps: 1) formation of a reactive chemical species; 2) reaction of these species with the surface to form products containing the substance to be etched; and 3) removal of those products from the surface. The excimer laser may play a role in any or all of the steps, and may do so via thermal or non-thermal means.

Metals, semiconductors, and selected insulators have all been etched using excimer lasers in conjunction with an appropriate chemical. Because halogens (F, Cl, Br, I) react rapidly with many substances, pure halogen gas or gaseous molecules containing a halogen atom are commonly used as the etchant chemical. An extensively studied system is excimer laser etching of copper in the presence of chlorine gas [13]. The Cl_2 spontaneously reacts with copper to form a $CuCl_x$ complex having a very low room-temperature vapor pressure. Excimer laser photons remove the $CuCl_x$ layer through non-thermal photodesorption, and thus produce etching. Because halogen atoms diffuse deeply into polycrystalline copper, large etch rates of up to several hundred Å/pulse can be achieved.

Etching of a crystalline silicon using Cl_2 vapor proceeds at a much slower rate ($\leq$1 Å/pulse) since only surface monolayers of $SiCl_x$ can form [14]. Figure 3 depicts an etch rate curve for silicon and the evidence for thermal and non-thermal etching regimes. Fluorine-containing compounds can produce silicon etch rates as high as several Å/pulse.

Weakly absorbing insulators such as quartz undergo etching only at relatively high fluences (hundreds of mJ/cm^2) and at small etch rates of no more than 1 Å/pulse [15]. Such low etch rates are usually a clear indication that the chemical interaction is limited to just the top surface layers.

Engineering Aspects

Semiconductor processing, microelectronic device fabrication, and treatment of human vision problems, offer some of the most exciting applications of excimer laser ablation and etching. But, as with any engineering effort, there exist a number of issues that must be addressed prior to the successful implementation of an excimer laser processing tool. Ideally, an excimer laser ablation or etching technique would possess the following attributes:

- Relatively low pulse fluences producing large removal rates.
- Use of near-UV wavelengths (308 or 351 nm) for simpler optical requirements.
- For etching systems, process initiation by photo-activation of a surface-absorbed species for the highest possible spatial resolution. (Gas or liquid phase photoactivation will permit diffusion of the reactive species and may lower resolution.)
- Volatile products to avoid redeposition and debris formation.
- High material selectivity (the ability to ablate or etch one material in the presence of a second material.)
- Ablation and etching products, and etching reactants, that are non-toxic and safe for handling and disposal.

As expected, few materials possess even a majority of these qualities. In addition, the excimer laser must fulfill requirements of its own, such as long gas lifetimes, reliable operation, pulse-to-pulse output energy stability, and spatially uniform beam intensities. With the increasing demand for "industrial" systems, excimer lasers meeting many of these requirements are now offered commercially.

Efficient material removal depends on the characteristics of the material to be processed. The absorption spectrum of the material is one of the chief factors in assessing ablation efficiency. For etching applications, the spectrum of the chemical medium must also be considered. If the excimer laser wavelength matches a region of strong absorption, there will be efficient coupling of light to the solid. Weakly absorbing materials will be difficult to etch or ablate at acceptable rates or quality.

Once light is absorbed, the chemistry and physics of the process must be such that an acceptable removal rate is achieved. This rate can be determined by a plot of depth of material removed per pulse versus incident fluence (such as Fig. 3). The rate usually increases with increasing fluence if all other parameters are held fixed. Details of rate versus fluence allows one to choose a suitable operating fluence that will maximize the amount of material removed without causing substrate damage. Similarly, there is a tradeoff concerning laser repetition rate. Maximizing the amount of material ablated or etched per unit time without inflicting unacceptable heating levels on the substrate requires careful optimization. Commercial excimer lasers can fire at several hundred pulses per second, enough to induce intense time-averaged heating under the proper conditions.

The fate of the removed material must be considered. A significant fraction of the etched or ablated products usually ends

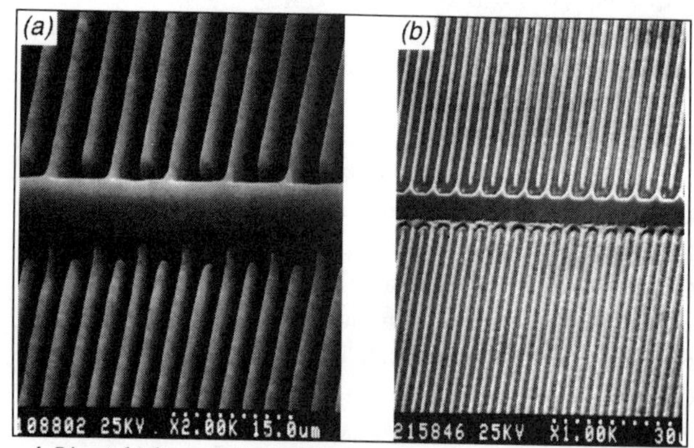

4. Direct, high-resolution patterning (approximately 1 μm lines and spaces) of surfaces by excimer laser image projection using a standard Kohler illumination system and a 10x UV microscope objective. a) a partially fluorinated organic polymer synthesized by plasma deposition from C_2F_4 feed gas; b) crystalline copper etched using Cl_2 vapor.

up on the processed surface, on the vacuum-cell wall, or on the cell window. There are very few materials that provide volatile etch or ablation products that leave the surface free of debris. Exceptions are fluorine etching of silicon and ablation of polyimide in the presence of helium gas [16]. Usually the process designer must select a suitable post-laser cleaning step. This may be as simple as washing the surface with an appropriate solvent, or as complex as plasma-ashing the surface in the case of polymers.

Finally, the toxicity level of the chemical by-products is always an issue. Gas-phase products may be poisonous and must be exhausted via an appropriate vacuum system. Solid products may also be unsafe, requiring strict handling procedures.

Patterning Technology

One of the great advantages of using directed light for processing is that the beam can be patterned. For materials that can be etched or ablated with acceptable quality, direct patterning of the surfaces is feasible. This type of "resistless" processing can offer enormous savings in cost and time over standard patterning techniques such as wet-chemical or plasma etching that require photolithographic processing [17].

Excimer laser patterning is almost always performed using a masking technique, either projection or proximity, rather than by scanning a focused beam. One reason is due to the high divergence and poor coherence of the laser, with the resultant poor focusing characteristics. A second reason is the low duty cycle of the laser—only about 10^{-4}%. Lastly, parallel processing is simply more efficient than serial processing.

The highest resolution yet obtained by excimer projection ablation is 0.13 μm lines and spaces in hard carbon films [17]. This is impressive, but it was accomplished under controlled laboratory conditions and does not represent achievable industrial manufacturing resolutions. More realistic are feature sizes of 1 μm or greater, which can be produced over field sizes of 1-2 cm^2. Figure 4 displays high resolution, excimer laser projection patterning in both a metal and a plastic. In the past few years high-resolution patterning has also been achieved by interferometric methods. This method has produced feature sizes of sub-100 nm in a polymer material [18].

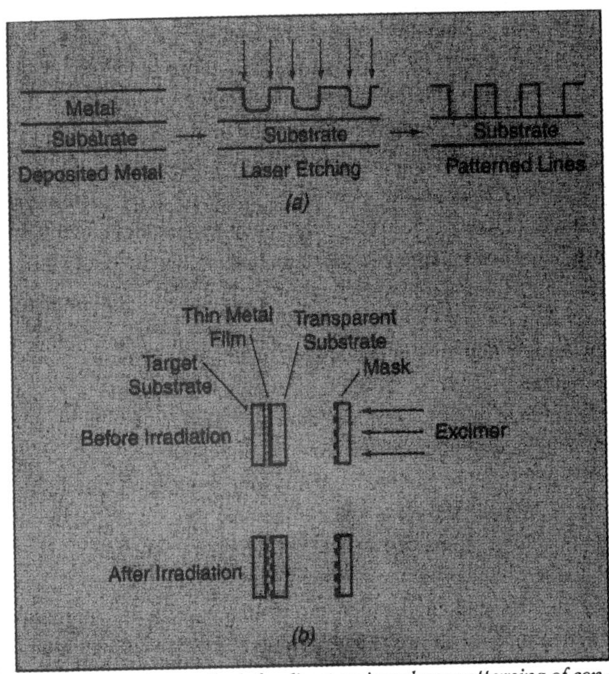

6. Shown are two methods for direct excimer laser patterning of conductive metal lines. a) In subtractive etching, a thin film is deposited and then pattern etched using the excimer projection technique. b) In the LIFT method, a thin metal film on a transparent substrate is ablated using patterned laser light. The pattern is transferred to a target substrate in proximity to the ablated film.

5. Via hole created by photoablation of Kapton-type polyimide at 308 nm. The film thickness is 5 μm, and the diameter is approximately 100 μm. The silicon substrate can be seen in the open via.

Applications

Microelectronics

There are a variety of microelectronic applications that can utilize excimer laser ablation and etching. Among the most straightforward is the creation of via holes, i.e., small openings (1-100 μm diameter) that are usually fabricated in dielectric materials to permit electrical connections between metal films on either side of the dielectric. Insulators such as quartz, polymers, and silicon nitride often serve as the dielectric material. Figure 5 displays a via opening fabricated in the organic polymer polyimide. Excimer lasers can provide clear and sharp fabrication of vias, and image projection technology can create a pattern of vias over an entire surface in just one exposure cycle [19]. Of course, via fabrication can be expanded to accommodate the many scientific needs for small holes in metal films, such as nozzles and apertures for electron beams.

A more intriguing application is the patterning of electrical circuitry by subtractive etching/ablation, or by the laser-induced-forward-transfer (LIFT) technique, a shown in Fig. 6 [20]. In subtractive etching, patterned excimer laser light is used to spatially etch a thin film of deposited metal to create the desired pattern of conductive lines. LIFT accomplishes the same goal by ablating a thin metal film in the desired pattern off of a

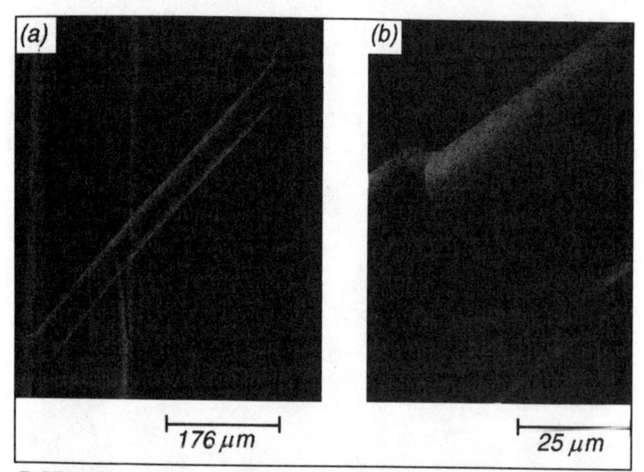

7. SEM photographs of an excimer laser stripped wire. a) Low magnification view showing entire 0.5 mm strip length. b) High magnification view displaying the interface between the stripped metal and the remaining plastic insulation.

transparent substrate and having the material deposit onto a nearby target substrate. The fidelity of the pattern can be retained to a resolution of 10-20 μm in the transfer process.

Pulsed Laser Deposition

The concept behind pulsed laser deposition (PLD) is simple enough. An excimer beam is focused onto a target with the result that a high flux of ablated material is directed onto a nearby deposition substrate. Depending upon a variety of conditions, adherent and high-quality films can be achieved with growth rates that may approach 0.5 μm per minute. PLD is an old technique that has been studied since the development of high-power lasers over 30 years ago [21]. The resurgence of interest in this method in recent years is clearly due to its success in producing high-quality thin films of YBCO High-T_c superconductors. Excimer lasers are the current laser of choice for PLD since the ultraviolet emission provides increased ability to break up the target into its constituent atoms and molecules, and for its ability to penetrate surface plasmas formed during the ablation. Both of these attributes are necessary to form high-quality deposited films. Other important qualities of excimer laser PLD include "congruent" evaporation, and the ejection of ablated material into a very narrow angular distribution toward the deposition substrate. The former permits deposition of films with the correct stoichiometry, while the latter allows for relatively efficient deposition rates [21].

Particle generation is a problematic feature of PLD that is difficult to solve. During the process, particles ranging in size from 0.1 to 50 μm may be deposited, causing degradation of the growing films. Various techniques have appeared to combat particle generation [21]. Variations in thickness uniformity also presents problems for PLD. Due to the forward-directed nature of the ablation plume, there is inherent non-uniformity in the way films are deposited. Researchers have resorted to beam scanning, target rotation, increasing the target to substrate distance, and a host of other measures to produce films of more uniform thickness.

The success of PLD growth of superconductors has spurred a renewed interest in depositing films of other substances. A variety of substances and materials have been deposited by this technique: PZT and other ferroelectrics, barium titanate ($BaTiO_3$), zinc ferrite ($Zn_xFe_{3-x}O_4$), alumina (Al_2O_3), silica (SiO_2), amorphic diamond, boron nitride (BN), and metal oxides such as MgO and ZrO [2, 22].

Micromachining

Most people are familiar with the ability of lasers to drill, cut, and weld bulk materials, and for these applications high power CO_2 and YAG lasers are ubiquitous. Yet there exist an increasing number of applications where the layer-by-layer ablation aspect of excimer lasers proves ideal for machining operations. Particularly amenable are applications in which surface features in the 1-100 μm range are required. The short pulse duration coupled with the ultraviolet output results in a minimal heat-affected zone. This leads to excellent edge definition and a high degree of precision. YAG and CO_2 infrared lasers are usually not capable of precision edge definition on a small scale. Intricate patterns of holes, lines, curved areas, etc., can be fabricated by an excimer laser machining tool. The polyimide-via-ablation process mentioned above is but one example of laser micromachining. Other polymer micromachining applications include fabrication of optical fiber alignment slots, patterning of total internal reflection mirrors in optical waveguides, and creation of other waveguide devices [2].

Wirestripping

Compared to mechanical, thermal, electric, or chemical means of removing plastic insulation from small diameter wires, laser removal offers precision and speed without the necessity of contacting the material. Utilizing the technique of excimer laser ablation of organic polymers, 248 nm removal of polyurethane wire insulation proceeds with much higher precision and cleanliness than does pulsed CO_2 laser wirestripping [23]. The strong UV absorption of most polymer insulators permits layer-by-layer removal, while the short pulse duration minimizes heat flow. Experiments performed on small-diameter magnet wire (~50 μm) have demonstrated the high quality of this technique. Figure 7 displays the high spatial precision and cleanliness of excimer laser wirestripping.

Corneal Sculpting

The success of excimer laser processing of synthetic polymers and plastics has opened up intense research into ablation of biological materials. Excimer laser radiation, in particular, the 193 nm ArF laser, has produced precise, clean incisions in the cornea, skin, and aorta [24]. These results suggest a potential surgical role for the excimer laser, and several companies now offer commercial excimer lasers properly modified for use in medical environments.

The use of excimer lasers for photorefractive keratectomy, more popularly known as corneal sculpting, holds considerable interest for people who wear eye-glasses or contact lenses [25].

Corneal sculpting has been used to correct severe astigmatism in individuals, but it is in the correction of ordinary myopia (nearsightedness) and hyperopia (farsightedness) that excimer laser sculpting holds its biggest promise. Myopia exists when the cornea curves too sharply, focusing the light rays in front of the retina so that distant objects appear blurred. When the cornea is too flat, hyperopia causes the light rays to focus behind the retina. This results in nearby objects being blurred. For myopia, the 193 nm laser is used to precisely ablate the central portion of the cornea, slightly flattening its curvature. In hyperopia, the laser would hone away the periphery of the lens to increase the curvature. In order to properly perform this surgery, the pulse-to-pulse energy variation must be very small, and the beam intensity must be spatially tailored to provide the correct ablation pattern on the lens. Needless to say, such requirements make medical excimer lasers costly. But the optimism that surrounds a technique potentially capable of freeing millions of people from having to wear corrective lenses more than outweighs the cost factor. In the U.S., FDA approval of this technique occurred in 1995.

James Brannon is a member of the research staff at the IBM Almaden Research Center in San Jose, California.

References

1. J. Ewing: "Excimer Lasers" in *Laser Handbook* (M. Stitch, Ed.), vol. 3, p. 135, North Holland Publishing Co., 1979.
2. J. Brannon: "Excimer Laser Ablation and Etching," *American Vacuum Society Monograph*, vol. M-10 (1993) The data in Tables 1 and 2 were obtained from this reference.
3. C.I.H. Ashby: "Laser-Induced Etching," in *Physics of Thin Films* (M. Francombe and J. Vossen, Eds.), vol. 13, p. 151 Academic Press 1987.
4. D. Abergali, L.T. Perelman, G.S. Janes, C. von Rosenberg, I. Itzkan, and M. Feld: *Lasers in the Life Sciences* (Journal), **6**(1), 55 (1994).
5. R. Kelly et al.: *Nucl. Instr. Meth. Phys. Res. B*, **9**, 329 (1985).
6. M. von Allmen: *Laser Beam Interaction with Materials*, New York: Springer-Verlag, 1987.
7. .E. Andrew, P.E. Dyer, R.D. Greenough, and P.H. Key: *Appl. Phys. Lett.*, **43**, 1076, (1983).
8. R.S. Dingus and R.J. Scammon: *SPIE Proc.*, Vol. 1427, p. 45 (1991).
9. R. Srinivasan and B. Braren: *Chem. Rev.* **89**, 1303 (1989).
10. R.W. Dreyfus, F.A. McDonald, and R.J. von Gutfeld: *J. Vac. Sci. Technol. B*, **5**, 1521 (1987).
11. J. Brannon, J.R. Lankard, A.I. Baise, F. Burns, and J. Kaufman: *J. Appl. Phys.*, **58**, 2036 (1985).
12. S. Kuper, J. Brannon, and K. Brannon: *J. Appl. Phys. A*, **56**, 43 (1993).
13. J. Brannon and K. Brannon: *J. Vac. Sci. Technol. B*, **7**, 1275 (1989).
14. W. Sesselmann, E. Hudeczek, and F. Bachmann: *J. Vac. Sci. Technol.* B, **7**, 1284 (1989).
15. G.L. Loper and M. Tabat: *SPIE Proc.*, Vol. 459, p. 121 (1984).
16. S. Kuper and J. Brannon: *Appl. Phys. Lett.*, **60**, 1633 (1992).
17. M. Rothschild and D. Erlich: *J. Vac. Sci. Technol. B*, **6**, 1 (1988).
18. H.M. Phillips, D.L. Callahan, R. Sauerbrey, G. Szabo, and Z. Bor: *Appl. Phys. A*, **54**, 158 (1992).
19. J.R. Lankard and G. Wolbold: *Appl. Phys. A*, **54**, 355 (1992).
20. R.J. Baseman, A. Gupta, R. Sausa, and C. Progler: in *Laser and Particle Beam Chemical Processing for Microelectronics*, vol. 101 pg. 237 Materials Research Society, 1988.
21. H. Sankur and J. Cheung: *Appl. Phys. A*, **47**, 271 (1988).
22. D.H. Lowndes, et al: *Science*, **273**, 898 (1996).
23. J. Brannon, A. Tam, and R. Kurth: *J. Appl. Phys.*, **70**, 3881 (1991).
24. G.H. Pettit: *IEEE Cir. Dev.*, **8**, 18 (1992).
25. M. Freudenheim: *The New York Times*, 13 May, p. C1 (1992).

CD■

Laser-Fabrication for Solid-State Electronics*

A variety of new laser-processing techniques improve device yield and performance

Richard M. Osgood, Jr.

Laser beams, particularly those from UV lasers, can cut, etch, and deposit materials in ways not previously considered possible. Among the best examples of these optical techniques for manipulating materials properties are those developed for fabricating advanced solid-state electronics circuits and devices [1]. Here, the ultrasmall dimensions of the individual devices and the complexity of the materials assembly provide an unparalleled challenge to the fabrication engineer.

This challenge has inspired many new techniques over the last decade, including immediately practical methods for drilling in printed circuit boards—the first application of excimer lasers on a production line [2]. More subtle fabrication techniques arepromised for the future. Optically patterned epitaxy, for example, may allow direct growth of laser or optical-device structures [3].

*This work was supported by the Defense Advanced Research Projects Agency/Air Force Office of Scientific Research, the Army Research Office, and The Semiconductor Research Corporation.

Reprinted from *IEEE Circuits and Devices Magazine,* Vol. 6, No. 5, pp. 25-31, September 1990.

Laser processing has generally been based on initiating a phase change or a chemical reaction that modifies the structure or chemical composition of a solid surface. Rapid heating of an insulator or metal, for example, can melt the surface and cause a desired reflow or smoothing of the surface texture. In semiconductors, laser heating can anneal or cause a dopant to be incorporated in the heated area. Among the many possible chemical operations are deposition of thin films, gas-phase etching, and even electrochemical deposition or etching in solutions. These reactionscan be driven either thermo- or photochemically.

Three general fabrication procedures are used in laser processing. In the first (Fig. 1a), a laser is focused to a single spot on the substrate to be worked. This spot can be rastered across the wafer to generate a pattern, a procedure termed direct writing because the spatial location of the pattern is not determined by a preexisting mask. Direct writing is relatively slow because each pixel must be generated serially and the processing speed is low. Nonetheless, it is an ideal approach for making custom or discretionary changes.

Focused laser beams are also the basis for nonplanar processing (Fig. 1b), which typically uses equipment very similar tothat used for direct writing. Nonplanar processing is essential when the topography of the substrate to be processed is so variable that a fixed-focus optical system can not be used. Forming via holes in a silicon wafer is a good example. Here, the typical wafer thickness (about 250 μm) is much greater than the depth of field for a typical optical system. But with a simple focused laser beam, one can compensate for changes in the height of the structure by smoothly adjusting the height of the lens above the substrate during processing [4].

For both of these focused laser-processing techniques, a cw laser is generally preferred because it allows the highest number of pixels to be processed per unit time. If a pulsed laser is used instead, its average power is severely limited because the single-pulse energy cannot exceed the threshold for air breakdown.

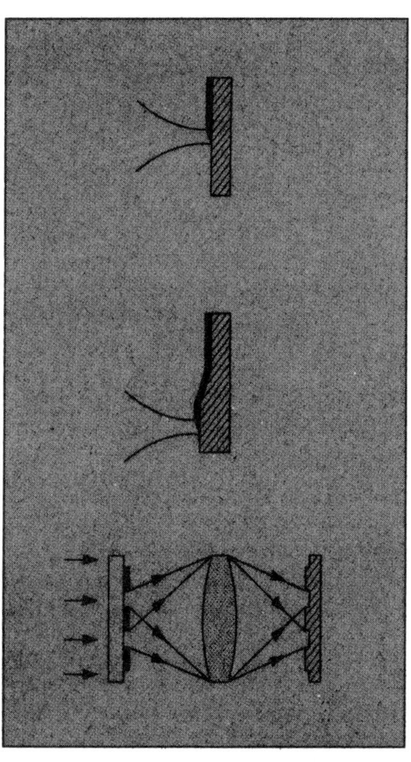

1. There are three types of commonly used laser chemical processing. Direct writing (a) rasters a narrowly focused laser beam across the substrate. Non-planar processing (b) is similar, but has provision for varying the beam's focus when topographic changes in the z direction exceed the beam's depth of field. In laser-projection processing (c), a wide-area beam is projected onto the surface. A beam that is prepatterned with a mask can produce direct patterning of the substrate.

In a third technique, a large-area beam is projected onto a surface. If the beam is prepatterned with a mask, direct patterning of the substrate is possible. Without a mask, there is uniform, i.e., flood, illumination. Because of the large area being illuminated, pulsed lasers (such as excimer lasers) can be used efficiently. Laser projection processes many pixels in parallel, and so it is compatible with conventional semiconductor fabrication. In fact, since patterning is done directly with the laser beam, this type of processing can be thought of as resistless lithography.

What makes laser sources useful in these fabrication techniques? For the focused-beam techniques, the laser's very high brightness is key. With a simple lens system a laser source can project an optical flux into a micrometer-sized spot that is many orders of magnitude greater than that from a conventional source. This higher flux translates into a correspondingly higher processing rate, and also makes possible processes that rely on nonlinear phenomena, such as surface melting.

A second useful characteristic is the short pulse length available from many types of lasers. The short pulses can strongly irradiate a substrate, yet only heat its surface. This makes it possible, for example, to melt only the upper layer of a semiconductor wafer.

Third, lasers generally make better ultraviolet light sources than do lamps in terms of average power, monochromaticity, and intensity. UV wavelengths are essential formany materials processing techniques because 1) they can dissociate molecular bonds in gases and condensed molecules by a simple single-photon-absorption process that produces no heat; 2) conversely, they can produce very controlled ablative processing because of the short absorption depths common in solid materials at UV wavelengths [5]. (In laser-induced ablation, energy is deposited in the substrate faster than thermal conduction can remove it from the directly irradiated volume. When this energy is sufficient to vaporize the volume, ablation occurs. A short absorption length, which is typical of UV light, allows well-defined ablation at modest laser energy).

Finally, laser beams are highly collimated and travel in straight lines. Therefore, lasers are convenient sources for coupling to a remote target.

The development of laser processing has been accelerated by a substantial research effort in the fundamental science of laser-surface interactions. This basic research has pointed out, for example, the physical limitations on proposed processing techniques and has suggested ways of improving existing methods of processing. This

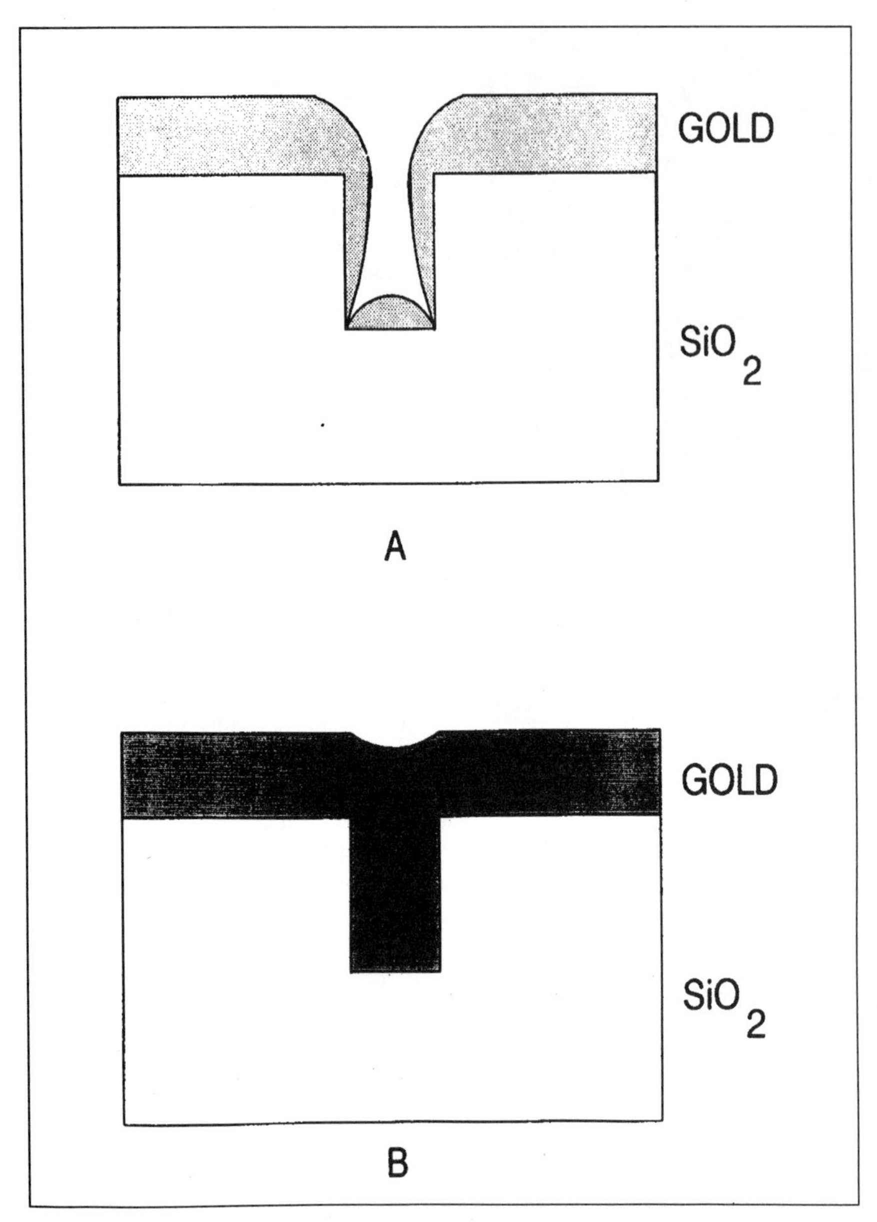

2. In laser planarization, the laser beam melts a surface made uneven by processing (a), and surface tension restores the plane surface (b), even over a high-aspect-ratio deep via.

interaction of research and practical development has produced a variety of successful techniques that are impressive in their own right, and also serve to illustrate the powerful capabilities of laser processing.

Laser Metallization for VLSI

Packing electrical interconnects at sufficiently high density to accommodate large numbers of on-chip devices is one of the most crucial and challenging areas of advanced VLSI chip fabrication. The main difficulty for designers of standard memory and logic chips is developing processes that allow multiple levels of metallization separated by dielectrics films, along with interlevel conducting vias. To prevent unacceptable resistive losses in the interconnections, highly conductive noble metals or copper must be used. Unfortunately, these metals are difficult to pattern in fine-line geometries.

For designers of ASICs and other custom chips, an important challenge is to develop flexible methods that permit rapid customization of the upper-level interconnection "wires." If such techniques were available, they could also be used for limited repair of IC metallization when failures could be corrected in the upper levels of the interconnects. Both ofthese areas of interconnect technology have now been addressed by laser techniques.

Many of the problems encountered when fabricating high-density interconnects are associated with the topography of the chip surface. In particular, because we require highly conducting lines of small transverse dimension, the line cross-section must be thin but "tall." Further, these lines must be formed by depositing metal within their patterned high-aspect cavities in layers of insulating material. The interconnects are typically formed by blanket deposition on the patterned insulator. But the structures are extremely difficult to fill by conventional means, which often leave the cross section half-filled or containing voids, particularly at the bottom of the structure (Fig. 2a). Even after a successful deposition has been made on the patterned, thick dielectric, the surface is left with a "hilly" topography. Any attempt to do high-resolution lithography on such a surface will be limited because of the short depths of field obtained with high numerical-aperture optics. Therefore, the surface must be planarized after deposition and prior to lithographic patterning.

These problems can be resolved by recently-developed laser techniques [6,7]. In one of these techniques, a short-pulse laser does transient melting of a substrate containing a conventionally deposited metal film (Fig. 2a). If the laser pulse is short enough and the optical energy is absorbed in the top layers of the deposited film, then the optically induced melting will be confined to the upper layer. After melting, the surface tension on the molten surface will establish a uniform horizontal plane, even filling features in the surface such as high-aspect vias and trenches (Fig. 2b). (This planarization process is best achieved with nonrefractory metals like gold and aluminum that have moderate melting points.) The laser planarization process also improves other characteristics, such as metal-dielectric adhesion and grain size. Laser planarization has been achieved with several laser sources, but the excimer laser is the most promising because of its large-area beam and significant pulse power, which allows single-pulse planarization. Excimer lasers are also attractive because their ultraviolet light couples comparatively well into most metals. And with short pulse lengths available from most commercial excimer lasers, it may be possible to planarize metals on top of polymeric materials, which have low threshold for thermal damage.

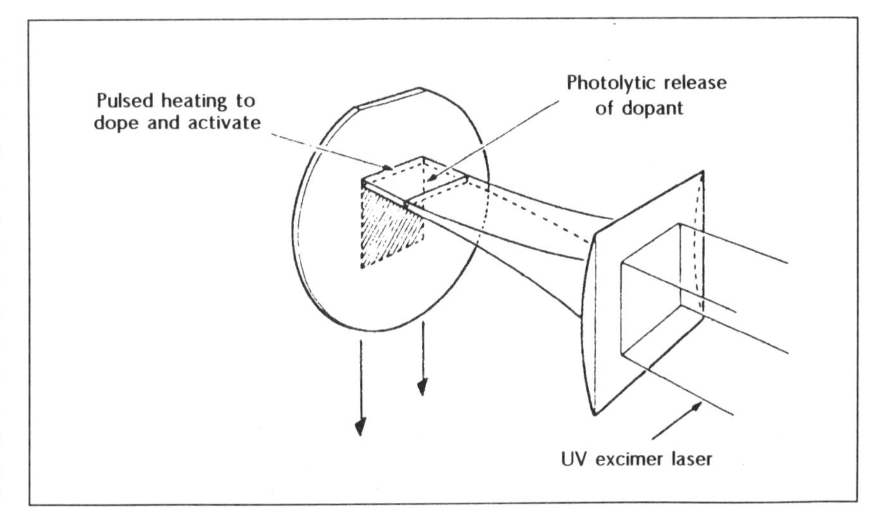

3. In laser doping of silicon, a UV-laser pulse photodissociates a gas or an adlayer dopant and simultaneously melts the silicon surface. The dopant diffuses into the melted surface.

Ultrashallow Doping

The rapid heating associated with a strongly absorbed, nanosecond laser can be used for other forms of processing. As silicon IC technology advances, device dimensions shrink not only laterally, but also in the direction normal to the surface. While current research suggests that laser chemistry may provide a useful route to controlled growth of ultrathin oxides (less than or equal to 100Λ), it is clear that rapid surface heating with excimer lasers is already a practical method for forming ultrashallow doping regions in silicon. It would be an attractive alternative to the current method, a complex, multistep, ion-implantation process.

Only several years ago, Deutsch and his coworkers at Lincoln Laboratory discovered that highly doped, ultrashallow regions could be formed with exposure to only a few pulses from an excimer laser [8]. An ultraviolet laser beam was used to produce transient near-surface melting (Fig. 3). When this is done, a gas atmosphere photodissociation, thermal cracking of the gas, or surface-adsorbed molecules can supply doping atoms for incorporation into the solid. (Suitable precursor molecules include phosphorous trichloride and diborane.) Incorporation of the dopant atoms then proceeds by diffusion into the liquid.

Deutsch and his coworkers used this doping process to make large-area p-n junctions for solar cells. Subsequently, Sigman and his group at Stanford showed that the shallow, highly doped regions formed with the addition of a modest low-temperature anneal excellent source-drain contacts for a MOSFET circuit [9]. They have modeled and characterized the technique to the point where it forms a viable alternative to ion-implantation in modern IC production.

A critical challenge to perfecting the Stanford group's technique was the development of a compatible surface masking process, which is needed because an unpatterned exposure was used for the doping process. Without a proper process margin for the laser exposure, melting or ablation of the mask occurs. Recently, Sugioka and Toyoda have shown that patterned doping in GaAs is possible with the use of a projected patterned laser beam. Such an approach would further simplify the laserdoping process.

In-situ Processing

Processing advanced VLSI chips relies increasingly on doing multiple processing steps in a single vacuum envelope. This approach, called "in-situ processing," attempts to eliminate atmospheric transfer steps, thus reducing airborne-introduced impurities. Some degree of patterning typically is among the semiconductor fabrication operations that must be performed. Unfortunately, lithography provides at present the sole method of patterning in IC fabrication. And since lithography is based on a sequential set of aqueous operations, it is not useful for vacuum operation. Fortunately, laser-projection processing provides an alternate, resistless approach to patterning. In addition, the low-temperature and ion-free nature of laser chemistry also provides processing advantages which that may be useful even for flood exposure.

Patterned projection processing has been demonstrated for several of the techniques already mentioned; in particular, patterned laser doping of semiconductors and patterned deposition of aluminum for metallization [11, 12]. Another example of laser projection patterning is patterned etching of GaAs, a process first demonstrated by irradiating a GaAs sample with ultraviolet light in a cell containing halide gases [13]. The pattern was defined by desorption of adlayers of product molecules, typically gallium or arsenic halide from the surface. Once these were removed, the surface was bare and more easily attacked by etchant atoms. Recently, this approach has been carefully refined and operated in an in-situ MBE

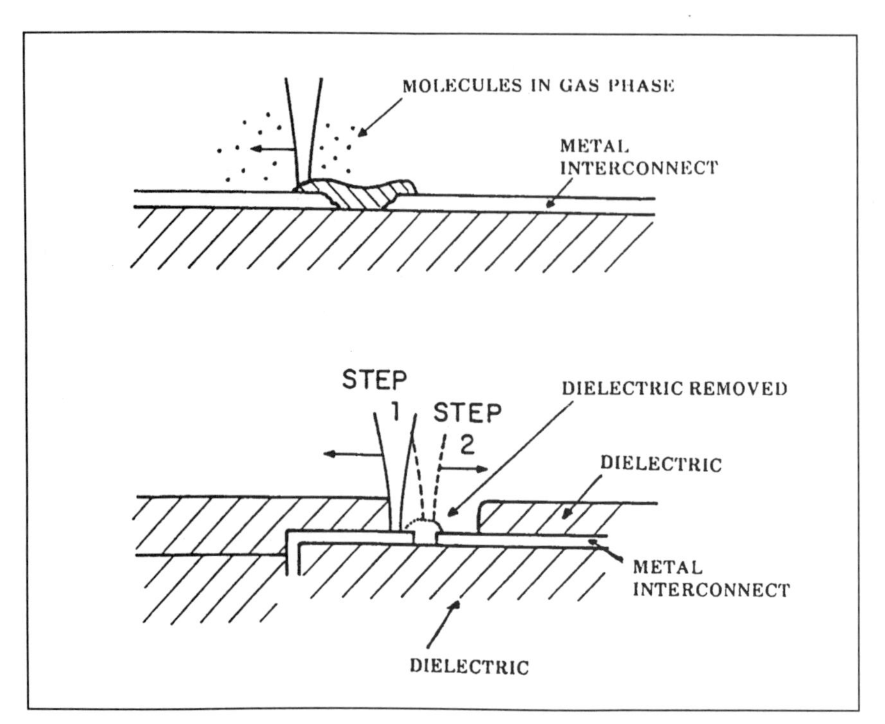

4. A multichip module (MCM) package can be repaired by laser. A break in a surface interconnect can be repaired by deposition (top). A second approach first removes the polyimide dielectric (solid beam, bottom). The break in the interconnect is then filled (dotted beam, bottom).

laser-processing system [14]. With this system, monolayers of etched product can be removed in one laser pulse, allowing very precise etch definition. In addition, it is possible to regrow material over the etched area.

Another series of experiments has used patterned laser illumination to define regions for selective epitaxial growth [3]. Here, laser light is used to dissociate a single molecular layer adsorbed on a singl-e-crystal substrate. Since the laser-induced chemistry only occurs where the beam impinges on the solid surface, the crystal growth is patterned by the laser light.

Electronic Packaging

An important trend in integrated electronicsis the adoption of multichip module (MCM) packages; i.e., a carrier that holds the bare chip and permits their interconnection via multiple layers of dielectric and metal thin films [15]. These packages, which have been used in mainframe computers for many years, increase the interconnection density and signal speed between chips. The MCM is an intermediate step between the older printed-circuit board and the as yet-to-be-realized wafer-scale integration.

Fabrication of MCMs, however, is extremely challenging because they are produced from alternating layers of normally incompatible materials¨typically, metals such as copper and dielectrics such as ceramics and polymers. These modules are now being patterned at dimensions approaching those used in integrated circuits, yet their total surface area is approximately one order of magnitude greater than that of most IC's. Further, unlike most integrated circuits, the modules rely heavily on the use of a third dimension¨depth into the package¨so dimensional tolerances must remain constant over significant changes in the "z" dimension. Finally, the complexity and size of MCMs significantly increases the cost per unit—approximately two orders of magnitude greater than the chips—so a loss in manufacturing yield carries a heavy financial penalty.

Excimer laser processing has had significant impact on the manufacture of MCM and MCM-like packaging. Perhaps the most important result has been in the etching of the organic materials used in many package designs. At ultraviolet wavelengths, precise submicrometer-defined etching can be obtained by high-power ablative removal of the polymeric material [5]. While the exact details of the ablative mechanism are still being explored, the basic approach is to couple sufficient energy into the molecular surface prior to thermal diffusion. Molecular bonds are then broken and the material is removed by supersonic ejection. Ultraviolet wavelengths are particularly effective because the ultraviolet absorption depth in polymeric compounds can be extremely short, approximately hundreds of angstroms at 193 nm in polyimide; thus the threshold for ablative decomposition can be achieved with relatively modest incident fluence (see the accompanying article by James Brannon.) This debris from the ablative process generally appears as gaseous products. Excimer-laser ablation provides a simple, dry method of patterning interlayer dielectrics. Pattern definition can be achieved either by patterning the laser beam or by using a proximity mask. An important issue to be resolved is that during ablation of dielectrics some carbonaceous debris is generated. This debris forms particles that may reduce the yield of subsequent processing steps.

Writing With a Focused Beam

By using a tightly focused laser beam, one can initiate and control a micrometer-scale chemical-reaction zone. Depending on the details of the reaction chemistry, this process can be used to dope, etch, or deposit on the substrate. Used in conjunction with a scanning system, the focused beam can provide a set of microscopic "tools" for circuit processing.

Repair is one of the most elementary operations to be done in conjunction with

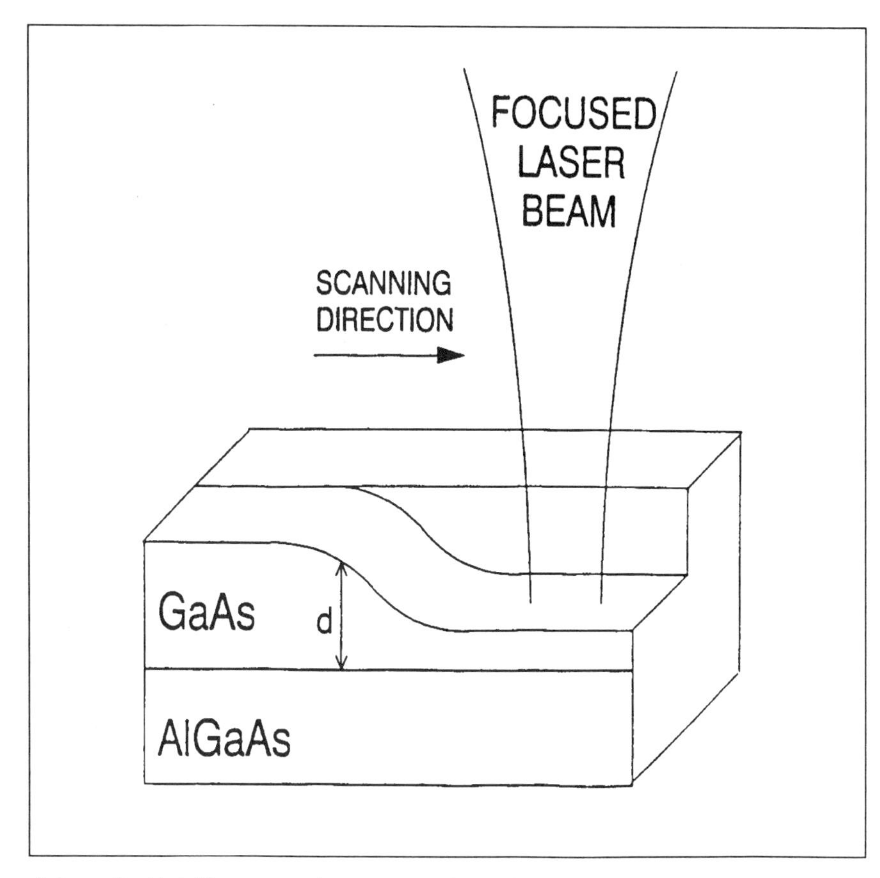

5. Laser liquid etching can produce an etch of smoothly varying depth.

laser writing. It is important in mask-making, integrated circuits, and packaging, as well as in fabricating some active-matrix liquid-crystal displays. Since repair typically involves a modest total processed area, even relatively slow writing processes can be utilized. Laser repair has long been a critical element in manufacturing masks and IC's because of the cost of the individual component or the otherwise unacceptably low manufacturing yield. Processing with focused-laser-controlled reactions adds an important new element to repair operations because low-temperature nonthermal operations become possible, as does simple additive deposition [16]. Instruments for both IC and mask repair are commercially available.

The recent growth in the use of sophisticated multichip modules for packaging has spurred interest in package repair, since the cost of each module is so high (Fig. 4). Depending on the technology envisioned, either refractory metal (for fired ceramic) or noble metal (for polyimide-copper) must be used as the repair material. Generally, the resolution requirements are easily met¨lines typically approximate 1 mil. However, step coverage over the broken lines imposes a real challenge on the deposition technology.

Laser writing for generalized use in integrated circuits (or MCM manufacture) has also been demonstrated and commercialized [17]. In this case, as in electron-beam writing, writing speed limits the extentof processing. However, the unparalleled ability to make computer-controlled or discretionary operations makes this approach extremely useful for custom or low-volume circuits.

Laser writing is most easily demonstrated in systems with chemical gain. Thus, laser writing in resist has been used as a compact tool for producing the final metallization layer for gate arrays. In addition, a series of such photoresist operations has been used in conjunction with an interleaved stack of spin-on polyimide layers to produce a semicustom MCM by a group at General Electric Laboratories [18]. In addition, laser direct writing of metal using direct deposition of a metal film from an organometallic precursor can be used as an alternate method for ASIC connection. It can also be used effectively as a design tool for prototyping both GaAs and silicon integrated circuits [17, 19].

Because focusing of a laser is extremely simple, it is easy to adjust the focal point in order to deal with gross changes in surface topography. Thus, lines of constant width have been written on the beveled edges of chips positioned on a silicon wafer substrate [20]. This technology makes an all-silicon MCM possible.

Focused-laser processing has also been used to produce devices for integrated optical devices in lithium niobate and III-V materials [22]. One method uses a form of aqueous laser microelectrochemistry to etch features in III-V materials such as GaAs and InP [22]. As a result of the microscopic laser-beam diameter very high local etch rates have been obtained, allowing etching to be done over large areas of the chip. When UV radiation is used for etching, very-high-aspect-ratio vertical walls can be formed on the etched features [23]. Thus far, laser microelectrochemistry appears to be an extremely useful method for making custom-waveguide structures for routing light signals across a semiconductor chip. Further, the ability to vary the etch rate as the beam traverses the semiconductor surface provides a useful fabrication tool for making low-loss, integrated, vertical and horizontal optical tapers and compact couplers (Figs. 5, 6).

Outlook

The development of new laser-processing techniques have heightened the interest in industrial UV laser sources. For example,techniques for generating intense CW ultraviolet light for laser direct

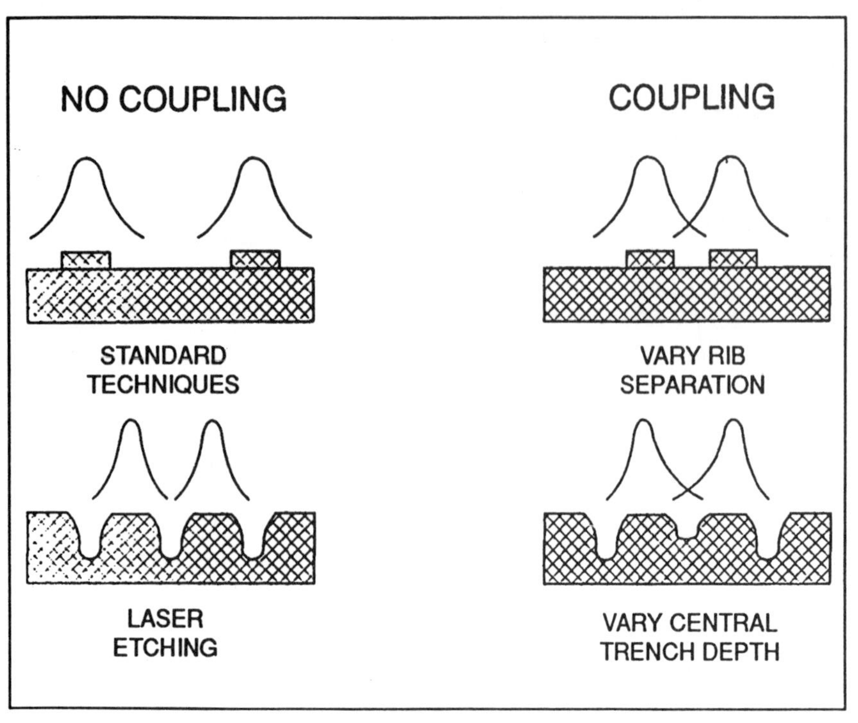

6. These crossections compare standard and laser, trench-defined parallel optical waveguides (left). Coupling in the laser-written structure can be achieved by reducing the depth of the central trench to allow leakage of the evanescent waveguide modes (right).

writing are constantly being improved. The industrial use of excimer lasers to ablate holes in epoxy boards has demanded development of an improved, reliable excimer-laser tool to insure adoption on the production line. The existence of such truly industrial tools will enable widespread uses of lasers in solid-state electronics manufacturing.

Acknowledgment

I wish to thank my colleagues in the Microelectronics Sciences Laboratories at Columbia University for their comments and suggestions.

References

1. K.G. Ibbs and R.M. Osgood, Jr. (eds.), Laser Chemical Processing for Microelectronics," Cambridge University Press, Cambridge, 1989; D.J. Ehrlich and J.Y. Tsao (eds.), Laser Microfabrication: Thin Film Processes and Lithography, Academic Press, Boston, 1989; I.P. Herman, Chem. Rev. 89, p. 1323, 1989.
2. F.G. Bachmann, "In Situ Patterning: Selective Area Deposition and Etching," 1989 MRS Fall Meeting, Paper B6. 1.
3. S.M. Bedair, M.A. Tischler, T. Katsuyama, and N.A. El-Masry, Appl. Phys. Lett., 47, p. 51, 1985; A. Doi, Y. Aoyagi, and S. Namba, Appl. Phys. Lett., 49, p. 785, 1986; V.M. Donnelly, V.R. McGrary, A. Appelbaum, D. Brasen, and W.P. Lowe, J. App. Phys., 61, p. 1410, 1987.
4. D.J. Ehrlich, R.M. Osgood, Jr. and T. F. Deutsch, Appl. Phys. Lett., 38, p. 1018, 1981.
5. R. Srinivasan and V. Mayne-Banton, Appl. Phys. Lett., 41, p. 576, 1982.
6. M. Delfino and T. Reifsteck, IEEE Elec. Dev. Lett., EDL-3, p. 116, 1982.
7. D. B. Tuckerman and A.H. Weisberg, IEEE Elec. Dev. Lett., EDL-7, p. 1,1986; R. Mukai, N. S. Asaki, and M. Nakano, IEEE Elec. Dev. Lett., EDL-8, p. 76, 1987.
8. T. F. Deutsch, J.C.C. Fan, G.W. Turner, R.L. Chapman, D.J. Ehrlich, and R.M. Osgood, Jr., Appl. Phys. Lett., 38, p. 144, 1981 and references cited in reference 1.
9. P. G. Carey, T.W. Sigman, R.L. Press, and T.S. Fahlen, IEEE Elec. Dev. Lett., EDL- 6, p. 291, 1985.
10. K. Sugioka and K. Toyoda, J. Vac. Sci. Technol., B6, p. 850, 1988.
11. K. Sugioka, K. Toyoda, Y. Gomi and S. Tanaka, Appl. Phys. Lett., 55, p. 619, 1989.
12. G.S. Higashi and C.G. Fleming, Appl. Phys. Lett., 48, p. 1051, 1986; G. Blonder, G.S. Higashi, and C.G. Fleming, Appl. Phys. Lett., 50, p. 768, 1987.
13. P.D. Brewer, D. McClure and R.M. Osgood, Jr., Appl. Phys. Lett., 49, p. 803, 1986.
14. P.A. Maki and D.J. Ehrlich, Appl. Phys. Lett., 55, p. 91, 1989.
15. R.R. Tummala and E.J. Rymaszewski, Eds., Microelectronics Packaging Handbook, Van Van Nostrand (1989).
16. D.J. Ehrlich, T.F. Deutsch, D.J. Silversmith and R.M. Osgood, Jr., IEEE Elec. Dev. Lett., EDL-1, p. 101, 1980.
17. J.Y. Tsao, D.J. Ehrlich, D. Silversmith, R. Mountain, IEEE Elec. Dev. Lett., EDL 3, p. 164, 1982; T. Cacouris, R.R. Krchnavek, H.H. Gilgen, R.M. Osgood, Jr., S. Kulick and J. Schoen, Proceedings of the IEDM, 1985.
18. See, for example, Proceedings, 2nd Dupont Symposium on High Density Interconnect Technology, (October, 1988).
19. J. Black, S. Doran, M. Rothschild, and D.J. Ehrlich, Appl. Phys. Lett., 5050, p. 1016, 1987.
20. D.B. Tuckerman, IEEE Elec. Dev. Lett., EDL-8, p. 540, 1987.
21. J.Y. Tsao, R.A. Becker, D.J. Ehrlich and F.J. Leonberger, Appl. Phys. Lett., 42, p. 559, 1983, and T. Krauss, A. Speth, M.M. Oprysko, B. Fan, and K. Grebe, Appl. Phys. Lett., 53, p. 947, 1988.
22. See, for example, A.E. Willner, M.N. Ruberto, D.J. Blumenthal, D.V. Podlesnik and R.M. Osgood, Jr., Appl. Phys. Lett., 54, p. 1839, 1988.23. D. Podlesnik and H. Gilgen in reference 1.

Phase-Shifting Masks Gain an Edge

Thanks to developmental gains in the PSM technology, optical lithography is poised to cut finer feature shapes and sizes

B.J. Lin

One area in which semiconductor technology has made significant progress pertains to the increase in packing density and chip size. Lithography has been the key among the many technologies required for such scaling, because it facilitates early down-sizing of experimental devices and aids the development of other down-sizing technologies. More important, lithography is used in every masking step, and thus it has a significant impact on the cost of semiconductor manufacturing. Although exotic techniques such as the direct writing of e-beams have been used to fabricate exploratory devices, optical lithography remains the workhorse of the industry because of its production readiness and economy. X-ray, e-beam, and ion-beam techniques have been explored for decades, but none has been able to keep up with the pace of packing density and chip size or to show feasibility in building complete functioning circuits. Fortunately, phase-shifting mask technology is now at hand to propel optical lithography to the next level of performance.

Many avenues have been pursued to improve optical microlithography. It is not difficult to see from the Rayleigh equation on resolution,

$$W = k_1 \frac{\lambda}{NA} \qquad (1)$$

that λ, the actinic wavelength; and NA, the numerical aperture of the imaging lens, have to be reduced and increased, respectively, to improve resolution. The latter has started from 0.17, through 0.28, 0.35, 0.45, 0.55, and is approaching 0.6; while the former has been reduced from the mercury g-line of 436 nm, briefly to the 405 nm h-line, then 365 nm i-line, and the KrF excimer laser wavelength of 248 nm. Even exploratory works on the 193 nm ArF excimer laser wavelength have been pursued. Using a k_1 value of 0.8, which is commonly accepted for manufacturing practice, a resolution of 0.3 μm can be reached by stretching λ to 248 nm, and NA to 0.6. This is the feature size that can support manufacturing of 64M-256M DRAMs and 16M-64M SRAMs. However, the depth of focus (DOF) follows the Rayleigh equation on DOF, i.e.,

$$DOF = k_2 \frac{\lambda}{NA^2} \qquad (2)$$

The depth of focus inevitably decreases when λ or NA is changed to improve resolution. With $\lambda = 248$ nm, $NA = 0.6$, and $k_2 = \pm 0.5$, the DOF becomes ±0.34 μm. Values of DOF smaller than that are not usable, because of resist thickness, focus sensing errors, device topography, chucking error and absence of wafer flatness. Some other techniques have been pursed to improve the resolution and preserve the DOF (or even improve it) for optical lithography. None of these techniques rely on decreasing the λ/NA ratio. Some, such as reducing stray light and vibration, restore lost contrast. Removing reflections in the photoresist layer reduces the large exposure latitude to manageable ones, because optical interference in the resist film induces a large variation in the exposure intensity. Proximity corrections facilitate a common exposure-defocus window for an arbitrary combination of small feature sizes and shapes.

However, the phase shifting mask (PSM) technique, as well as illumination optimization, can improve the contrast of the optical image to improve the resolution and DOF. The potential improvement is a reduction of k_1 by between 0.15 and 0.3, with a tolerable reduction in k_2. That is, the DOF can be maintained at ±0.34 μm while the resolution is improved to 0.18 μm, with $\lambda = 248$ nm, $NA = 0.5$, $k_1 = 0.35$ and $k_2 = \pm 0.35$, assuming that the other improvements have already reduced the usable k_1 from 0.8 to 0.65.

Operating Principles

Phase shifting masks take advantage of the interference effect in a coherent or partially coherent imaging system to reduce the spatial frequency of a given object, to enhance its edge contrast, or to achieve both. This technique results in a combination of higher resolution, larger exposure latitude, and larger DOF. The shifting of phase is ac-

Reprinted from *IEEE Circuits and Devices Magazine*, Vol. 9, No. 2, pp. 28-35, March 1993.

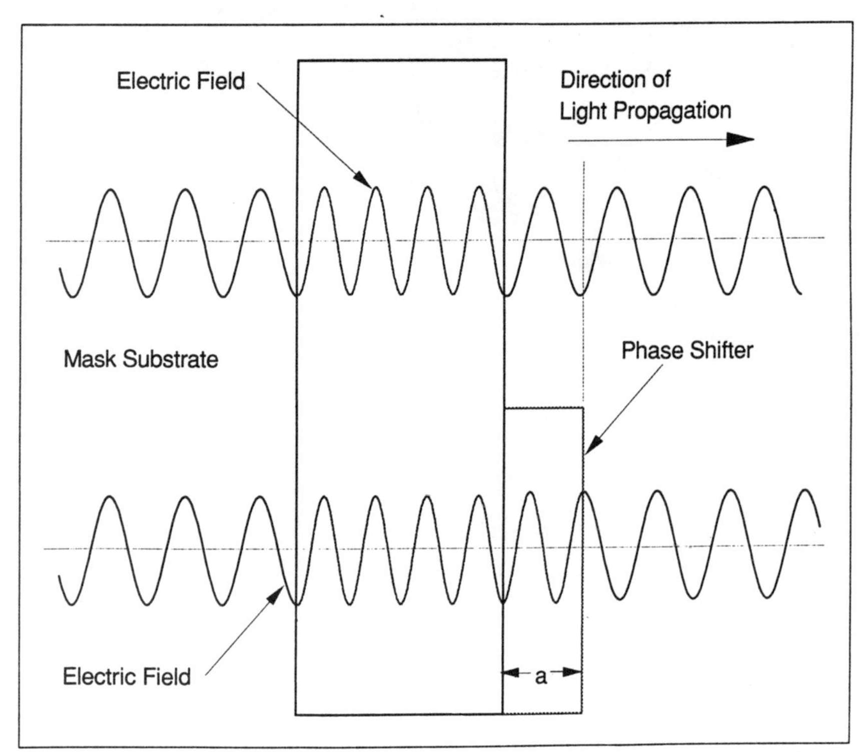

1. The phase shifting layer introduces a difference in optical path. Here, the difference is $\lambda/2$.

complished by adding an extra patterned layer of transmissive material on the mask (Fig. 1). As light propagates through the substrate and the extra layer, its wavelength is reduced (from that in air) by the refractive index of the substrate and the extra layer. If we compare the optical path through the extra material, and the path without it, there is a difference of $(n-1)a$, where n is the refractive index of the extra layer, and a is the layer's thickness. The phase difference, θ, is

$$\theta = 2\pi\frac{a}{\lambda}(n-1) \quad (3)$$

Usually, a phase shift of π is desirable, not only because the maximum difference in amplitude, namely from 1 to -1, is achieved, but also because this is the region of the sine wave where amplitude differences are least affected by small phase variations. In other words, $\partial\cos\theta/\partial\theta$ is minimum at $\theta = \pi$, under which conditions Eq. 3 yields,

$$a = \frac{\lambda}{2(n-1)} \quad (4)$$

This is the thickness required to induce a π shift. Because one popular phase shifter, SiO_2, has a refractive index of approximately 1.5, a is now approximately equal to the actinic wavelength. Note that the phase shift can be any odd number of π, i.e., $(2m+1)\pi$, where $m = 0, 1, 2, \ldots$ and that phase shifting is relative. The extra higher refractive index material can be viewed as the phase shifter, or the lower refractive index air path can be considered as such. For simplification of discussion, the extra layer is often called the phase shifter unless specifically defined otherwise. The imaging improvements due to PSM can be easier understood by examining each individual phase shifting approach.

PSM Approaches

There are many ways to introduce the phase shifters to the mask pattern. Their configurations and working principles will be discussed as follows. Reference of these approaches can be found in [1].

Alternating Phase Shifting (ALT PSM)

This system is characterized by phase shifting every other element in a closely packed array. The Alt PSM approach is shown in Fig. 2b, in comparison to the binary intensity mask (BIM) in Fig. 2a. In the former case, the electric field amplitude is -1 at the shifted areas on the mask. This -1 amplitude effectively reduces the spatial frequency of the electric field so that it is less inhibited by the lens transfer function and forms a higher-contrast amplitude image at the wafer plane. When this electric field is recorded by the photoresist, only the intensity, which is proportional to the square of the electric field amplitude, can be recorded. Hence, the reduced spatial frequency is doubled back to the original frequency but the image has much more contrast. Edge contrast is also improved because the electric field must pass through a zero to -1 amplitude, which assures a zero intensity at the wafer.

Subresolution-Assisted Phase Shifting (SA PSM)

The Alt PSM method demonstrates the potential of phase shifting. However, it requires closely packed patterns to be effective. In actual circuit layouts there are many situations where critical dimensions are sufficiently far away from any adjacent patterns to provide phase shifting. In order to provide phase shifting for isolated openings such as contact holes and line openings, subresolution phase shifters are introduced (Fig. 2c). The dimension of these phase shifters is below the resolution limit of the optical imaging system, thus, images cannot be printed. Their sole function is to enhance the edge contrast of the pattern of interest.

Rim Phase Shifting (Rim PSM)

SA PSM and Alt PSM are still limited by the inability to provide phase shift to opaque patterns. Rim PSM (Fig. 2d), overcomes such a problem and can be applied to an arbitrary mask layout. Here, phase shifting only takes place at the rim of the mask patterns. The center of the patterns is blocked by the absorber to prevent large areas of negative amplitude from producing bright areas where they are supposed to be dark. Again, bright areas result from negative or positive field amplitudes because the photoresist can only detect intensity which is proportional to the square of the electric field. Note that edge contrast enhancement is now the sole imaging improving function of these phase shifters.

Attenuated Phase Shifting (Att PSM)

The Att PSM is another PSM approach that applies to arbitrary mask layouts. It can be implemented on either a transmissive mask or a reflective one. The dark areas of the mask can be phase shifted to π, but with an attenuated amplitude to prevent producing

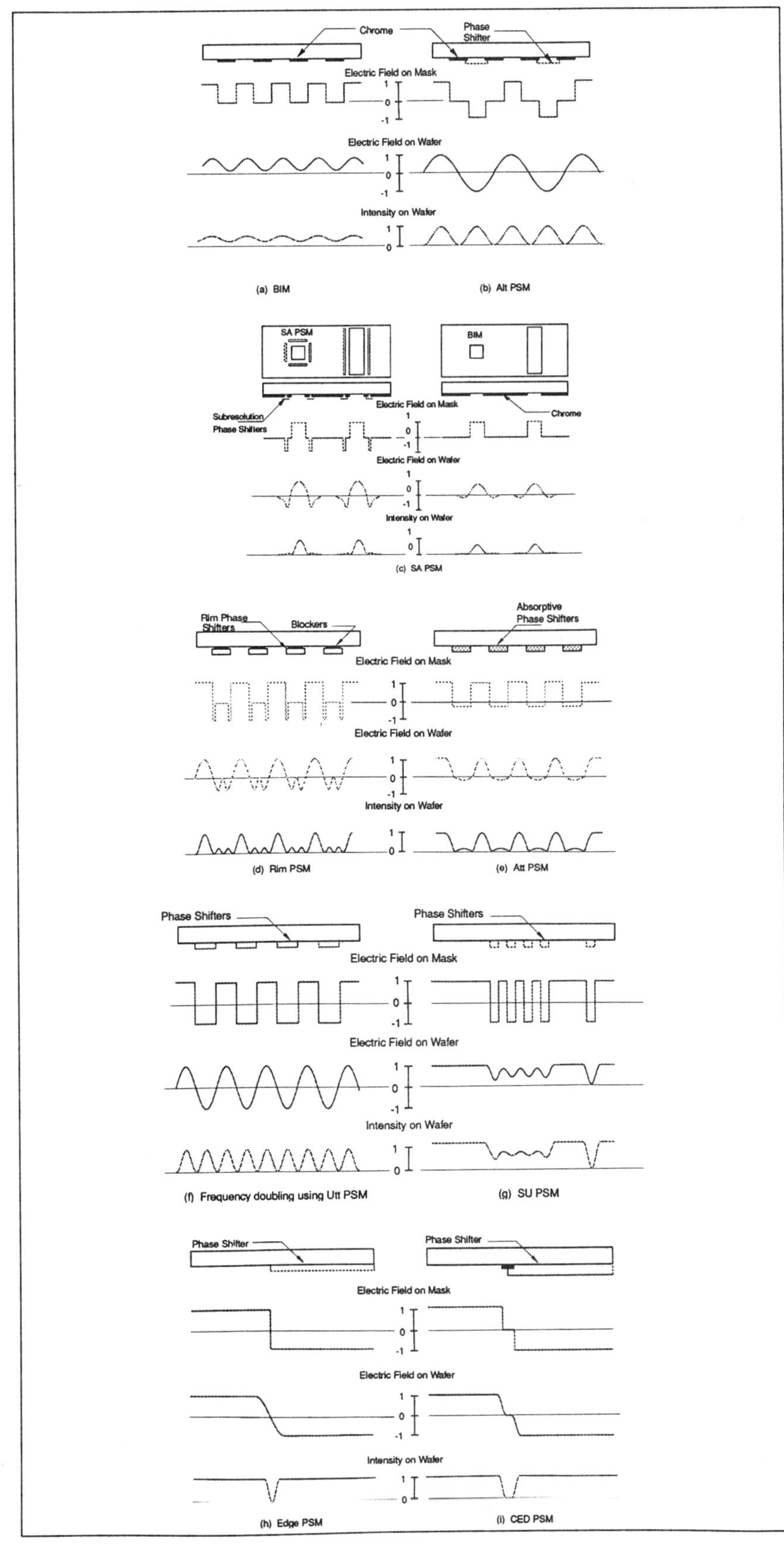

2. Binary intensity mask and PSMs. BIM (a), Alt PSM (b), SA PSM (c), Rim PSM (d), Att PSM (e), frequency doubling using Utt PSM (f), SU PSM (g), edge PSM (h), CED PSM (i).

too much light in these areas (Fig. 2e). The negative amplitude provides the desired improvement in image edge contrast, and the attenuation prevents the negative amplitude from becoming too large and subsequently exposing the resist.

Unattenuated Phase Shifting (Utt PSM)
Transparent phase shifters can also be used to improve the optical image without having to use a mask absorber. The transparent characteristics can be taken advantage of in spatial frequency doubling, or to make an opaque line image from the phase shifted edge (edge PSM) as shown in Figs. 4f, and 4h, respectively. Sometimes the phase shifted edge is covered with an absorber stripe (Fig. 2i) to reduce the dependence on the edge profile for narrow images and to control the linewidth of larger images. This covered edge (CED) PSM is not an Utt PSM but shares the same working principle as edge PSM. Packing CED PSM closely returns the structure to Alt PSM.

Small phase shifter patterns in Utt PSM become opaque patterns because the opaque edges of these patterns are not resolvable from each other. Hence, as shown in Fig. 2g, small opaque features can be formed by this sub resolution Utt Psm (SU PSM) and large opaque areas can be synthesized with many SU features.

Pros and Cons of PSM Approaches

Alt PSM
The Alt PSM is the only known PSM approach that can achieve $k_1 = 0.35$ for closely packed patterns. The DOF is also very impressive at such low k_1. Therefore, much attention has been given to this approach. However, one should be aware of the following characteristics:

(1) Requirements of reflection-controlled resist systems: Even though the exposure latitude is improved by Alt PSM at low k_1, it is still smaller than that given by unshifted systems with high k_1, and cannot support the large exposure variation caused by multiple reflections in the resist, as previously discussed. A resist system is required that suppresses multiple reflections in the resist, such as a top-surface imaging or multi-layer resist system. Otherwise, the critical dimension control tolerance cannot be maintained.

(2) Unwanted dark lines with light-field mask: Alt PSM works best with dark-field masks where the absorber background

can cover the phase shifter edges. In a light-field mask, the phase shifter edges become unwanted opaque images because they are edge phase shifted, just as edge PSM (Fig. 3).

(3) Uneven phase shifting because of non-uniform packing: When the distance between line edges becomes different, the phase shifting improvement becomes non-uniform. At large distances, the improvement can be completely lost.

(4) It is mathematically impossible to achieve perfect alternation for arbitrary 2-dimensional layouts.

SA PSM

SA PSM can readily be combined with Alt PSM to improve the image of isolated openings, but its limitations are:

(1) The resolution potential is not as high as Alt PSM; the minimum printable k_1 is larger than 0.5.

(2) Mask writing requires resolution much higher than that of BIM.

(3) The mask grid size has to be smaller, causing problems in high pattern count, thus high data volume.

(4) Much room is required; an area approximately 0.4 - 0.6 λ/NA to either side of a feature has to be reserved for SA phase shifters, making a total extension of 1.6 λ/NA in either x or y.

Rim PSM

Rim PSM applies to any arbitrary layout and is most effective for line and hole openings. Unlike Alt and SA PSM, self-aligned fabrication of rim PSM is possible, making a second aligned mask writing unnecessary. However, its limitations are:

(1) Exaggerated proximity effect; that is, features of different shapes and sizes do not share a common exposure window. Even though each individual feature has a large exposure-defocus latitude, none exists for combination of features when proximity effects are severe.

(2) Much room is required. Feature openings have to be larger than the desired image by at least 0.2 λ/NA and each rim is of the order of 0.15-0.2 λ/NA, making a total extension of 0.5-0.6 λ/NA in either x or y.

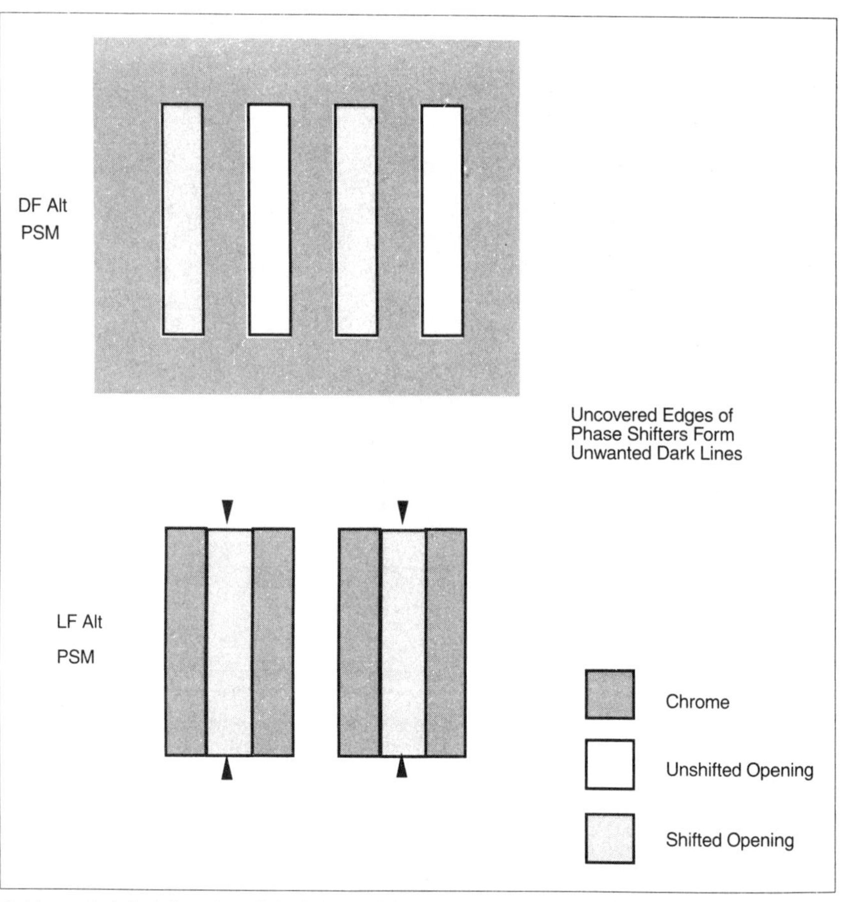

3. Unwanted dark lines in a light-field Alt PSM.

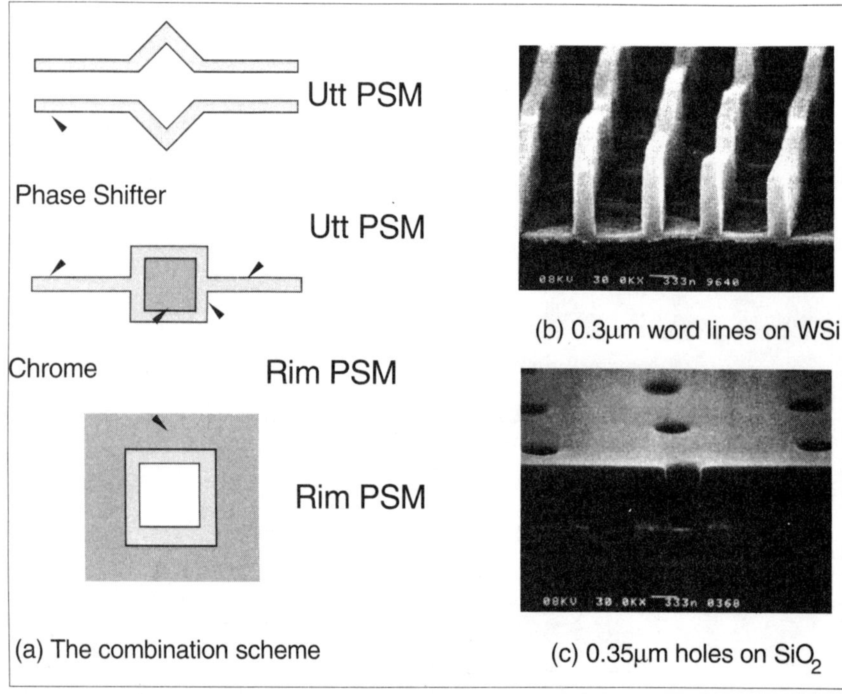

4. A combination of rim PSM and Utt PSM (Courtesy by K. Nakagawa et al., Fujitsu Ltd.) The combination scheme (a); 0.3 μm resist word lines on WSi, k_1 = 0.41 (b); 0.35 μm holes on SiO_2, k_1 = 0.48 (c).

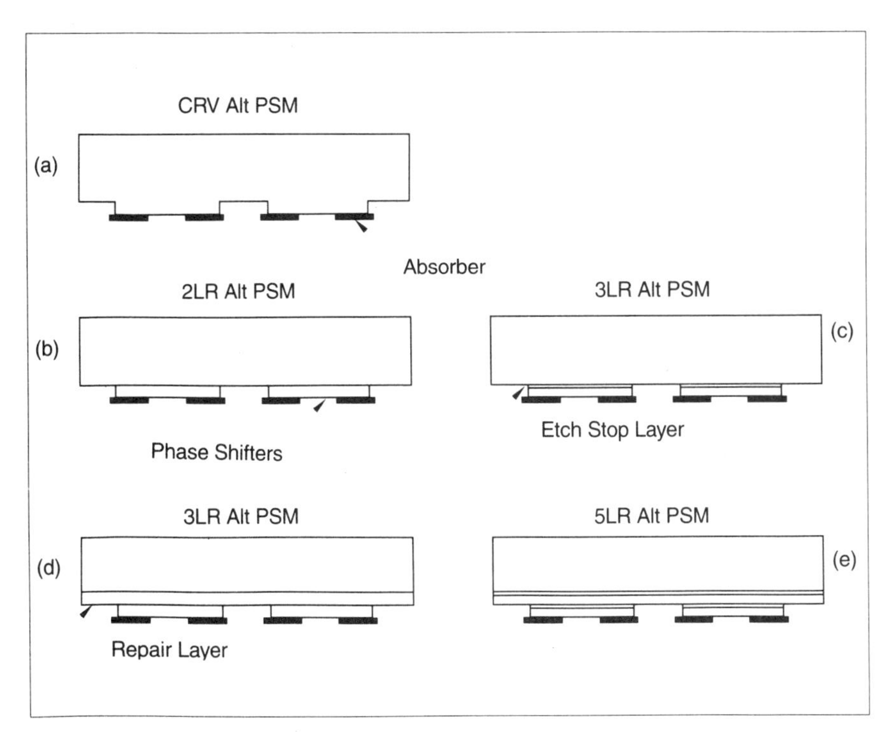

5. Crossection of CRV and MLR PSMs. CRV Alt PSM (a). 2LR Alt PSM (b). 3LR Alt PSM structure to facilitate repair (d). 2LR becomes 3LR PSM to introduce etch stop layer (c). 3LR PSM becomes 5LR PSM to introduce etch stop layer (e).

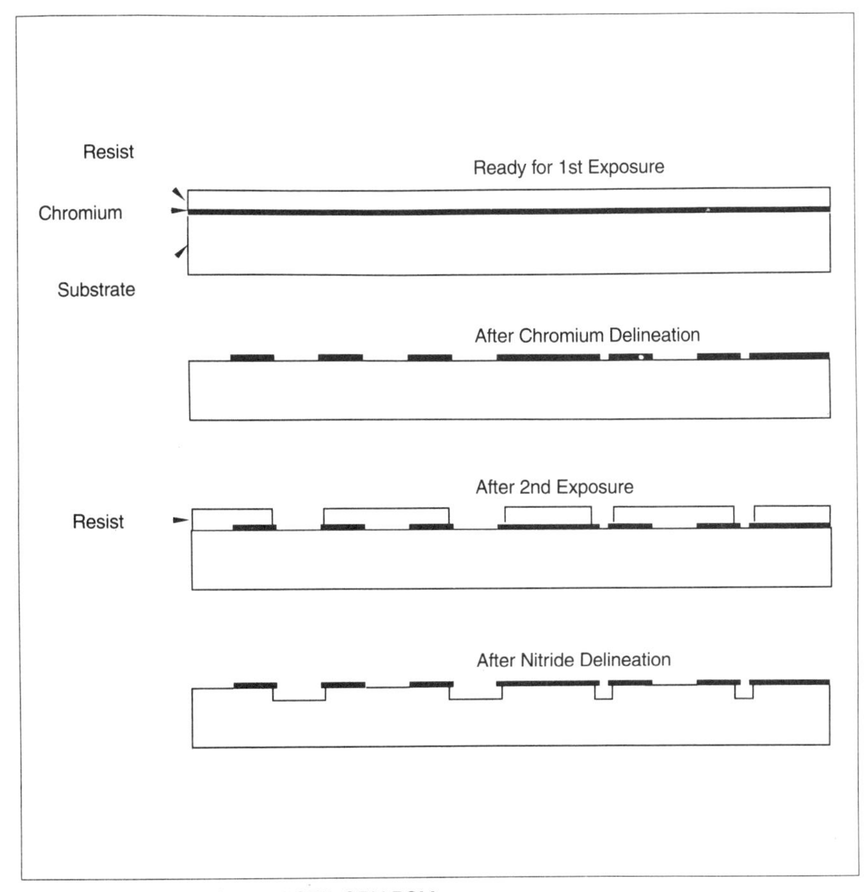

6. Fabrication of Alt, SA, and CED CRV PSMs.

(3) Improvements of line-space pairs at low k_1 are negligible.

(4) The linewidth variation is extremely sensitive in size variation of the rim shifter.

Att PSM
Similar to rim PSM, the Att PSM is most effective for line and hole openings, applies to arbitrary patterns, can be self-aligned, has proximity effects, and needs larger openings that the desired images. However, it is better than rim PSM in all aspects; Namely, the *DOF* is larger, exposure dosage lower, proximity effects less significant, and the extra room required per feature is approximately 0.1 λ/NA instead of 0.5-0.6 λ/NA.

The difficulties with Att PSM are:

(1) The mask blanks are not commercially available at large quantity and guaranteed quality.

(2) New absorber patterning processes may have to be developed.

(3) Perfect repair of defects at the pattern edges is difficult. However, Att PSM lends itself to less defect-prone fabrication processes.

Utt PSM
Utt PSM can best be used in combination with rim PSM (Fig. 4). In fact, exploratory 64M DRAM circuits have been fabricated with such a combination [2]. Its improvements on small opaque features is better than rim PSM, and it automatically results from the diminishing absorber areas as the feature size gets smaller for opaque features.

However, composing an entire mask using SU PSM is not practical. Similar to SA PSM, the subresolution elements can be extremely burdensome in terms of pattern count and high resolution mask making.

Edge or CED PSM are desirable when the layout is looped because one single edge form by a phase shifter cannot self-exist. However, when the circuit does not call for loops and edge PSM is used, an additional masking level is required to cut the loops.

It is not obvious that a single PSM approach can best deal with specific patterns. In order to accommodate an arbitrary design, the proper combination of approaches (composite PSM) is inevitable. In the foreseeable future, more and better combinations can be expected.

PSM Fabrication, Repair and Inspection

The fabrication process is strongly dependent on the type of PSM chosen, the availability of a uniform phase-shifter etch process, the choice of phase shifter materials, alignment capability of a second-level exposure, and the repair requirements. Two major substrate types are now considered. First, there is the carved PSM (CRV PSM), with which the phase shifter is part of the quartz substrate (Fig. 5). Phase shifting is produced by selectively removing a predetermined thickness of quartz to facilitate the optical path difference.

Second, there is the multi-layer PSM (MLR PSM), which has a substrate containing a separate phase shifting layer and, most likely, an etch stop layer. An additional layer to the so-called 2LR PSM has been proposed to facilitate PSM repair [3]. If the phase shifting layers cannot be differentially etched with respect to the quartz substrate, additional etch stop layers have to be used, resulting in 3LR and 5LR PSMs, as shown.

Fabrication of CRV PSM

With a uniform quartz etching process, the PSM can be carved from a conventional chrome-on-quartz substrate for BIM. Figure 6 shows the fabrication steps for Alt, SA, and CED PSMs. All the chrome patterns are delineated conventionally, followed with an aligned second exposure which opens the areas designed for phase shifting so that the quartz can be etched to the desired thickness for phase shifting. The alignment need not be extremely precise because the edge of the phase shifters is covered by chrome patterns.

For the simple Att PSM consisting of a single layer of attenuated phase shifter that satisfies both the π-shift and the attenuation requirements, the single exposure, development and etching steps are similar to those of BIM. When the transmissive absorber does not provide the desired amount of phase shift, it is etched first, and is followed with self-aligned etching of the quartz substrate to produce the CRV Att PSM. No second-level exposure is needed. Fabrication of the CRV Utt PSM is even simpler. One starts with an uncoated quartz blank, images with resist and etches the quartz into the depth for a π-shift.

Many different methods can be used to fabricate the CRV rim PSM. It is preferable to adopt a self-aligned process to avoid the

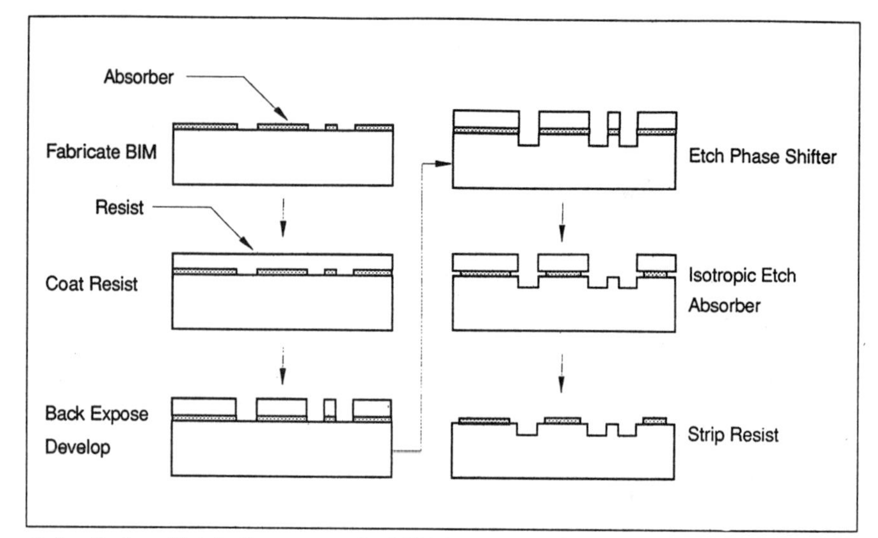

7. A self-aligned fabrication process for CRV rim and Utt PSMs.

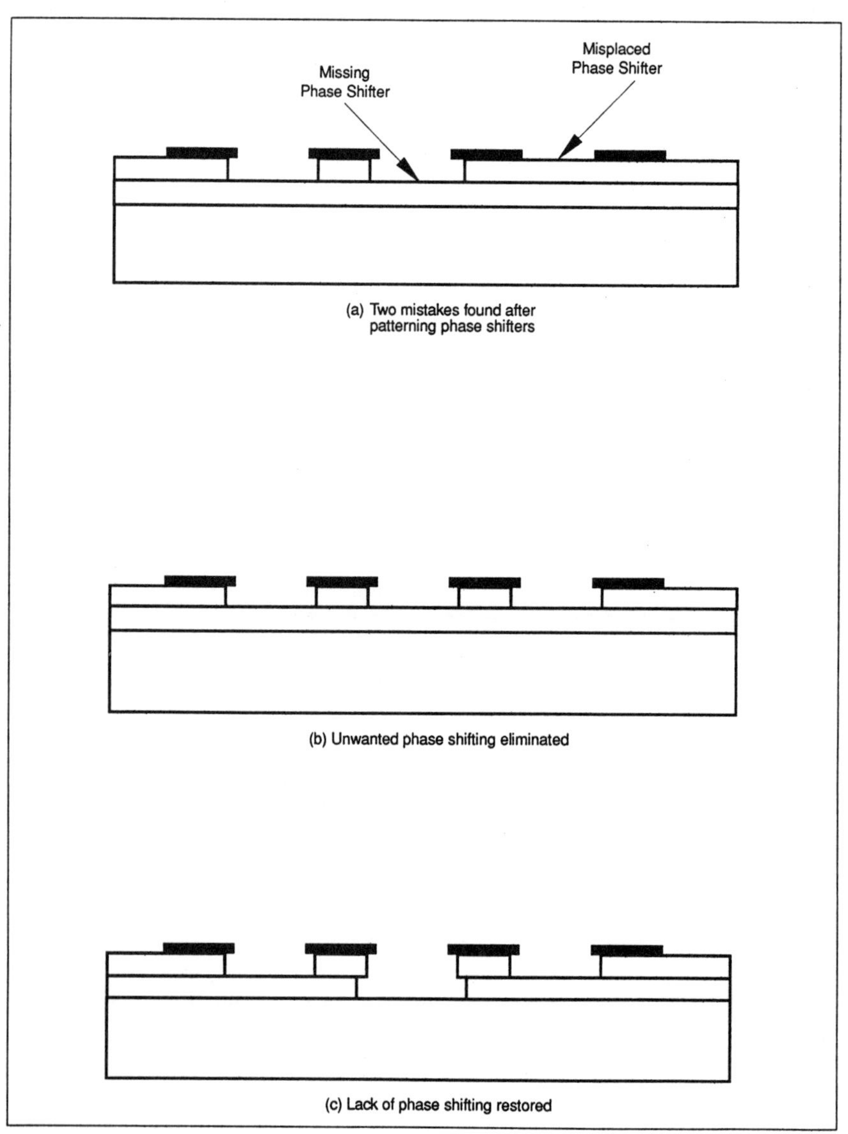

8. Repairing a double-π MLR PSM.

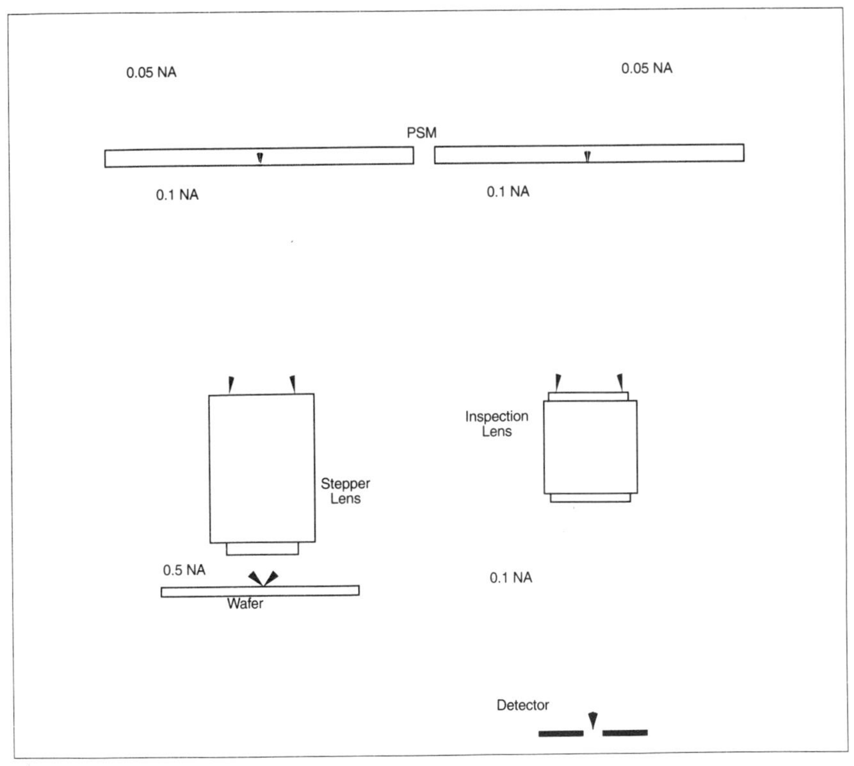

9. *Functional PSM inspection.*

need of an aligned second exposure. In Fig. 7, the absorber and the desired thickness of phase shifter is delineated just as in the case of a CRV Att PSM. before the resist etch mask is stripped, an isotropic chrome etch creates the undercut in chrome to expose the phase shifter rim [4]. Stripping the resist completes the process.

Fabrication of MLR PSM

A straightforward way to design a MLR PSM is to have a separate phase shifter between the absorber and the quartz substrate (Fig. 5b). The phase shifter materials must have a high etch selectivity with respect to quartz. Except for Utt and Att PSMs, etch selectivity from the absorber to the phase shifter is also required. If no such phase shifter ia available, an additional etch stop layer has to be inserted, resulting in the configuration shown in Fig. 5c. The fabrication steps are similar to those in of the CRV PSM, except that the tight uniformity tolerance of the etch process is now relaxed. The tradeoffs are lower yield and higher cost of substrate manufacturing.

For most PSM approaches, it is preferable to have the absorber above the phase shifter. Not only does this allow the CRV configuration, but is also permits the mask blank to be fabricated at the blank manufacturer and saves the mask maker the task of depositing a defect-free phase shifting layer in a less suitable environment. There are also optical advantages in reducing scattering and multiple reflections caused by variation of the refractive index and imperfect interfaces, However, the shifter-on-absorber configuration is suitable for the Att PSM. Here, the absorber serves as an inherent etch stope for the phase shifter either during fabrication or repair.

Repairing PSMs

Except for the shifter-on-absorber Att PSM and Utt PSM, the absorber is not covered by any other layer, and the repair of missing or misplaced absorber is similar to repairing BIM. However, when a focused gallium-ion beam is used for repair [5], the gallium ions implanted into the mask substrate cause transmission loss. The subsequent gallium stain removal process may induce phase shifting errors. Laser repair is not susceptible to staining but the edge definition accuracy has to be improved in order to repair defects at critical areas. In either case, a differential etching process to remove the phase shifter against chrome and vice versa has to be developed. The lack of an additive process for phase shifter may be overcome by taking advantage of the fact that a shift of 2π is equivalent to a θ shift. Therefore, subtraction can be used to create the equivalent of addition. However, light scattering from a shallow edge is different from that for a deep edge. Repairing a missing phase shifter defect by etching the location to shift 2π can leave dark lines at the boundary.

With either existing focused ion beam or laser repair equipments, only the defect edge can be viewed. The capabilities to evaluate the amount of phase shifting and to detect the etch end point are needed. One has to overcome these deficiencies of the repair equipment or has to avoid repair by developing low-defect fabrication processes.

MLR PSM alleviates the problem of repair end-pointing, just as it does the etch end point problem in fabrication. The layers may be configured to facilitate PSM repair. The repair steps of a double π-shifter configuration [3] are shown in Fig. 8. This further trades off ease of repair with cost and difficulties in substrate manufacturing.

PSM Inspection

In additional to inspecting defects similar to BIM, the inspection of PSM defects include missing or misplaced phase features, placement and acuity of phase shifter edges, and phase errors. Existing equipment can inspect, with little modification, phase shifter edges not covered with the absorber. This is possible despite the inability of either bright-field or dark-field illumination to turn phase changes into detectable intensity changes; any uncovered phase shifter edge scatter light to change the intensity there.

In order to measure the amount of phase shift and the errors hereof, some means to convert phase information into intensity information has to be used. Phase contrast imaging or interference contrast can convert phase distribution into intensity distribution to be detected.

Ultimately, a functional inspection can benefit PSM inspection, and perhaps even PSM repair. An example of functional viewing is shown in Fig. 9, where a small-field 1X objective using the actual imaging wavelength is the key component. The *NA* of this objective is *nX* less than that of a *nX* stepper lens to simulate the mask side of the imaging process. The projected image is now *nX* larger than the wafer image, which makes it easier to examine. Care should be taken to ensure that the illumination mimics the actual partial coherence and pupil distribution.

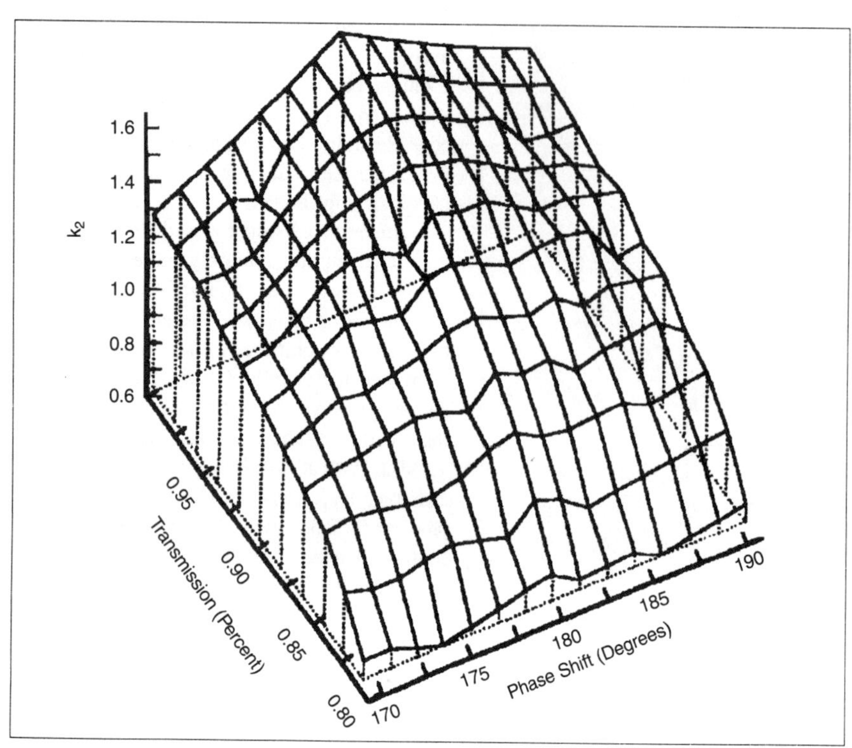

10. Variable k_2 as a function of phase shift and phase shifter transmission.

Tolerances

In addition to substrate transparency, absorber edge location, smoothness, and opacity in BIM, the variation of phase and tranmissivity of the phase shifter are of concern with PSM. For Utt and rim PSM, the edge location, profile, and smoothness of the phase shifter are additional concerns. The tolerance of these parameters has to be evaluated separately and interactively to specify the mask writing, measurement, inspection, and repair equipments, as well as the mask fabrication processes.

Consider an example for the Alt PSM approach operating at $k_1 = 0.5$ (Fig. 10). The normalized *DOF*, k_2, is evaluated as a function of phase shift and transmission of the phase shifter. Assume a ±10 percent tolerance of the minimum feature size at an exposure latitude of 10 percent, which can be supported by a reflection-controlled resist system such as top-surface-imaging or multi-layer resists. When a single layer resist system is used, the exposure latitude requirement increases to 30 percent. Using an anti-reflection coating with the resist system reduces the latitude to 20 percent. Then, one evaluates his particular *DOF* requirement, which consists of such factors as wafer flatness, chucking errors, resist thickness, topography, focussing errors, and lens field curvature. Typically, a ±0.5 μm *DOF* is required. An ultimate *DOF* goal of ±0.35 μm has been proposed. The tolerance can now be determined using Eq. 2 for a given set of λ and *NA*. Even though the actual tolerance has to be determined by the particular PSM approach, the linewidth tolerance and the exposure latitude requirements of a best, worse and medium case is given here for Alt PSM. A more detailed treatment of Alt PSM tolerance can be found in [6].

(1) At $k_1 = 0.5$, and a *DOF* requirement of ±0.5 μm, λ = 248 nm, and *NA* = 0.7, the minimum feature size attainable is 0.18 μm from Eq. 1. If the exposure latitude requirement is 10 percent, the transmission tolerance is 15 percent and the phase tolerance is ±10 percent. When a resist system other than top-surface-imaging or multi-layer resists is used, the tolerance becomes tighter because the exposure latitude is much larger than 10 percent.

(2) At $k_1 = 0.35$, and a *DOF* requirement of ±1 μm, λ = 365 nm, and *NA* = 0.5 to make 0.26 μm minimum features, the transmission tolerance is 10 percent and the phase tolerance is ±2°. Tightening the transmission tolerance to zero only relaxes the phase tolerance to ±3°.

(3) At $k_1 = 0.35$, and a *DOF* requirement of ±0.5 μm, λ = 248 nm, and *NA* = 0.5 to make 0.18 μm minimum features, the transmission tolerance is 5 percent and the phase tolerance is ±5°.

Summary

The potential, working principles, and approaches in phase shifting masks have been shown. The tradeoffs of each approach, then fabrication, inspection, repair, and tolerances have been discussed. It is feasible to use the phase shifting technology to improve optical lithography to 0.18 μm feature size with $k_1 = 0.35$, λ = 248 nm, and *NA* = 0.5. Further resolution improvements are still possible, but much development is required to making phase shifting masks a manufacturing reality. **CD**

Dr. B. J. Lin joined the IBM T.J. Watson Research Center in 1970, where he served as a research staff manager in charge of developing single- and multi-layer resist processes for optical and e-beam lithography. He joined IBM GTD in 1985 to manage the advanced optical lithography department. In 1989, he moved to the IBM advanced Technology Center, then took an assignment at Sematech with the phase-shifting mask team. He founded Linnovation, Inc. in August 1992 to expand his research in optical lithography.

References

1. B.J. Lin, "Phase Shifting and other challenges in optical mask technology," *SPIE BACUS Proceedings*, **1496**, p. 54, 1990.

2. K. Nakagawa, M. Taguchi, and T. Ema, "Fabrication of 64M DRAM with i-line phase-shift lithography," *IEDM Technical Digest*, p. 817, 1990.

3. N. Hasegawa, "Repair technology of phase shift masks," *Semiconductor World*, p. 103, 1991.

4. Y. Yanagishita, N. Ishiwata, Y. Tabata, K. Nakagawa, and K. Shigematsu, "Phase-shifting photolithography applicable to real IC patterns," *SPIE Proceedings*, **1463**, p. 207, 1991.

5. L. Harriot, "Focused ion beam repair of phase-shift photomasks," *SPIE Proceedings*, **1671**, 1992.

6. A.D. Katnani and B.J. Lin, "Phase and transmission error study for the alternating-element (Levenson) phase-shifting mask," *SPIE Proceedings*, **1674**, 1992.

7. Y. Funaki and Y. Mochizuki, "Lithography for 64M developments centered around phase shift," *Nikkei Microdevices*, p. 46, 1991.

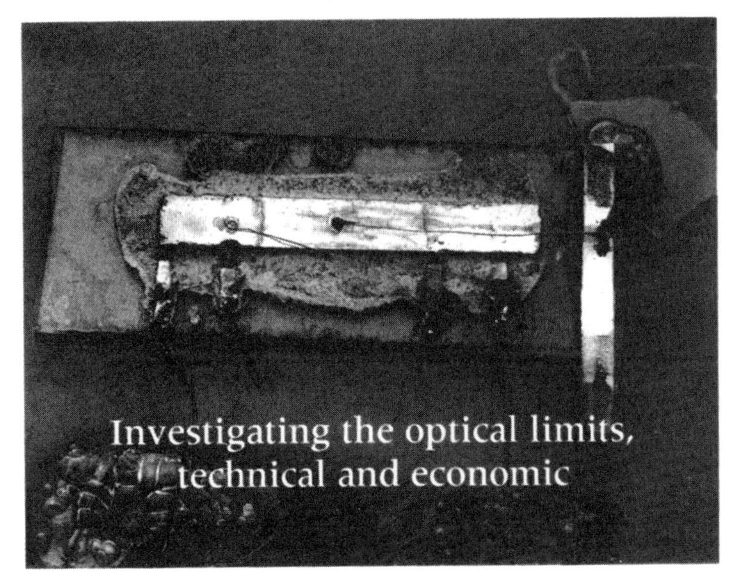

The first integrated circuit, a phase shift oscillator. Courtesy of Texas Instruments.

Advanced Lithography for ULSI

J. Bokor, A. R. Neureuther, and W. G. Oldham

Progress in lithography has always been the key pacing item in the drive toward smaller transistors, higher chip density, high speed, and increased performance. In the early 1970s, contact and proximity printing were used to print the first large scale integrated (LSI) circuits with feature size in the 10 micron range. In 1975, the Perkin-Elmer projection aligner was introduced, an event that can be viewed as the birth of practical optical projection lithography and the very large scale integrated (VLSI) circuit. Today, high performance steppers, using ultraviolet light and the world's most advanced optics, are being used to manufacture 64 Mbit DRAMs and advanced microprocessors with minimum feature sizes of 350 nm.

On numerous occasions during this 20+ years of history, the apparent limits of optics have been identified, and a halt to the progress in optical lithography has been predicted to be imminent. However, the reports of the death of optical lithography have consistently been exaggerated (apologies to Mark Twain). The tremendous commercial incentive for continued improvement of IC technology has been the driver for a series of innovations in optics that have time and again overcome the predicted limits. What's more, increasingly powerful computers made possible by the improvements in IC technology have played an important role in the progress in optics, via sophisticated computer aided lens design. Another noteworthy factor in the development of advanced lithography optics has been improved metrology for characterizing lens aberrations using optical interferometry techniques.

From today's perspective, we see that lens technology has advanced to the point where lithographic feature sizes equal to, or even smaller than, the wavelength of the light used in the stepper can be achieved in production. How far optical lithography can go from here depends both on how short a wavelength is possible and how close we can come to the absolute limit of diffraction. In this article, we will examine the current thinking on these questions, and discuss what might happen if and when optical lithography really can no longer be used.

Characterizing Optical Performance

Figure 1 shows a schematic diagram of the optical system in a lithography stepper. Critical parameters for lithography are the minimum printable feature size, S, and the focus range over which the image is adequately sharp (depth of focus, DOF) of the projection lens. These depend in turn on the numerical aperture, NA, of the lens and the

Reprinted from *IEEE Circuits and Devices Magazine*, Vol. 12, No. 1, pp. 11-15, January 1996.

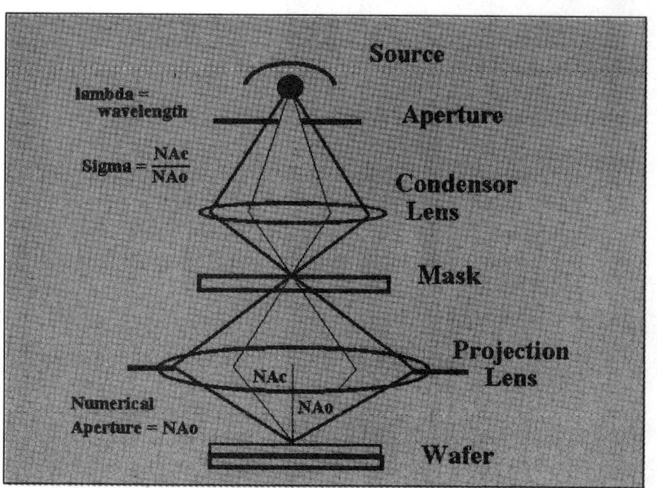

1. Schematic diagram of an optical projection printing system showing the source of illumination, the collection of the source rays by the condenser lens, the mask which to convey its pattern diffracts rays at new angles, and the projection lens, which collects ray out to a maximum acceptance angle and focuses them onto the wafer.

actinic wavelength, λ, and for diffraction limited (ideal) optics are described by the simple relations:

$$S = k_1 \lambda / NA$$

$$DOF = k_2 \lambda / (NA)^2$$

where k_1 and k_2 are empirical constants (both in the range of 0.5 to 1.0). They can be theoretically calculated under special conditions. For example, Lord Rayleigh showed that the resolving power of microscope objectives is described by these equations with $k_1 = 0.61$ and $k_2 = 1$. These values are commonly referred to as the "Rayleigh limit." In practical lithography, k_1 and k_2 are generally dependent on a large number of tool, resist, and process parameters, the type of pattern being imaged, as well as the requirements of the shape and allowed size range of the developed resist profile. A k_1 value of unity and a k_2 value of 0.7 were achieved within a few years after the high-volume introduction of optical projection lithography. Steppers used in production today use the 365 nm "i-line" produced by mercury vapor lamps, and have NA as high as 0.65. Using k_1= 0.6, a resolution of 350 nm is achieved in production. Further improvement of k_1 down to below 0.5 may be expected in the future with improvements owing to reduction of aberrations, improved resists, and resolution enhancement methods including the use of phase masks.

Lithographic feature size is set by the resolution, so further reduction of feature size could be obtained by reducing k_1, reducing λ, and/or increasing NA. As shown in Fig. 2, all three strategies have been pursued simultaneously in the past, and are projected to continue for the foreseeable future. The severe reduction in DOF associated with this trend becomes a formidable problem, owing to the fact that real wafers are not perfectly flat to begin with, and deposition and etching of thin films creates topography on the wafer surface. To cope with this problem, planarization of the wafer across the image field must be used. Reduced DOF also puts additional demands on the precision of the stepper focusing system.

Wavefront Engineering

Methods for making significant practical improvements in the working resolution (low k_1) and depth of focus (high k_2) are being investigated using what are collectively referred to as "resolution enhancement" techniques. These enhancements include modified illumination, phase-shifting masks, and in-lens filtering. Each deliberately emphasizes the high spatial frequency components that form the image. This process may be thought of as wavefront engineering to improve image fidelity at the wafer.

Producing sharp images of small features in projection printing is analogous to generating a crisp strike of a snare drum in a sound system. The highest frequency sound waves are best at resolving the strike of the drum, just as the most off-axis waves in a projection printer (which, when they interfere at the wafer plane produce the highest spatial frequencies in the image) are best at resolving a small feature. The sound system is fundamentally limited by the highest frequency it passes, and so, too, is the projection printer. While the total bandwidth of a sound system may be limited, improvement in overall fidelity is possible by emphasizing certain sub-bands of the allowed frequencies using a graphic equalizer.

The resolution enhancement techniques in optical lithography are basically different approaches for implementing equalization, in that they emphasize off-axis rays (high frequencies) relative to near-axis waves (low frequencies). The modifications to the

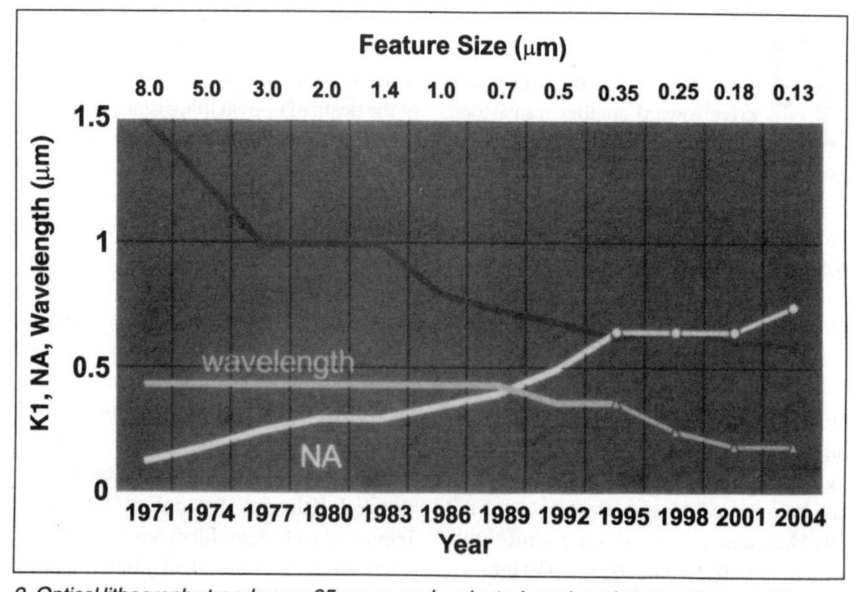

2. Optical lithography trends over 25 years, and projected continuation over the next 10 years, illustrating the contributions to progress from wavelength reduction, increase of numerical aperture (NA) and the k_1 parameter.

spatial frequency spectrum that they introduce are depicted in Fig. 3. The horizontal axis, labeled frequency, describes the spatial frequency, and is high for off-axis rays and low for near-axis rays. In modified illumination, such as quadrupole illumination, the high frequencies are increased and the low frequencies are reduced by simply blocking unwanted rays before they strike the mask. The mask, which diffracts the incident light, is assumed to be periodic. It basically acts like a mixer and shifts the source energy into weighted copies at each of the diffracted orders. This mixing of the source spectrum with the mask means that for particular combinations of the pattern period and the off-axis illumination, a very dramatic emphasis of the high frequencies can be obtained. Modified illumination is now available on commercial steppers and can enhance both k_1 and k_2 for certain geometries.

Adding a phase-shifting function to the normally on-off properties of the mask increases its ability to mix more of the energy up to high frequencies. In fact, a 180 degree phase-shift between alternate openings on a periodic mask can even eliminate the energy remaining in the low frequencies. The use of phase-shifting masks (PSM) has been reviewed recently [1]. A weak form called attenuating PSM, which uses a single layer of 6-10% transmitting material with 180 degree phase shift, is ready for production use at a k_1 of about 0.5. Another form of weak PSM using edge (rim) shifters is also emerging. Also under investigation is a strong PSM approach in which alternating openings are given phases differing by 180 degrees (known as the Levenson mask) with potential for k_1 of about 0.4.

An in-lens filter can also produce a similar effect by attenuating the near-axis low frequency components. In-lens filtering is difficult to implement with existing exposure tools because it requires access to the pupil plane of the projection printer. Thus far, it is receiving attention only at the laboratory level.

Other improvements include optical proximity correction (OPC). The task of OPC is to precompensate the layout features on a mask to help counteract nonideal effects in mask making, imaging, resist processing, pattern transfer, and so on. For example, in imaging at high resolution, the low pass filtering of the lens results in a spatial spreading of the light at low intensity levels into neighboring features. This spillover coherently interacts with the light from the desire feature which, in turn, affects the feature size and even feature location. Very elegant edge movement and intensity calculation schemes have been introduced based on singular value decomposition methods from communication theory to speed up the OPC process from what was once thought to be years to a few hours per square mm.

How much farther is it possible to enhance image properties, and what kind of difficulties are being encountered? The ultimate resolution of an optical system is related to the smallest spatial period (one full line and space) that can be printed by interfering two laser beams at the extreme off-axis angles through the lens, and corresponds to 0.5 λ/NA. While the sharpness of the interference nulls might be used to make one feature type (line or space) 2 to 5 times smaller than the other, the average value for the minimum k_1 is 0.25 (half of the minimum period). But this is for a very simple and systematic pattern. The basic rub in practice is that designers want and need a richness of layout pattern sizes, types and orientations. In fact, enhancing the resolution of minimum sized features actually degrades features about 1.5 times larger. While a slight improvement in resolution for the small feature types is useful, the more significant gains will come from IC designers and lithographers working together to establish restricted sets of layout pattern shapes, which will benefit from resolution enhancement techniques to a greater degree.

Process Related Effects

The resist reaction to the incident image and the pattern transfer process must be consistent with creating well defined resist profiles and device features. In effect, the optics

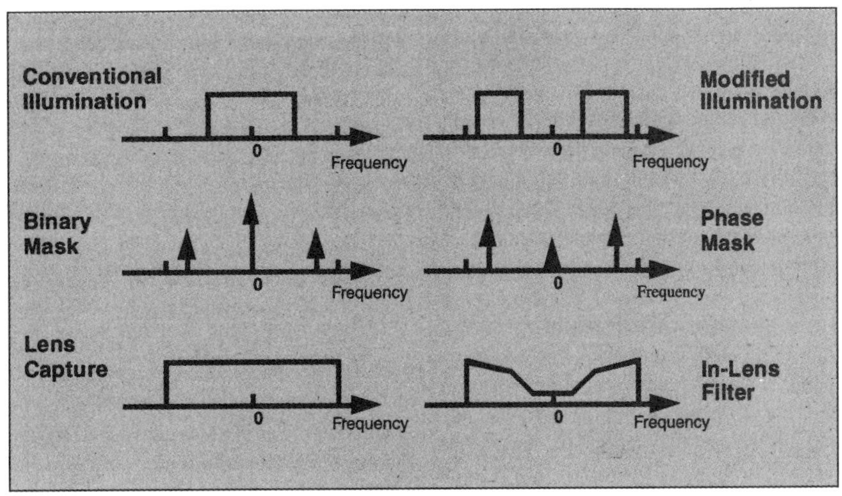

3. Filtering functions versus spatial frequency utilized by the resolution enhancement techniques of modified illumination, phase-shifting masks, and in-lens filtering to emphasize high frequencies.

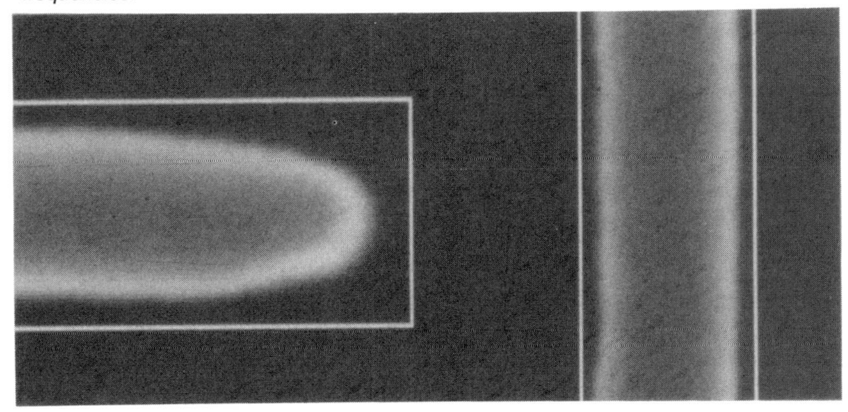

4. Comparison of a scanning electron micrograph of a projection printed resist image with the design layout, emphasizing important differences such as line end shortening.

deliver a sine wave approximation to the openings in the mask and the chemistry must then convert these sine waves into square resist profiles. An important final step is the pattern transfer of the resist profile into the device structure. A discussion of many of the resist material and related etching issues can be found in [2] and [3]. There is a very complex interrelationship between the lithography exposure tool, the resist material, its processing, and pattern transfer.

Integrating knowledge of resolution enhancement effects, optical proximity effects, resist dissolution, and pattern transfer into an overall view point which spans both IC design and process development is one of the principle challenges in extending the limits of optical lithography. For example, Fig. 4 compares a scanning electron micrograph (SEM) of a printed resist image with the initial mask pattern. The resist pattern in the SEM differs significantly from the design, especially in that the terminating line is foreshortened. The foreshortening in this particular example is particularly severe because both the optics and the resist processing contributions add. While IC design and process development have been successfully partitioned through the introduction of design rules, the savvy designer still looks over the fence to find out what can be exploited and what should be avoided.

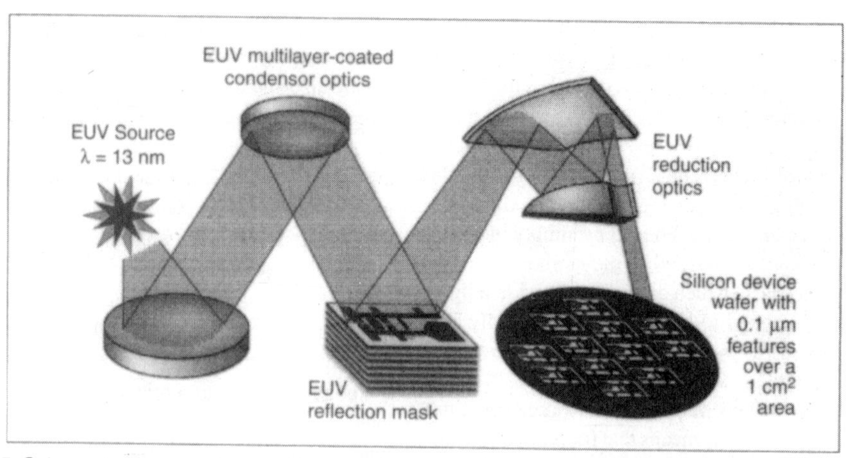

5. Schematic diagram of an EUV lithography system showing a laser-plasma source of 13 nm radiation, a 2-mirror condenser system, a reflection mask with patterned absorber on top, and a simple 2-mirror projection system that images the mask absorber pattern onto the wafer.

Towards Shorter Wavelengths

Intense research is underway to continue to reduce the actinic wavelength for lithography. The wavelength region below about 300 nm is generally referred to "deep UV" (DUV). Efficient, powerful, ultraviolet lamps that operate in the DUV are not available. Fortunately, gas-discharge pumped excimer lasers offer high powers at a selection of DUV wavelengths, including 248 nm and 193 nm.

A serious issue for DUV optics is the severely restricted choice of transparent optical materials. Fused silica is the optical material of choice because of its homogeneity and its desirable combination of mechanical and thermal properties, allowing it to be polished to high perfection. It is suitable for use at 248 nm, and probably at 193 nm. However, below 193 nm, absorption in fused silica becomes prohibitive. Refractive optical systems made from only a single lens material have large chromatic aberrations, requiring special efforts to narrow the linewidth of the laser source. This task adds complexity and expense to the laser source, and tends to reduce the available power. The mostly-reflective approach of reduction step-and-scan optics, pioneered by Silicon Valley Group in the commercial Micrascan system, is far more color tolerant. A significant effort is underway to explore 193 nm lithography with such a system. This optical system incorporates some fused silica elements.

One of the important issues that still must be addressed is that radiation damage of the fused silica is caused by long-term exposure to 193 nm radiation. This damage takes several forms, including enhanced absorption due to the formation of color center defects, and radiation induced densification (compaction), which can induce lens aberrations. Efforts to develop formulations of fused silica that are more resistant to radiation damage are in progress.

Another very challenging problem in the DUV is the loss of transparency in photoresist materials. IC manufacturing today is based on "volume exposure" of resists, i.e., nearly uniform exposure throughout the full resist thickness of about 1 micron. Wet development of the resist material results in nearly ideal vertical walled patterns that can be used to precisely transfer the desired pattern into underlying regions through etching or implantation processes. A resist thickness of about 1 micron is required for many of these pattern transfer operations. Special resists have been developed in recent years which have acceptable transparency at 248 nm, but the problem is significantly more difficult for 193 nm. If the combination of wavelength and resist chemistry do not permit adequate transparency, it is still possible in principle to perform pattern transfer with resist thickness exceeding the absorption depth. Such a process is called "surface imaging," since the image forming region is shallow compared to the full resist thickness. After chemical treatment, typically using an organo-silicon compound, the exposed region is differentiated from the unexposed regions because it contains 10-15% silicon. The exposed and treated silicon containing regions are then used as an etch mask for an anisotropic plasma etching step, thus transferring the pattern into the full thickness of the resist. Excellent, vertical resist wall profiles can be achieved, suitable for subsequent high definition pattern transfer. The combination of chemical treatment and plasma etch step is known as "dry development." Surface imaging chemistry is in its infancy, and such processes will undergo much further development before they are widely used in manufacturing.

Research on the properties of optical materials in the ultraviolet suggests that 193 nm is close to the short wavelength limit for refractive optics. Absorption becomes too high at shorter wavelengths. If we project to an NA of 0.75, and a k_1 of 0.5 using resolution enhancements, we arrive at a resolution of 130 nm. With a k_2 of 1.2, the depth of focus will be only 400 nm, which will put significant demands on wafer planarity, as well as stepper engineering. Nevertheless, many workers in advanced IC lithography believe that this is a plausible, even likely scenario. Chastened by the history of overly pessimistic estimates in this field, most workers are now reluctant to *ever* count out optical lithography.

Beyond Ultraviolet Optics

What about 100 nm and beyond? A fundamental question some are asking is whether 100 nm lithography will ever be needed. The issue here is whether the cost of the entire IC manufacturing process at such fine feature sizes can be low enough so that the increased functionality can be delivered at an attractive price. We remain optimistic that CMOS technology will advance to the 100 nm generation and beyond. The market for semiconductor electronics is exploding and may be expected to continue to grow for quite some time. World-wide competition to supply this market will be a tremendous incentive for innovation. In fact, just recently, Japan's MITI (Ministry for International Trade and Industry) announced its new Super Advanced Electronics Technology Program, with one of its goals being the development of technology for making circuit patterns less than 100 nm.

The question is how to do the lithography for 100 nm and smaller feature sizes. While history shows that it is dangerous to count out the continued extension of optical lithography, the wavelength limit for refractive lenses is 193 nm; hence, by the time we reach 100 nm features, it seems clear that something new will be needed.

A number of options are presently under investigatation at various laboratories. The logical evolutionary path is to continue to scale optical projection lithography to shorter wavelength. This entails moving to all-reflective optics. Incrementally shrinking the wavelength into the vacuum ultraviolet (VUV) range (100-200 nm) would only be effective if the *NA* of the reflective optical projection system could be made as large as the ultimate 193 nm lenses, perhaps as high as 0.75 or so. It appears to be an extremely difficult challenge to design a purely reflective system that can also meet stringent distortion requirements over the large field size needed for IC lithography.

However, a short wavelength optical projection system might be realized by making a somewhat larger jump in wavelength, down to 13 nm, the extreme ultraviolet (EUV) region of the spectrum. Significant effort in the US, Japan, and Europe is in progress on EUV lithography [4]. Figure 5 shows a schematic diagram of such a system. The primary reason this particular wavelength is attractive is that optical coating technology using Si/Mo multi-layer films has been developed that can yield relatively high reflectivity at normal incidence for this wavelength. Reflectivity slightly higher than 65% at this wavelength has been demonstrated, and could conceivably be improved to 70 percent. At this wavelength, feature sizes less than 100 nm can be achieved using an *NA* of only 0.1, and 4X and 5X reduction optical designs involving only 3 or 4 mirrors that also meet projected distortion and field-size specifications have been reported. The proposed light source is a radiating hot plasma heated by a high power pulsed laser. An interesting new feature in this technology is the introduction of a reflection mask. Since transparent optical materials are unavailable, the idea is to use a flat reflector as a mask blank, with a patterned absorber film on top of the mirror coating. Preliminary experiments at AT&T Bell Laboratories, Sandia National laboratories, and NTT laboratories in Japan have demonstrated better than 100 nm resolution over small fields using prototype 2-mirror optics.

There are significant difficulties that must be overcome before practical EUV lithography can be realized. One of the most difficult is that, in order for a reflective optical projection system to achieve diffraction limited performance, each of the mirror surfaces must be figured with an absolute accuracy of a small fraction of the operating wavelength. For λ = 13nm, the required surface figure accuracy is less than 1 nm, which is well beyond the current state of the art in terms of measurement, let alone fabrication technology. However, this challenge has stimulated a great deal of innovation in the areas of optical fabrication and metrology recently, and there has been tremendous progress towards meeting the requirements.

Research is also being conducted on a number of other approaches for lithography at 100 nm and below. Earlier predictions of the death of optical lithography gave birth to several technologies, which have by now been under investigation for many years, including x-ray proximity, direct-write e-beam, projection e-beam, and ion-beam lithographies. More recently, dramatic advances in micro-electromechanical (MEMS) technology have inspired a whole new set of ideas for patterning based on massively parallel scanning arrays. Some of these include arrays of electron beams, x-ray beams, and proximal probes (atomic force microscope-type tips). All of these approaches represent a significant shift away from the optical projection paradigm. Each has advantages and disadvantages, as well as very vigorous proponents and doubters. At this point, it is unclear which of the contending solutions for 100 nm lithography and beyond will make it into production. These are certainly exciting times for research in advanced lithography. ***CD***

J. Bokor, *A. R. Neureuther*, and *W. G. Oldham* are faculty members of the Department of Electrical Engineering and Computer Sciences at the University of California at Berkeley. They are actively involved in research on lithography, as well as other aspects of integrated circuit technology.

References

1. B. J. Lin, "Phase-shifting masks gain an edge," *Circuits and Devices* **9,** March, p. 28, 1993.

2. D. H. Ziger, "Understanding IC lithography: resist issues affecting IC design and manufacture," *Circuits and Devices* **8,** September, p. 42, 1992.

3. A. R. Reinberg, "Etching and lithography running neck and neck," *Circuits and Devices* **9,** January, p. 24, 1993.

4. F. Zernike and D. T. Attwood, eds., *Extreme Ultraviolet Lithography*, Optical Society of America, Wash. DC, 1995.

CHAPTER 4

OPTOELECTRIC INTEGRATED CIRCUIT DEVICE TECHNOLOGY

Phase-locked Laser Arrays Revisited

An Updated Review of Phase-locked Arrays of Semiconductor Diode Lasers

Dan Botez and Luke J. Mawst

In January 1986 a review article on phase-locked arrays of diode lasers was published in this magazine. Now, more than a decade later, major breakthroughs have occurred, both in theory and experiment, that have allowed phase-locked arrays to meet their promise of reliable, high-continuous-wave (CW) power (≅ 0.5W) operation in diffraction-limited beams as well as multiwatt (5-10 W) near-diffraction-limited peak-pulsed-power operation. As an update to the 1986 article, this article describes a corrected picture for the array modes, arrays of antiguides and the concept of resonant leaky-wave coupling, and relevant recent results.

Overview

By comparison with other types of high-power, coherent semiconductor-based sources ("broad area"-type master oscillator power amplifier (MOPA), unstable resonator), phase-locked arrays have some unique advantages: graceful degradation; no need for optical isolators; no need for external optics to compensate for phasefront aberrations due to thermal-and/or carrier-induced variations in the dielectric constant; and, foremost, intrinsic beam stability with drive level due to a strong, built-in, real-index profile. The consequence is that, in the long run, phase-locked arrays are bound to be fundamentally more reliable than either MOPAs or unstable resonators.

At the time the original review article [1] was written, three major types of phase-locked arrays had been investigated: evanescent-wave coupled, diffraction-coupled, and Y-junction coupled (see Fig. 5 in [1]). Up to 1988 the results were not at all encouraging: maximum diffraction-limited, single-lobe powers of ≈ 50 mW or coherent powers (i.e., fraction of the emitted power contained within the theoretically defined diffraction-limited-beam pattern) never exceeding 100 mW. Thus, the very purpose of fabricating arrays (to surpass the reliable power level of single-element devices) was not achieved. The real problem was that researchers had taken for granted the fact that strong nearest-neighbor coupling implies strong overall coupling. In reality, as shown in Fig. 1, nearest-neighbor coupling is "series coupling," a scheme plagued by weak overall coherence and poor intermodal discrimination [2]. Strong overall interelement coupling occurs only when each element equally couples to all others (so-called "parallel coupling") [2]. In turn, intermodal discrimination is maximized and full coherence becomes a system characteristic. Furthermore, parallel-coupled systems have uniform near-field intensity profiles, and are thus immune to the onset of high-order-mode oscillation at high drive levels above threshold.

Parallel coupling can be obtained in evanescent-wave-coupled devices, but only by weakening the optical-mode confinement, and thus making the devices vulnerable to thermal- and/or injected-carrier-induced variations in the dielectric constant. For both full coherence and stability it is necessary to achieve parallel coupling in structures of strong optical-mode confinement (i.e., built-in index steps ≥ 0.01). As shown below, only strongly guided, leaky-wave-coupled devices can meet both conditions [3-5]. A sta-

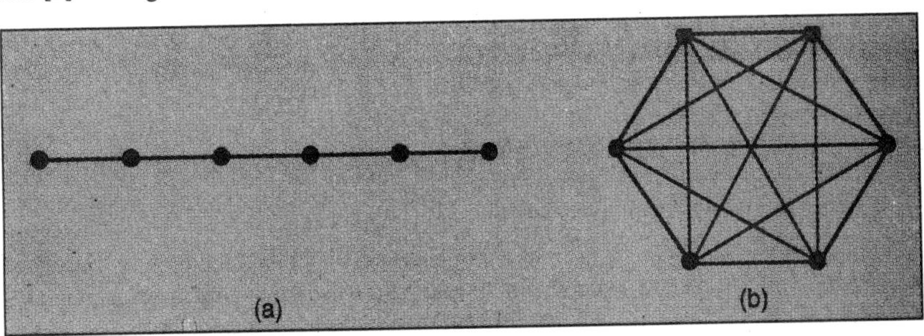

1. Types of overall interelement coupling in phase-locked arrays: (a) series coupling (nearest-neighbor coupling, coupled-mode theory); (b) parallel coupling.

Reprinted from *IEEE Circuits and Devices Magazine,* Vol. 2, No. 1, pp. 25-32, November 1996.

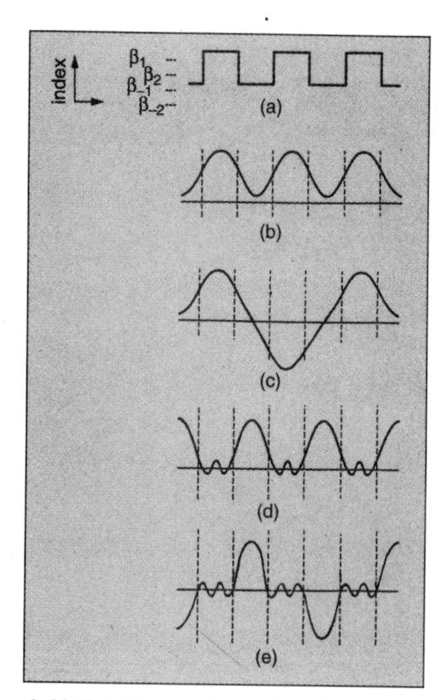

2. Modes of array of periodic real-index variations: (a) index profile; (b) in-phase evanescent-wave type; (c) out-of-phase evanescent-wave type; (d) in-phase leaky-wave type; (e) out-of-phase leaky-wave type. Respective propagation constants are β_1, β_2, β_{-1} and β_{-2}.

ble, parallel-coupled source is highly desirable for systems applications since it has graceful degradation; that is, the failure or obscuration of some of its components does not affect the emitted beam pattern. Furthermore, it has been recently shown both theoretically [6] and experimentally [7] that, by contrast to series-coupled systems, parallel-coupled systems are fundamentally stable against coupling-induced instabilities.

In 1978 Scifres et al. [8] reported the first phase-locked array: a five-element gain-guided device. It took eight years before Hadley [9] showed that the modes of gain-guided devices are of the leaky type. Gain-guided arrays have generally operated in leaky, out-of-phase (i.e., two-lobed) patterns with beamwidths many times the diffraction limit due to poor intermodal discrimination. Furthermore, gain-guided devices, being generated simply by the injected-carrier profile (Fig. 1a in [1]), are very weakly guided and thus vulnerable to thermal gradients and gain spatial-hole burning. The first real-index-guided, leaky-wave-coupled array (i.e., the so-called antiguided array) was realized in 1981 by Ackley and Engelmann [10] While the beam patterns were stable, the lobes were several times the diffraction limit, with in-phase and out-of-phase modes operating simultaneously due to lack of a mode-selection mechanism. Positive-index-guided arrays came next in array research (1983-8). In-phase, diffraction-limited-beam operation could never be obtained beyond 50 mW output power. Some degree of stability could be achieved in the out-of-phase operational condition such that, by 1988, two groups [11, 12] reported diffraction-limited powers as high as 200 mW.

In 1988 antiguided arrays were resurrected and, right from the first attempt, researchers obtained close to 200 mW diffraction-limited in-phase operation [3]. Hope for achieving high coherent powers from phase-locked arrays was rekindled, although there was no clear notion as to how to controllably obtain single-lobe operation. The breakthrough occurred in late 1988 with the discovery of resonant leaky-wave coupling [4], which allowed parallel coupling among array elements for the first time, and thus the means of achieving high-power, single-lobe, diffraction-limited operation. The experimental coherent powers quickly escalated such that to date, up to 2 W has been achieved in a diffraction-limited beam [13], and up to 10 W in a beam twice the diffraction limit [14].

Array Modes Revisited

As shown at the top of Fig. 2, a monolithic array of phase-coupled diode lasers can be described simply as a periodic variation of the real part of the refractive index. Two classes of modes characterize such a system: evanescent-type array modes (Figs. 2b and c), for which the fields are peaked in the high-index array regions; and leaky-type array modes, for which the fields are peaked in the low-index array regions (Figs. 2d and e). Another distinction is that while evanescent-wave modes have effective-index values between the low and high refractive-index values, leaky modes have effective-index values below the low refractive-index value [5, 15]. For both classes of modes the locking condition is said to be "in-phase" when the fields in each element are cophasal, and "out-of-phase" when fields in adjacent elements are a π phase-shift apart.

As described in the original paper, evanescent-type modes were the first to be analyzed, simply because it could be assumed that they arose from the superposition of individual-element wavefunctions, and thus could readily be studied via the coupled-mode formalism [16, 17]. While coupled-mode theory proved quite useful in understanding early work on phase-locked arrays, it has severe limitations: it does not apply to strongly coupled systems, and it does not cover leaky-type array modes. Ironically, control of leaky modes turns out to be the key for high-power, phase-locked operation. Thus, after many years of use, coupled-mode theory has suddenly become obsolete as far as the design and analysis of high-power coherent devices is concerned.

One major reason why leaky array modes had been overlooked for so long is that they are not solutions of the popular coupled-mode theory. The other reason is that for few-element (up to five) arrays, leaky modes play a minor role since they are very lossy [15]. However, for high-power devices (10 or more elements), leaky-mode

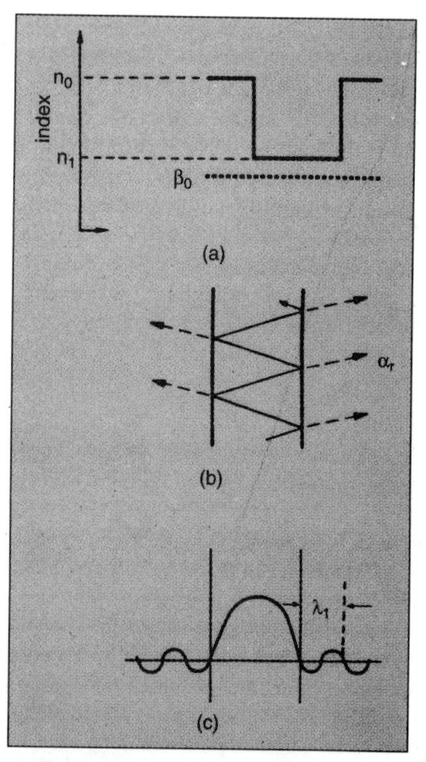

3. Schematic respresentation of real-index antiguide: (a) index profile; (b) ray-optics picture (α_R is edge radiation loss coefficient); (c) near-field amplitude profile (λ_1 is the leaky-wave periodicity in the lateral direction).

operation is what ensures stable diffraction-limited-beam operation to high drive levels.

Evanescent-type array modes are of importance only when the modal gain in the high-index regions is higher than the modal gain in the low-index regions. Many workers [18-20] have shown that excess gain in the high-index array regions generally favors oscillation in the evanescent out-of-phase mode, in close agreement with experimental results. There is, however, one major limitation: the built-in index step, Δn has to be below the cutoff for high-order (element) modes [19, 21]. For typical devices, $\Delta n \leq 5 \times 10^{-3}$. In turn, the devices are sensitive to gain spatial hole burning [22] and thermal gradients.

Leaky-type modes are favored to lase when gain is preferentially placed in the low-index regions [4, 23, 24]. Unlike evanescent-type modes there is no limitation on Δn; that is, no matter how high Δn is, the modes favored to lase comprise fundamental element modes coupled in-phase *or* out-of-phase. This fact has two important consequences. First, one can fabricate single-lobe-emitting structures of high-index steps (0.05-0.20), that is, stable against thermal- and/or carrier-induced index variations. Second, it becomes clear why predictions of coupled-mode theory that excess gain in the low-index array regions [19, 21] favor the in-phase evanescent-type mode have failed. That is, the so called net-gain-between-elements approach to obtain fundamental-mode operation (see Fig. 5b in [1]) was flawed. The reason is simply the fact that leaky modes could not be taken into account when using coupled-mode analysis. Ironically, the very array structures proposed by Streifer et al. for in-phase mode operation [19], when analyzed using exact theory by Fujii et al. [25], were found to primarily favor operation in an out-of-phase leaky mode. To analyze both array-mode types, for a given structure, one has to use either exact theory [15, 24, 25] or the Bloch-function method [23, 26, 27].

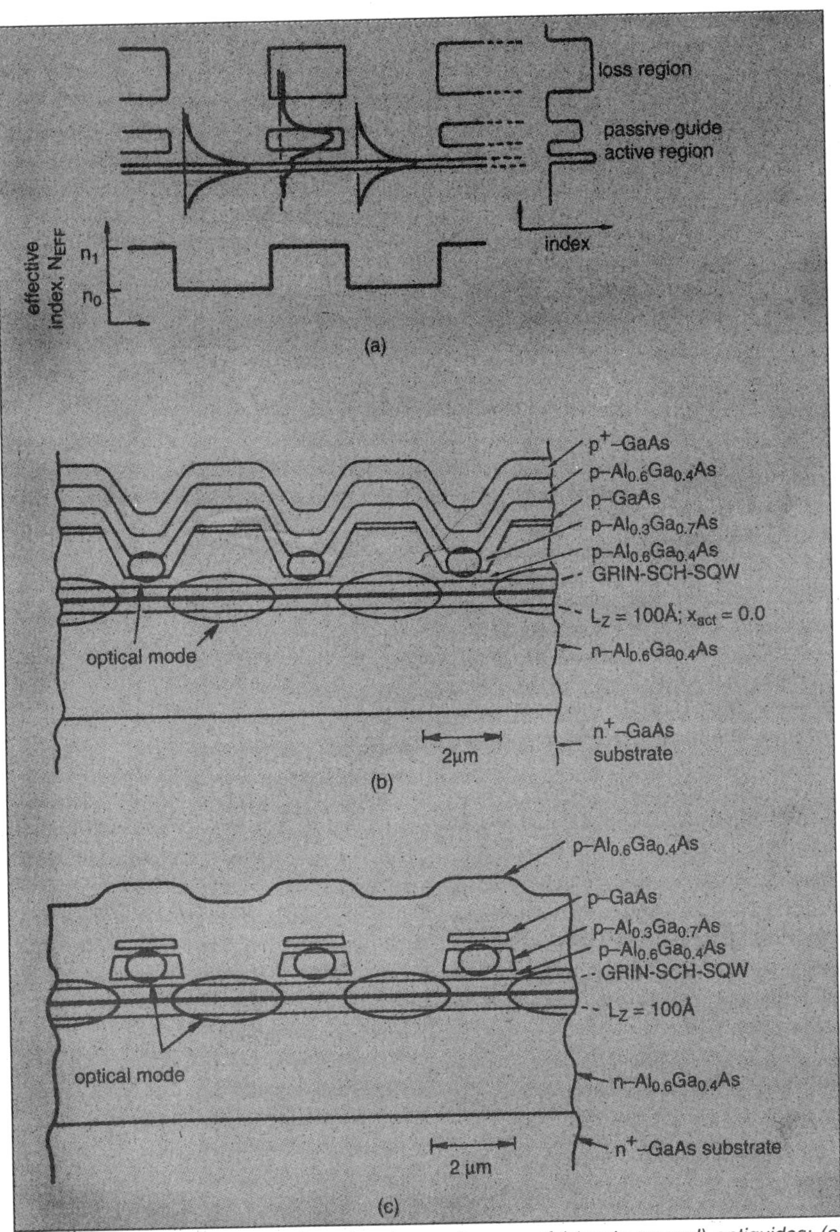

4. Schematic representation of modern types of arrays of (closely spaced) antiguides: (a) practical way of fabricating arrays of closely spaced antiguides; (b) complimentary-self-aligned (CSA) array type [15]; (c) self-aligned-stripe (SAS) array type [13, 14, 29, 30].

Arrays of Antiguides and Resonant Leaky-wave Coupling

The basic properties of a single real-index antiguide are shown schematically in Fig. 3. The antiguide core has an index, n_0,lower than the index of the cladding, n_1. The index depression, Δn, is (1-2) $\times 10^{-3}$ for gain-guided lasers and (2-5) $\times 10^{-2}$ for strongly index-guided lasers. The effective-index of the fundamental mode β_0/k is below the index of the core. The quantum-mechanical equivalent is thus a quasibound state above a potential barrier. By contrast, the quantum-mechanical equivalent of the fundamental mode in a positive-index guide is a bound state in a potential well. Whereas in a positive-index guide, radiation is trapped via total internal reflection, in an antiguide, radiation is only partially reflected at the antiguide-core boundaries (Fig. 3b). Light refracted into the cladding layers is radiation leaking outwardly with a lateral wavelength λ_1 (Figure 3c) [4, 15]:

$$\lambda_1 \cong \lambda \Big/ \sqrt{2n\Delta n + (\lambda / 2d)^2}, \qquad (1a)$$

and can be thought of as a radiation loss [28]:

$$\alpha_R = (l+1)^2 \lambda^2 / d^3 n \sqrt{2n\Delta n} \qquad (1b)$$

where d is the antiguide-core width, Δn is the lateral refractive-index step, n is the av-

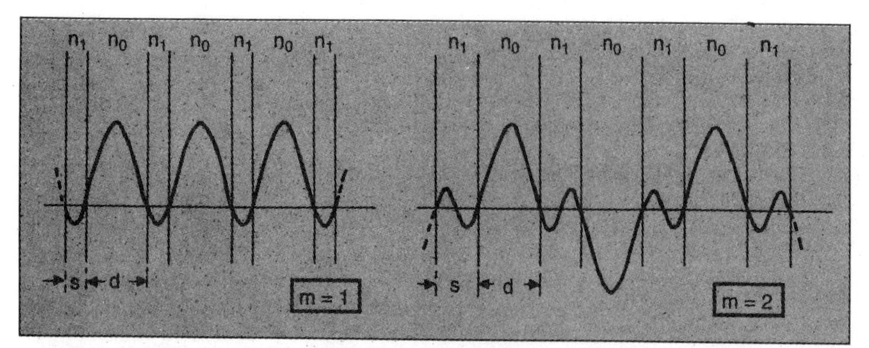

5. Near-field amplitude profiles in resonantly coupled arrays: (a) in-phase resonant mode; (b) out-of-phase resonant mode; m is number of interelement near-field intensity peaks. (After [5].)

Table 1. Performance to date of strong index-antiguided arrays

Drive condition	Maximum power	
	DL*	Narrow beam
CW	0.5 W	1 W; 1.7 DL
Pulsed	2.1 W	10 W; 2.0 DL

*DL = diffraction limited

erage index value, λ is the vacuum wavelength, and l is the (lateral) mode number. For typical structures ($d = 3$ µm, $\Delta n = 2\text{-}3 \times 10^{-2}$) at $\lambda = 0.85$ µm, typical λ_1 and α_R values are 2 µm and 100 cm^{-1}, respectively. Since $\alpha_R \propto (l + 1)^2$, the antiguide acts as a lateral-mode discriminator. For a proper mode to exist, α_R has to be compensated for by gain in the antiguide core [28]. Single antiguides have already been used for quite some time in CO_2 "waveguide" lasers.

Historically, the first arrays of antiguided lasers were gain-guided arrays [8] (Figure 1a in [1]), since an array of current-injecting stripe contacts provides an array of (carrier-induced) index depressions for which the gain is highest in the depressed-index regions. While for a single antiguide the radiation losses can be quite high [28], closely spacing antiguides in linear arrays significantly reduces the device losses [3-5] since radiation leakage from individual elements mainly serves the purpose of coupling the array elements.

The first real-index antiguided array was realized by Ackley and Engelmann [10]: an array of buried heterostructure (BH) lasers designed such that the interelement regions had higher refractive indices than the effective refractive indices in the buried active mesas. Since the high-index interelement regions had no gain, only leaky array modes could lase. The device showed definite evidence of phase locking (in-phase and out-of-phase) but had relatively high threshold-current densities (5-7 kA/cm^2) since the elements were spaced far apart (13-15 µm), thus not allowing for effective leaky-wave coupling.

For practical devices the high-index regions have to be relatively narrow (1-3 µm), which is virtually impossible to achieve using BH-type fabrication techniques. Instead, one can fabricate narrow, high-effective-index regions by periodically placing high-index waveguides in close proximity (0.1-0.2 µm) to the active region [3-5, 13] (see Fig. 4). In the newly created regions the fundamental transverse mode is primarily confined to the passive guide layer; that is, between the antiguided-array elements the modal gain is low. To further suppress oscillation of evanescent-wave modes an optically absorbing material can be placed between elements (Fig. 4a).

The first closely spaced, real-index antiguided array was realized by liquid-phase epitaxy (LPE) over a patterned substrate [3]. Devices made in recent years are fabricated by metal-organic chemical vapor deposition (MOCVD) and can be classified into two types: the complimentary-self-aligned (CSA) stripe array [15] (Fig. 4b); and the self-aligned-stripe (SAS) array [13, 14, 29, 30] (Fig. 4c). In CSA-type arrays preferential chemical etching and MOCVD regrowth occur in the interelement regions. For SAS-type arrays the interelement regions are built-in during the initial growth, and then etching and MOCVD regrowth occur in the element regions. A most recent approach, which does not involve regrowth, is the formation of low-index, high-gain regions via preferential *Zn*-diffusion disordering of a superlattice upper cladding layer [49].

Owing to lateral radiation, a single antiguide can be thought of as a generator of laterally propagating traveling waves of wavelength λ_1 (Fig. 3). Then, in an array of antiguides, elements will resonantly couple in-phase or out-of-phase when the interelement spacings correspond to an odd or even integral number of (lateral) half-wavelengths ($\lambda_1/2$), respectively (see Fig. 5). The resonance condition is

$$s = m\lambda_1 / 2$$
$$\begin{array}{ll} m = \text{odd} & \text{resonant in-phase mode} \\ m = \text{even} & \text{resonant out-phase mode} \end{array} \qquad (2)$$

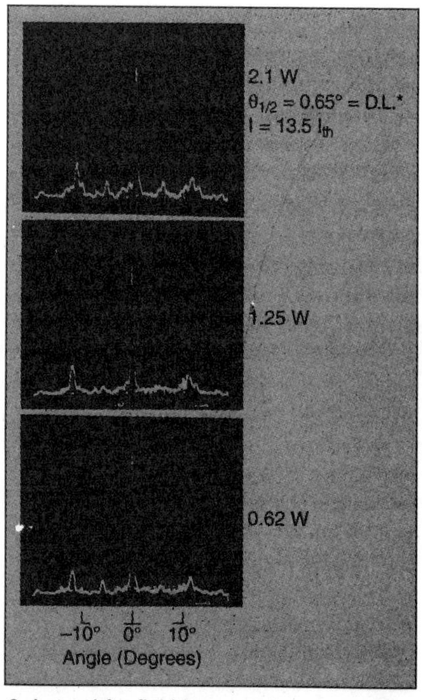

*6. Lateral far-field beam patterns at several power levels for a resonant array with a monolithic Talbot filter. A diffraction-limited beam is maintained above the 2 W output power level. (After [13].) *D.L. = diffraction limited*

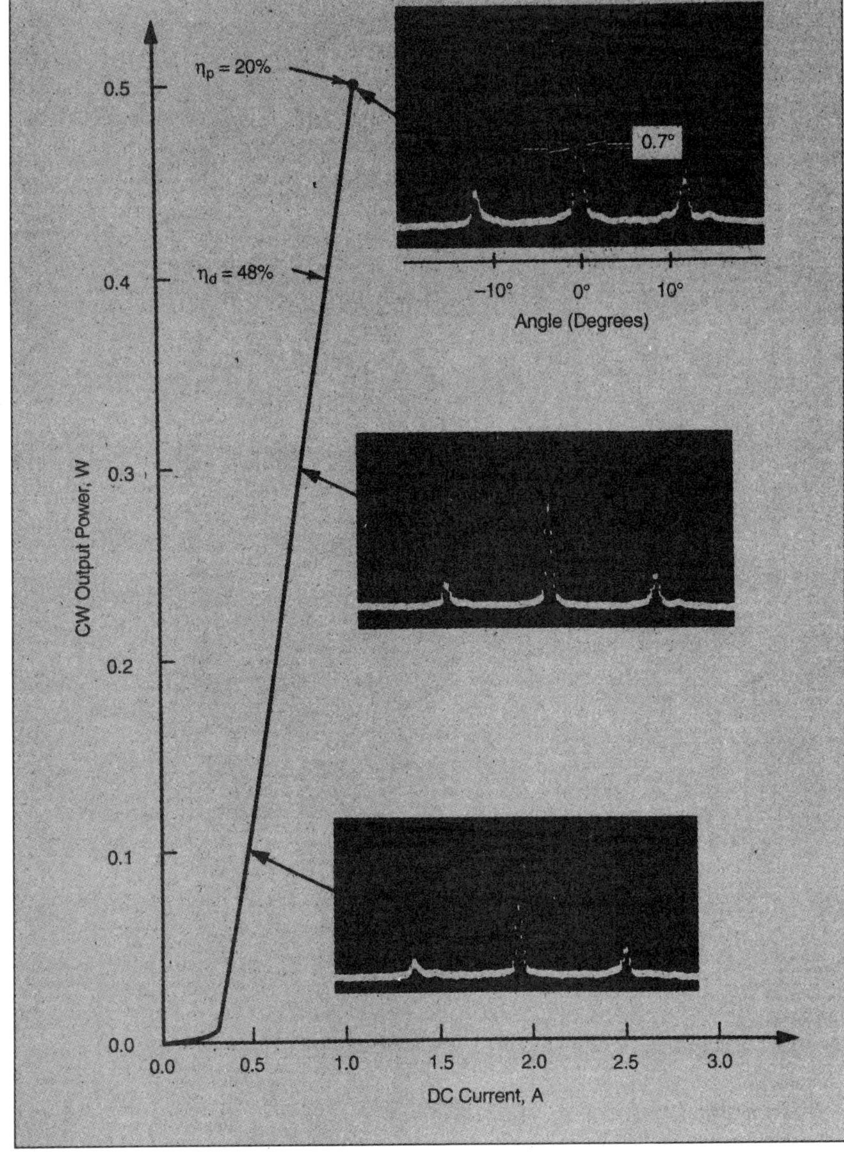

7. CW operation of optimized 20-element ROW array: light-current characteristic and lateral far-field patterns at various power levels. Beam pattern is diffraction-limited up to 0.5 W.

where s is the interelement spacing. Typical s values are 1 μm. Then, for in-phase-mode resonance, $\lambda_1 = 2$ μm. The lateral resonance effect in antiguided arrays is quite similar to the 2nd-order Bragg resonance in DFB-type structures [31]. In fact a resonant antiguided array is a *lateral* complex-coupled 2nd-order DFB structure.

When the resonance condition is met, the interelement spacings become Fabry-Perot resonators in the resonance condition [5]. Then each element can equally couple to all others, thus realizing, for the first time, the dream of researchers in the phase-locked-array field: parallel coupling (Fig. 1b). Resonant arrays of antiguides are called resonant-optical-waveguide (ROW) arrays [5]. As expected for paralleled-coupled devices, ROW arrays have uniform near-field intensity profiles [15], and maximum intermodal discrimination [5, 15].

Intermodal discrimination is provided by three effects: modal overlap with the gain region [15, 24], the so-called Γ-effect [15, 32]; edge radiation losses [15]; and interelement loss [15, 33]. Typical values for the discrimination between the (fundamental) in-phase mode and the next (high-order) array mode are 10-15 cm^{-1} for 10-element arrays [15]. Further intermodal discrimination can be achieved by employing intracavity Talbot-type spatial filters [34, 35]. It was in fact a 20-element array with Talbot-type spatial filters [36] that allowed, in 1990, the "breaking" of the 1 W coherent-power barrier for monolithic semiconductor diode lasers.

Relevant Recent Results

In pulsed operation, 20-element devices have demonstrated diffraction-limited beams to 1.5 W and 10.7 times the threshold [37], and to 2.1 W and 13.5 times the threshold [13] (Fig. 6). At the 2.1 W power level only 1.6 W is coherent uniphase power and the main lobe contains 1.15 W (so-called "power in the bucket"). Actually, close to 100% of the coherent power can be garnered in the main lobe by using aperture-filling optical techniques [38]. The fact that diffraction-limited beams can be maintained to very high drive levels is a direct consequence of the inherent self-stabilization of the in-phase resonant mode with increasing drive level [39].

When carrier diffusion is taken into account [40], it has been shown that for devices with $\Delta n \cong 0.025$, the adjacent mode reaches threshold (i.e., multimode operation starts) at more than seven times the threshold of the in-phase mode. Furthermore, if the index step is increased to ≈ 0.16, the effect of gain spatial hole burning is negligible, such that one can achieve single-(in-phase)-mode operation to drive levels > 15 times the threshold, powers of ≈ 3 W, in beams with ≅ 70% of the energy in the main lobe [40].

In CW operation, one is limited by thermal effects. The best result in a purely diffraction-limited beam is 0.5 W CW from a 20-element device of 80 μm-wide aperture (Fig. 7). For 1000 μm-long devices with HR and AR facet coatings (98 and 4% reflectivity), slope efficiencies of 48-50% are observed and the power conversion efficiency reaches values in the 20-25% range at 0.5 W output. Preliminary lifetests at 0.5 W CW output show room-temperature extrapolated lifetimes in excess of 5000 h [41]. Thus the prediction of the original article [1] (see "Conclusion" section) has been fulfilled.

Recently, by using large-aperture (120 μm), 20-element structures of 5 to 1 element/interelement width ratio, we obtained [42] 1 W CW operation in a beam 1.7 times the diffraction limit (Fig. 8). Close to threshold (0.1 W) the beam is diffraction limited, and 75% of the energy resides in the main

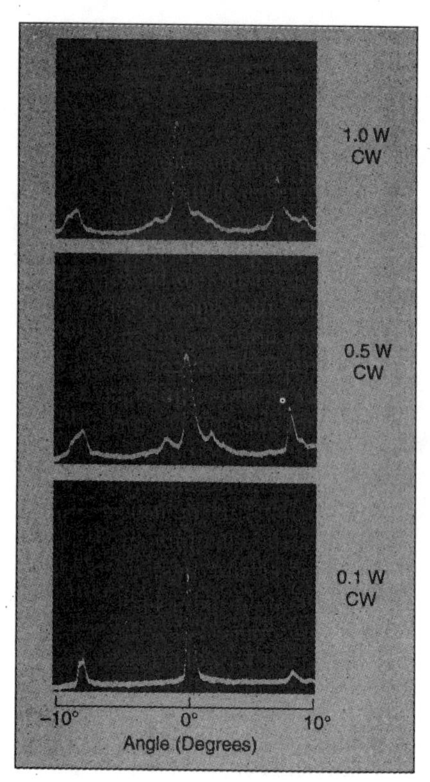

8. CW far-field patterns of 20-element ROW array with 5 μm-wide elements and 1 μm-wide interelements at various power levels. (After [42].)

lobe. The 1 W value represents the highest CW coherent power achieved to date from any type of fully monolithic diode emitters (i.e., without phase-correcting optics).

The results presented so far are for AlGaAs/GaAs structures (λ = 0.84-0.86 μm). Antiguided arrays have also been made from strained-layer quantum-well material (λ = 0.92-0.98 μm) in both the CSA [43] and the SAS-type array configurations: InGaAs/InGaP/GaAs [14] and InGaAs/AlGaAs/GaAs [29, 30]. Resonant devices were achieved only with the CSA configuration, and provided 1 W pulsed diffraction-limited-beam operation [43]. From SAS-type devices Major et al. [30] have demonstrated stable and efficient CW operation to 0.5 W in a beam 1.5 times the diffraction limit.

Nonresonant devices typically have beams 2 to 3 times the diffraction limit. First of all, the in-phase mode has a raised-cosine-shaped envelope and thus gain spatial hole burning is nonuniform across the device causing mode self-focusing [39, 40], similar to evanescent-wave coupled devices [22]. Then adjacent modes can readily reach threshold and the beamwidth increases to 2 to 3 times the diffraction limit. The combination of the in-phase mode and one or two adjacent modes uses the available gain efficiently, thus not allowing other modes to come in. This explains why stable beams can be maintained to very high drive levels and peak powers: 5 W to 45 times the threshold from 80 μm-wide aperture devices [36], 10 W from 200 μm-wide aperture devices [14] (Fig. 9), and 11.5 W from 185 μm-wide aperture devices [44]. Particularly notable is the achievement of a beam pattern 2 times the diffraction limit with 6W in the central lobe [14] (Fig. 9) from an Al-free structure (λ = 0.98 μm).

Bloch-function array analysis has been recently extended [45, 46] from infinite-extent arrays to finite-extent arrays. Analytical formulae can then be obtained for the resonant-array-mode loss [45] intermodal discrimination [46], and near- and far-field patterns [47]. Thus, we have now a *complete* analytical model that should allow for straightforward device design.

Finally it has been found both theoretically [6, 7] and experimentally [7] that ROW arrays do not suffer from the coupling-induced instabilities that generally plague series-coupled devices such as evanescent-wave coupled and Y-junction coupled arrays. Quiescent behavior up to 0.45 W cw power and 3.4 times the threshold, in near diffraction-limited beams, has been recorded from ROW arrays with negligible interelement loss [7]. (Too high a value for interelement loss causes saturable-absorption-induced self-pulsations) [7]. This is the *first time* that a phase-locked array has been found to be temporally stable to substantial powers and drive levels above threshold. Li and Erneux [6] have shown that the intrinsic stability of ROW arrays is due to parallel coupling.

Conclusions - A Decade Later

Arrays based on coupling of positive-index guides, although not successful for high-power coherent applications, may still be of use. One application could be controlled beam steering. For instance, one could create a phase ramp across the aperture of an evanescent-wave-coupled array by simply tailoring the injected-carrier profile. By varying currents through separate contact pads, the beam pattern could then be made to shift controllably.

It is now apparent that positive-index-guided devices were just one stage in array development. What has finally made phase-locked arrays a success has been the discovery of a mechanism for selecting in-phase leaky modes of antiguided structures. Parallel coupling in strong index-guided structures has allowed ROW arrays to reach stable diffraction-limited operation to powers 20-30 times higher than for other array types. This finally fulfills the phase-locked arrays promise of vastly improved coherent power by comparison with single-element devices. Table 1 summarizes the best results to date.

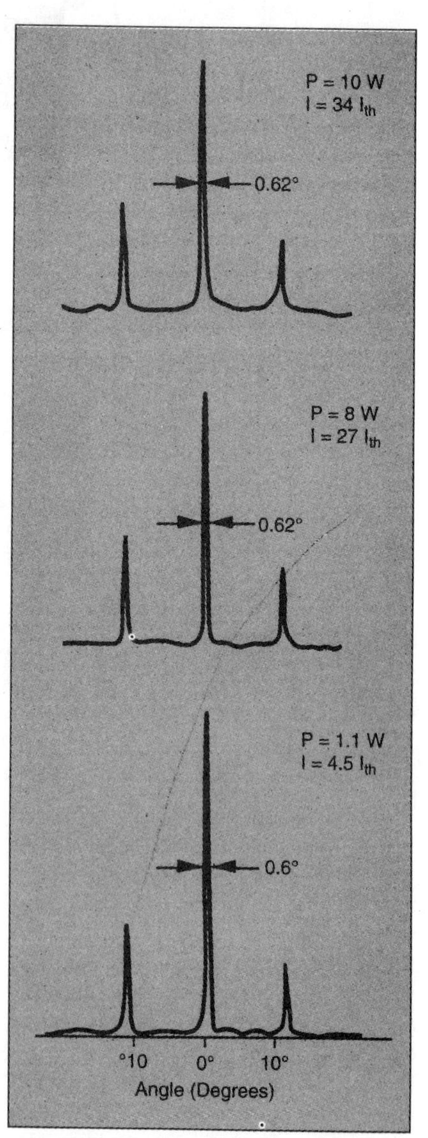

9. Pulsed far-field patterns of 40-element antiguided array of Al-free (λ = 0.98 μm) diode lasers. The beam is 2.0 times the diffraction limit up to 34 times the threshold (10 W). (After [14].)

In the future, for edge-emitting coherent arrays, the eventual limitation may just be thermal; that is , just as for incoherent devices, maximum emitting apertures would be 400-500 µm wide. Thus, the projected maximum diffraction-limited CW powers are in the 3-5 W range. Beyond 5 W, one will have to resort to 2-D surface-emitting arrays. A 2-D surface-emitting ROW array has already been demonstrated [48]. Interunit coupling occurs via resonant leaky-wave coupling, while radiation outcoupling is realized via 45° micromachined turning mirrors. Preliminary results from nine-unit arrays are quite encouraging: 3.9 W with 45% visibility and 6 W with 33% visibility [48]. For a very large number of units (≥100) there is always the concern that the device may operate in several mutually coherent regions and thus, to control hundreds of emitters, a master oscillator (i.e., injection locking) may be required. All in all we foresee that 2-D ROW arrays will eventually be capable of CW coherent powers of the order of 50 W.

Further Reading

In-depth reviews of work on monolithic phase-locked diode-laser arrays can be found in the article: D. Botez, "High-power Monolithic Phase-locked Arrays of Antiguided Semiconductor Diode Lasers," *Proc. Inst. Elec. Eng. Part J, Optoelectronics,* vol. 139, pp. 14-23, Feb. 1992.; and in Chapter I of the book *Diode Laser Arrays* eds. D. Botez and D.R. Scifres, Cambridge University Press, Cambridge, July 1994.

Dan Botez is the Philip Dunham Reed Professor of Electrical Engineering and Director of the Reed Center for Photonics at the University of Wisconsin-Madison. *Luke J. Mawst* is an Assistant Professor at the University of Wisconsin-Madison, where he is involved in the development of semiconductor laser structures using the InGaP-based material system. **C&D**

References

1. D. Botez and D.E. Ackley, "Phase-locked Arrays of Semiconductor Diode Lasers, " *IEEE Circuits and Devices Magazine*, vol. 2, pp. 8-17, Jan. 1986.

2. W.J. Fader and G.E. Palma, "Normal modes of *N* coupled lasers," *Opt. Lett.*, vol. 10, pp. 381-383, 1985.

3. D. Botez, L. Mawst, P. Hayashida, G. Peterson, and T.J. Roth, "High-power diffraction-limited-beam operation from phase-locked diode-laser arrays of closely spaced 'leaky' waveguides (antiguides)," *Appl. Phys. Lett.*, vol. 53, pp. 464-466, Aug. 1988.

4. D. Botez, L.J. Mawst, and G. Peterson, "Resonant leaky-wave coupling in linear arrays of antiguides," *Electron. Lett.*, vol. 24, pp. 1328-1330, Oct. 1988.

5. D. Botez, L.J. Mawst, G. Peterson, and T.J. Roth, "Resonant optical transmission and coupling in phase-locked diode-laser arrays of antiguides: The resonant-optical-waveguide array," *Appl. Phys. Lett.*, vol. 54, pp. 2183-2185, May 1989.

6. Ruo-ding Li and Thomas Erneux, "Stability conditions for coupled lasers: series coupling versus parallel coupling," *Opt. Comm.*, vol. 99, pp. 196-200, 1 June 1993.

7. S. Ramanujan, H.G. Winful, M. Felisky, R.K. DeFreez, D. Botez, M. Jansen and P. Wisseman, "Temporal Behavior of Resonant-Optical-Waveguide Phase-Locked Diode Laser Arrays," *Appl. Phys. Lett.*, vol. 64, pp. 827-829, Feb. 1994.

8. D.R. Scifres, R.D. Burnham, and W. Streifer, "Phase Locked Semiconductor Laser Array," *Appl. Phys. Lett.*, vol. 33, pp. 1015-1017, Dec. 1978.

9. G.R. Hadley, J.P. Hohimer and A. Owyoung, "High-order ($\upsilon > 10$) eigenmodes in ten-stripe gain-guided diode laser arrays," *Appl. Phys. Lett.*, vol. 49, pp. 684-686, 1986.

10. D.E. Ackley and R.W.H. Engelmann, "High Power Leaky Mode Multiple Stripe Laser," *Appl. Phys. Lett.,* vol. 39, pp. 27-29, July 1981.

11. D. Botez, P. Hayashida, L.J. Mawst and T.J. Roth, "Diffraction-limited-beam, high-power operation from X-junction coupled phase-locked arrays of AlGaAs/GaAs diode lasers," *Appl. Phys. Lett.*, vol. 53, pp. 1366-1368, Oct. 1988.

12. H. Hosoba, M. Matsumoto, S. Matsui, S. Yano and T. Hijikata, "Phased array laser diode – SAW-TOOTH channelled array," *The Review of Laser Engineering*, vol. 17, pp. 32-42, Jan. 1989 (in Japanese).

13. L.J. Mawst, D. Botez, C. Zmudzinski, M. Jansen, C. Tu, T.J. Roth, and J. Yun, "Resonant self-aligned-stripe antiguided diode laser array," *Appl. Phys. Lett.*, vol. 60, pp. 668-670, 1992.

14. H.Yang, L.J. Mawst, M. Nesnidal, J. Lopez, A. Bhattacharya, and D. Botez, "10 W Near-Diffraction-Limited Pulsed Power from Al-free Phase-Locked Antiguided Arrays," Tech. Dig. LEOS '96 Annual Meeting, Paper MM3, Nov. 18-20th 1996, Boston, MA.

15. D. Botez, L.J. Mawst, G.Peterson, and T.J. Roth, "Phase-locked arrays of antiguides: Modal content and discrimination," *IEEE J. Quantum Electron.*, vol. QE-26, pp. 482-495, March 1990.

16. J.K. Butler, D.E. Ackley, and D. Botez, "Coupled Mode Analysis of Phase-Locked Injection Laser Arrays," *Appl. Phys. Lett.*, vol. 44, pp. 293-295, Feb. 1984, see also *Appl. Phys. Lett.*, vol. 44, p. 935, May 1984.

17. E. Kapon, J. Katz, and A. Yariv, "Supermode Analysis of Phase-Locked Arrays of Semiconductor Lasers," *Opt. Lett.*, vol. 10, pp. 125-127, Apr. 1984.

18. S.R. Chinn and R.J. Spiers, "Modal Gain in Coupled-Stripe Lasers," *IEEE J. Quantum Electron.*, vol. QE-20, no. 4, pp. 358-363, Apr. 1984.

19. W. Streifer, A. Hardy, R.D. Burnham, and D.R. Scifres, "Single-Lobe Phased-Array Diode Lasers," *Electron. Lett.*, vol. 21, no. 3, pp. 118-120, Jan. 1985.

20. J.K. Butler, D.E. Ackley and M. Ettenberg, "Coupled mode analysis of gain and wavelength oscillation characteristics of diode laser phased arrays," *IEEE J. Quantum Electron.*, vol. QE-21, pp. 458-464, May 1985.

21. W. Streifer, A. Hardy, R.D. Burnham, R.L. Thornton, and D.R. Scifres, "Criteria for Design of Single-Lobe Phased-Array Diode Lasers," *Electron. Lett.*, vol. 21, no. 11, pp. 505-506, May 1985.

22. Kuo-Liang Chen and Shyh Wang, "Spatial hole burning problems in evanescently coupled semiconductor laser arrays," *Appl. Phys. Lett.*, vol. 47, pp. 555-557, Sept. 1985.

23. P.G. Eliseev, R.F. Nabiev and Yu. M. Popov, "Analysis of laser-structure anisotropic semiconductors by the Bloch-function method," *J. Sov. Las. Res.*, vol. 10, 6, pp. 449-460, 1989.

24. G.R. Hadley, "Two-dimensional waveguide modeling of leaky-mode arrays," *Opt. Lett.*, vol. 14, pp.859-861, Oct. 1989.

25. H. Fujii, I. Suemune and M. Yamanishi, "Analysis of transverse modes of phase-locked multi-stripe lasers," *Electron. Lett.*, vol. 21, pp. 713-714, Aug. 1985.

26. D. Botez and T. Holcomb, "Bloch-function analysis of resonant arrays of antiguided diode lasers," *Appl. Phys. Lett.*, vol. 60, pp. 539-541, Feb. 1992.

27. R.F. Nabiev and A.I. Onishchenko, "Laterally coupled periodic semiconductor laser structures: Bloch function analysis," *IEEE J. Quantum Electron.*, vol. 28, pp. 2024-2032, Oct. 1992.

28. R.W. Engelmann and D. Kerps, "Leaky modes in active three-layer slab waveguides," *Proc. Inst. Elec. Eng. Part I*, vol. 127, pp.330-336, Dec. 1980.

29. T.H. Shiau, S. Sun, C.F. Schaus, K. Zheng and G.R. Hadley, "Highly stable strained layer leaky-mode diode laser arrays," *IEEE Photonics Technol. Lett.*, vol. 2, pp. 534-536, 1990.

30. J.S. Major, Jr., D. Mehuys, D.F. Welch, and D. Scifres, "High power high efficiency antiguide laser arrays," *Appl. Phys. Lett.*, vol. 59, pp. 2210-2212, Oct. 1991.

31. C.A. Zmudzinski, D. Botez, and L.J. Mawst, "Simple description of laterally resonant, distributed-feedback-like modes of arrays of an-

tiguides," *Appl. Phys. Lett.*, vol. 60, pp. 1049-1051, March 1992.

32. D. Botez and L.J. Mawst, "Γ effect: Key intermodal-discrimination mechanism in arrays of antiguided diode lasers," *Appl. Phys. Lett.*, vol. 60, pp. 3096-3098, 1992.

33. G.R. Hadley, "Index-guided arrays with a large index step," *Opt. Lett.*, vol. 14, pp. 308-310, 1989.

34. L.J. Mawst, D. Botez, T.J. Roth, W.W. Simmons, G. Peterson, M. Jansen, J.Z. Wilcox, and J.J. Yang, "Phase-locked array of antiguided lasers with monolithic spatial filter," *Electron. Lett.*, vol. 25, pp. 365-366, 1989.

35. P.D. Van Eijk, M. Reglat, G. Vasilieff, G.J.M. Krijnen, A. Driessen, and A.J. Mouthaan, "Analysis of modal behavior of an antiguide diode laser array with Talbot filter," *IEEE J. Lightwave Technol.*, vol. LT-9, pp. 629-634, 1991.

36. D. Botez, M. Jansen, L.J. Mawst, G. Peterson and T.J. Roth, "Watt-range, coherent, uniphase powers from phase-locked arrays of antiguided diode lasers," *Appl. Phys. Lett.*, vol. 58, pp. 2070-2072, 1991.

37. L.J. Mawst, D. Botez, M. Jansen, T.J. Roth and J. Rozenbergs, "1.5 W diffraction-limited-beam operation from resonant-optical-waveguide (ROW) array," *Electron. Lett.*, vol. 27, pp. 369-371, 1991.

38. J.R. Leger, G.J. Swanson and M. Holtz, "Efficient side lobe suppression of laser diode arrays," Appl. Phys. Lett., vol. 50, pp. 1044-1046, 1987.

39. R.F. Nabiev, P. Yeh and D. Botez, "Self-stabilization of the fundamental in-phase mode in resonant antiguided laser arrays," *Appl. Phys. Lett.*, vol. 62, pp. 916-918, March 1993.

40. R.F. Nabiev and D. Botez, "Comprehensive Above-Threshold Analysis of Antiguided Diode Laser Arrays," IEEE J. Select. Topics Quantum Electron, vol. 1, No. 2, pp. 138-149, June 1995.

41. L.J. Mawst, D. Botez, M. Jansen, T.J. Roth, C. Zmudzinski, C. Tu and J. Yun, "Resonant-optical-waveguide antiguided diode laser arrays," *SPIE Proceedings OE-LASE '92 Meeting*, vol. 1634, pp. 2-12, 1992.

42. C. Zmudzinski, D. Botez, and L.J. Mawst, "Coherent, One-Watt Operation of Large-Aperture Resonant Arrays of Antiguided Diode Lasers," *Appl. Phys. Lett.*, vol. 62, pp. 2914-2916, 1993.

43. C. Zmudzinski, L.J. Mawst, D. Botez, C. Tu and C.A. Wang, "1 W Diffraction-limited-beam operation of resonant-optical-waveguide diode laser arrays at 0.98μm," *Electron. Lett.*, vol. 28, pp. 1543-1545, July 1992.

44. J.S. Major, Jr., D. Mehuys and D.F. Welch, "11.5 W, Nearly Diffraction-Limited Pulsed Operation of an Antiguided Laser Diode Array," *Electron. Lett.*, vol. 28, pp. 1101-1103, June 1992.

45. D. Botez and A.P. Napartovich, "Phase-locked Arrays of Antiguides: Analytical Theory," *IEEE J. Quantum Electron.*, vol. 30, pp. 975-980, April 1994; Erratum, vol. 32, Dec. 1996.

46. D. Botez, A. Napartovich and C. Zmudzinski, "Phase-Locked Array of Antiguides: Analytical Theory - II," *IEEE J. Quantum Electron.*, vol. 31, pp. 244-253, Feb. 1995.

47. A. Napartovich, D. Botez, "Analytical Theory of the Array Modes of Resonant Arrays of Antiguides," Kvantovaya Electronika (Russian Journal of Quantum Electronics), vol. 26, No. 8, pp. 670, August 1996.

48. L.J. Mawst, D. Botez, M. Jansen, C. Zmudzinski, S.S. Ou, M. Sergant, T.J. Roth, C. Tu, G. Peterson and J.J. Yang, "Two-Dimensional Surface-Emitting Leaky-Wave Coupled Laser Arrays," *IEEE J. Quantum Electron.*, vol. 29, pp. 1906-1918, June 1993.

49. J.M. Gray and J.M. Marsh, "850–nm antiguided laser array fabricated using a zinc disordered superlattice," Tech. Dig. Conference on Lasers and Electro-Optics '96, vol. 9, p. 78, Anaheim, CA, June 2-7 1996.

Quantum Well Semiconductor Lasers Are Taking Over

Amnon Yariv

Abstract

Semiconductor quantum well lasers are characterized by confinement of the electrons and holes to extremely thin (~70Å) regions. This leads to major and important improvement in all the operating characteristics of these lasers compared to conventional semiconductor lasers—specifically, to more than an order of magnitude reduction in threshold current and a much narrower spectral width.

Judged by economic impact, semiconductor lasers (SCLs) have become the most important class of lasers. This is due mostly to the key role they play in two areas of technology: compact disc players and long distance optical fiber communication. [1] Auspicious as this beginning is, it pales in comparison to what the future holds for these lasers. The state of affairs is probably not very different from those early days in the 1950s when the discrete transistor took its first tentative steps. Most of the applications today still depend on a single, relatively bulky and inefficient current-guzzling SCL. Already on the drawing boards are plans for new computer interconnect (within and between boards) circuits that will employ thousands of monolithic SCLs with their associated electronic circuits, grown and fabricated on single crystals of GaAs, InP and related semiconductor alloys.

SCLs will also figure in new applications ranging from two-dimensional display panels and erasable optical data storage, and invade new domains such as medical, welding and spectroscopic applications that are now the captives of solid state and dye lasers.

The main reason behind this major surge in the role played by SCLs is the joint progress in material growth technologies and the theoretical understanding of a new generation of SCLs—the quantum well lasers (QWLs). In what follows we will describe some of the recent progress that has led to these developments.

The main difference between a QWL and the conventional SCL is mostly in the thickness t of the active region (See Fig. 1). The active region of a semiconductor laser is the reduced energy gap layer into which the electrons and holes are injected from adjacent, higher gap n and p regions, respectively. These electrons and holes are induced to recombine by the laser light field. The energy released by this process (a photon for each electron-hole recombination) is added coherently to the optical wave. The active region is about 2000Å thick in a conventional SCL, while it is between 50Å and 100Å in a QWL. Such thin layers, corresponding to ~7-15 atomic layers, are grown exclusively using the techniques of molecular beam epitaxy (MBE) and metal organic chemical vapor deposition (MOCVD).

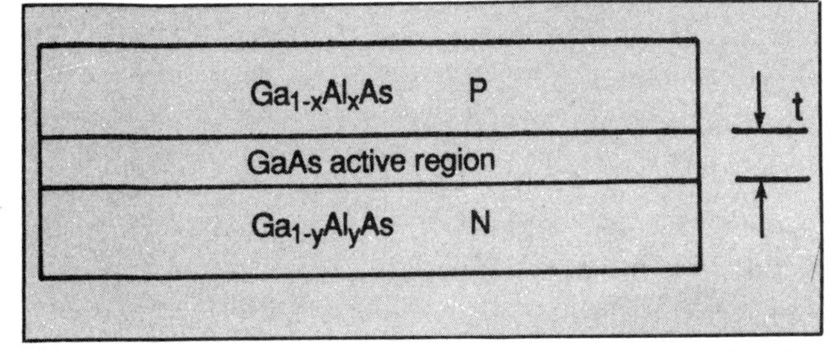

Fig. 1. The basic layered structure of a heterostructure GaAs/GaAlAs semiconductor laser.

The electrons and holes in such thin layers display quantum size effects. This is another way of saying that the electron-in-a-box (or hole-in-a-box) confinement energies, familiar from our first quantum mechanics course, due to the confinement of the carriers in the extremely thin active layer is large compared to kT at room temperature. This leads to major qualitative differences in the energy distribution of the electrons and holes and thus to major modification of all the basic optical properties of these new media.

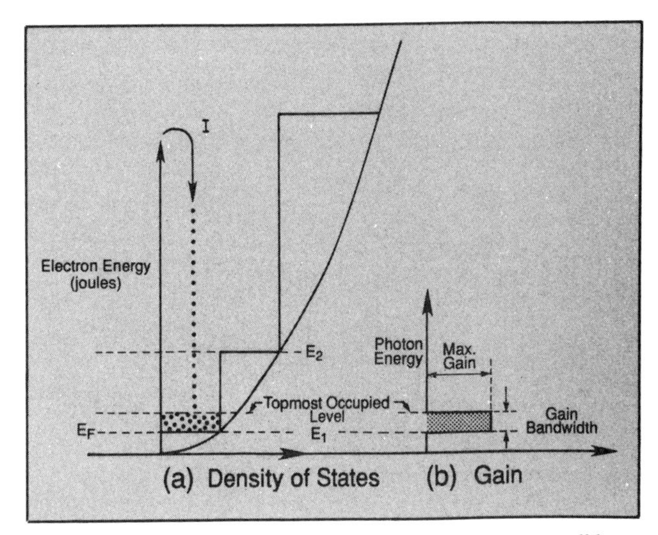

Fig. 2. (a) Density of states function vs. energy of a quantum well laser. Occupied states are shown at bottom. (b) Optical gain vs. photon energy corresponding to (a).

This new state of affairs is illustrated in Fig. 2. The staircase-like function depicts the density of available electron states, i.e., the number of available states per unit area—per unit energy of the quantum layer. The steps at E_1 and E_2 and beyond are the signature of the two-dimensional nature of the confined carriers. The smooth curve shows for comparison the density of states of a bulk semiconductor crystal. Note the lack of discontinuous steps in the bulk case. We can think of the density of states curve, Fig. 2(a), as the profile of a bathroom sink containing the electron fluid. The injection current I plays the role of the input water stream, while spontaneous electron-hole recombination, which removes electrons (and holes), takes the place of the bottom drain hole. At a given current I all the electron states up to some uppermost level E_F (the quasi-Fermi level) are filled up in the same way that the height of the

Reprinted from *IEEE Circuits and Devices Magazine*, Vol. 5, No. 6, pp. 25-28, November 1989.

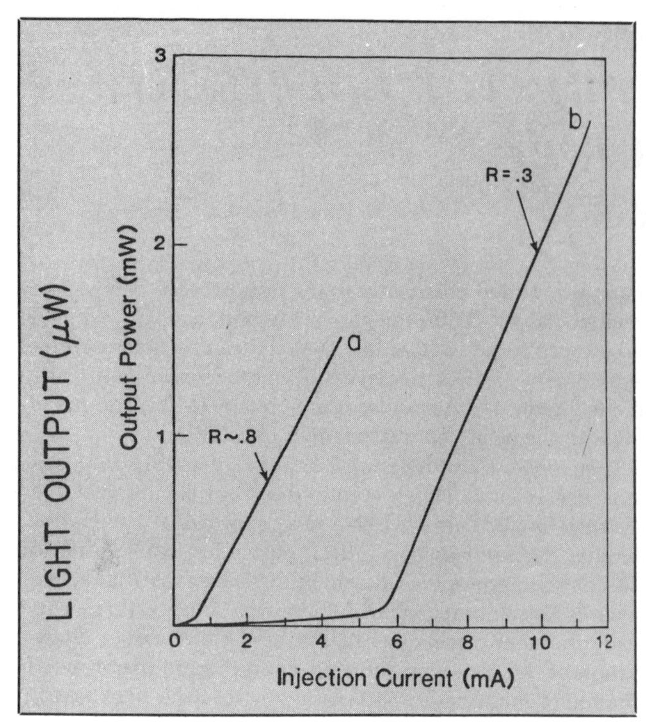

Fig. 3. Light output versus current curve for a 250μm long buried GRIN-SCH single QW laser with (a) high reflectivity dielectric coated cleaved facets, (b) uncoated cleaved facets. Both curves are for the same laser.

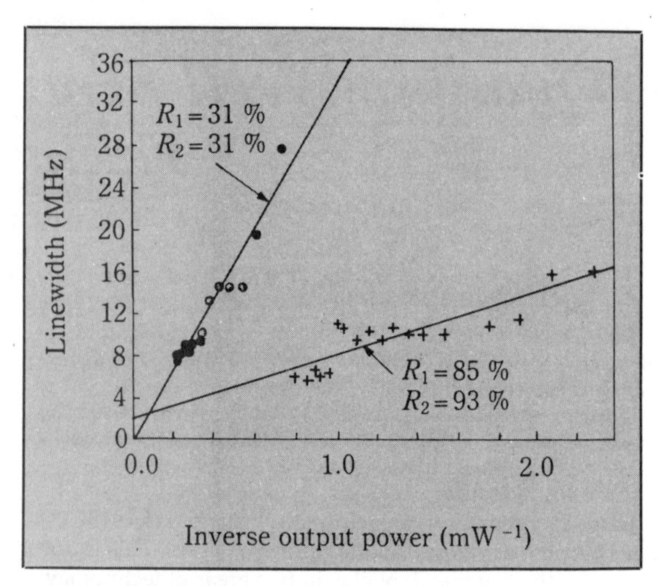

Fig. 4: Spectral linewidth as a function of the reciprocal of the output power at FACET with reflectivity R_1 for an uncoated 250μm long 1μm stripe buried heterstructure GRIN SCH SQW laser and for the same laser with coated facets reflectivities R_1 = 85 percent and R_2 = 93 percent.

water level in a sink with a drain depends on the input flow rate.

It turns out from basic laser theory [2] that a plot of the density (per unit area per unit energy) of filled electron states vs. electron energy (at some current I) mimics the plot of the optical gain vs. photon energy of the same semiconductor medium. (This is not surprising—having more electrons of a given energy translates into more induced power due to these electrons, which is added to that of the inducing optical wave and thus to a larger optical gain.) Such a curve is shown in Fig. 2(b). It follows immediately that because of the very flat nature of the bottom of our "sink," the maximum gain becomes available at very small input (injection) currents. And here lies one of the main advantages of the QWL—large gains at small currents. This effect becomes somewhat less dramatic than the figure indicates if one allows for rounding off due to finite temperatures and collison processes, but it still plays a key role. Semiconductor lasers with QW active regions exploiting this consequence have recently [3] been demonstrated with threshold currents as low as 0.55 ma which is over an order of magnitude less than typical commercial lasers. The experimental data is given in Fig. 3. The low threshold current is due not only to the large differential ($\partial g/\partial I$) gain, but also to the major reduction (X40) in active volume that occurs when one switches from conventional active regions ($t \gtrsim 400$Å) to quantum wells ($t \lesssim 100$Å). This reduces in a proportionate manner the number of electrons in the active region. This threshold current is sufficiently low that in many practical situations the laser may be considered by the optoelectronic designer as a thresholdless device converting, in a proportional manner, current to coherent light. This makes it possible to eliminate most of the optical power monitoring and much of the optical power stabilizing circuitry used in conjunction with present day lasers. These constitute a major stumbling block to the development of integrated optoelectronic circuits (IOECs). We can extend the same basic reasoning to "quantum wire" lasers in which the carriers are confined in two dimensions and the active region has cross-sectional dimensions of ~ 100 × 100 Å. A theoretical analysis indicates that lasers with threshold currents well below 0.1 ma should be achievable.[1] Attempts to fabricate such wires are now underway in a number of university and industrial laboratories.

The main loss mechanism of conventional SCL media is due to free carrier absorption by the injected electrons and holes. The resulting modal absorption coefficient, i.e., the absorption coefficient of the laser beam, is proportional to the total number of electrons (or holes) per unit length, along the direction of propagation, of the active region. It follows immediately that the QWL with its greatly reduced active volume contains proportionately fewer electrons and holes and thus is far less lossy than conventional lasers. One of the most important consequences of the greatly reduced loss has been the reduction of the spectral line-width of SCLs employing quantum wells (See Fig. 4.)

The spectral (Hz) width of the laser output is limited by spontaneous emission to a limiting value of [5]

$$(\Delta\upsilon)_{laser} = \frac{V_g^2 E_L\,(\alpha_i - (1/L)\ln R)\,(-1/L)\,(\ln R) n_{sp}\,(1+\alpha^2)}{8\pi P} \tag{1}$$

where the photon energy is E_L, V_g is the group velocity, P the output power at one of the facets n_{sp} in the spontaneous emission factor [5] which in most lasers is between 1.5 and 2.0. The modal absorption coefficient is α_i and the amplitude-phase coupling factor is α. [6] In the limit $\alpha_i << (1/L)$ $\ln R$ and $R \approx 1$ the dependence of $(\Delta\upsilon)_{laser}$ on R is proportional to $(1-R)^2$ so that one would expect the linewidth to approach zero in the limit of $R \rightarrow 1$; however, n_{sp} limits the reduction since it increases exponentially as the threshold current density decreases. [5] Values of $(\Delta\upsilon)_{laser}$ < 1MHz are required in today's coherent optical communication

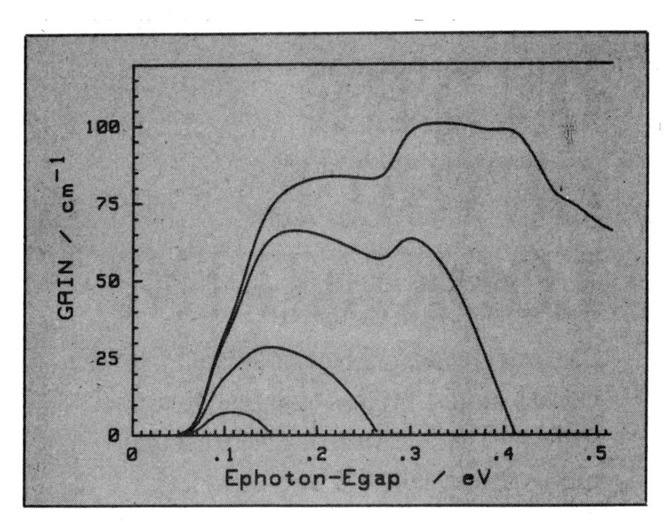

Fig. 5. Theoretical optical gain vs. photon energy at different injection currents.

systems and such values are most easily obtainable with advanced QWLs.

The narrow linewidth of the QWL benefits also from a reduced value of the α parameter. Theoretical analysis [5,6] backed by recent experiments indicates that in QWLs, values of α in the range of 2-3 are obtained as against $\alpha \sim$ 5-6 in conventional SCLs.

The low losses as well as the excellent uniformity of the epitaxial QW material has been pursued in obtaining efficient high power from QWLs. Scientists at Spectra Diodes, Inc. have recently reported semiconductor laser arrays based on GaAs/GaA1As, with cw power output exceeding 40 watts(!) [7] Even more impressive are the conversion efficiencies—wall plug power to optical power output—exceeding 50 percent. Differential quantum efficiencies, which are a measure of the ultimate achievable values, exceed 80 percent.

Since many of the applications envisaged for the QWLs involve high speed data streams, their modulation frequency response is of major concern. Recently K. Lau and collaborators of ORTEL Corporation determined experimentally that these lasers are extraordinarily fast with response times shorter than 19 psec reported.[8]

The last unique feature of QWL which we will describe has to do with their tunability. The QW medium, when pumped by sufficiently high currents, can provide optical gain over a large range of optical wavelengths. This range (the "gain bandwidth" of Fig. 2(b)) increases as the pumping current is increased, reflecting the gradual filling up of the density of states (our "sink") profile of Fig. 2. At sufficiently high currents, the electrons begin to occupy the second quantized level E_2. Fig. 5 shows calculated gain vs wavelength plots of a QW medium with the pumping current as a parameter. Of special interest is the extremely wide, nearly flat spectrum that is obtained at certain current levels. These characteristics are reminiscent of those of dye lasers, which are used mostly in applications that require tuning the output wavelength of a laser over a large range. It follows that QWLs are capable of similar wide tuning ranges. A plot of optical power vs. λ of a GaAs QWL employing an external grating as the wavelength selecting element is shown in Fig. 6. [9] The tuning range spans ~11 percent of the center wavelength. The prospect of replacing, in some applications, dye lasers with their very small efficiencies (~0.1 percent) and large areas (~1 m^2) with a matchbox sized device operating near 50 percent efficiency is real and challenging.

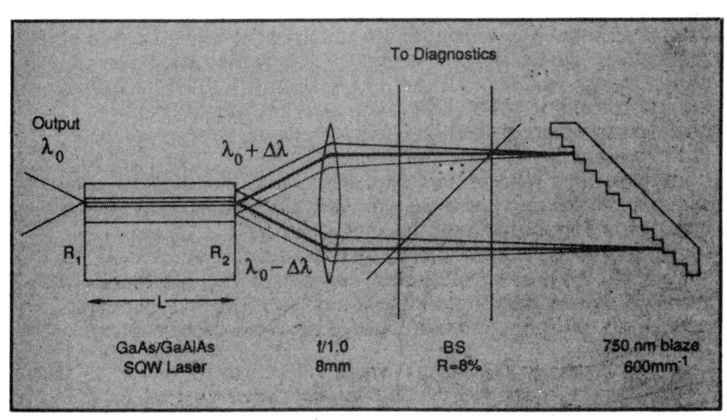

Fig. 6. The experimental configuration of a quantum well laser with an external grating.

The main challenges facing the QWL designer at the moment include:

1. the mode and coherence control of high power wide area lasers including laser arrays,
2. single channel lasers with mode-controlled multiwatt outputs,
3. monolithic tuning of QW lasers,
4. producing lasers with I_{th}<0.1 ma,
5. ever shorter wavelengths (important for optical desk computer data storage and video and audio disks of enhanced capacity).

The level of research on QWL at universities and industry here and abroad guarantees continued progress in this field and an increase of the pace at which these new upstart lasers will replace their older brethren.

Acknowledgement

The author acknowledges his colleagues (K. Lau, N. Bar-Chaim, P. Derry) and his students (M. Mittelstein, D. Mehuys) whose research efforts are responsible for much of the material described. The continuous support of the Defense Advanced Research Projects Agency, Office of Naval Research and National Science Foundation of the quantum well laser research at Caltech is gratefully acknowledged.

References

[1] See for example D. A. B. Miller, "Quantum Wells for Optical Information Processing," *Opt. Eng.*, May 1987, p. 368 and H. Okamoto, "Semiconductor Quantum-well Structures for Optoelectronics," *Jap. J. Appl. Phys., vol.* 26, p. 315, 1987.

[2] See for example A. Yariv, *Quantum Electronics*, 3rd ed., John Wiley and Sons, New York, 1989, p. 271.

[3] P. L. Derry, A. Yariv, and K. Lau, N. Bar-Chaim, K. Lee and J. Rosenberg, "Ultralow-Threshold Graded-Index Separate-Confinement Single Quantum Well Buried Heterostructure (Al,Ga)As Lasers with High Reflectivity Coatings," *Appl. Phys. Letts.*, vol. 50, no. 14, p. 874, 1987.

[4] A. Yariv, "Scaling Laws and Minimum Threshold Currents for Quantum Confined Semiconductor Lasers," *Appl. Phys. Lett.*, vol. 53, p. 1033, 1988.

[5] P. Derry, T. R. Chen, Y. Zhuang, J. Paslaski, M. Mittelstein, K. Vahala, A. Yariv, K-Y. Lau and N. Bar-Chaim, "Properties of Ultra Low Threshold Single Quantum Well (Al,Ga)As Lasers for Computer Interconnects," *Optoelectronics Devices and Tech.*, vol. 3, no. 2, p. 117, 1988.

[6] Y. Arakawa and A. Yariv, "Theory of Gain, Modulation Response and Spectral Linewidth in AlGaAs Quantum Well Lasers," *IEEE J. Quant. Elec.*, QE-21, p. 1666, 1985.

[7] D. Welch, D. Scifres, Spectra Diodes, Inc., private communication.

[8] K. Lau, N. Bar-Chaim, P. L. Derry and A. Yariv, "High-Speed Digital Modulation of Ultralow Threshold (<1mA) GaAs Single Quantum Well Lasers Without Bias," *Appl. Phys. Lett.*, vol. 51, no. 2, p. 69, 1987.

[9] M. Mittelstein, D. Mehuys and A. Yariv, "Broadband Tunability of Gain-Flattened Quantum Well Semiconductor Lasers with External Gratings," *Appl. Phys. Lett.*

[10] M. Mittelstein, D. Mehuys and A. Yariv, "Broadband Tunability of Gain-Flattened Quantum Well Semiconductor Lasers with an External Grating," submitted to *Applied Physics Letters*.

Organic-on-Inorganic Semiconductor Heterojunctions: Building Block for the Next Generation of Optoelectronic Devices?

S. R. Forrest

Abstract

We describe a new class of optoelectronic devices utilizing thin films of stable crystalline organic semiconductors layered onto inorganic semiconductor substrates. The electrical properties of these devices are determined by the energy barrier at the heterojunction contact between the organic and inorganic materials, and in many ways are similar to ideal diffused-junction inorganic semiconductor devices. The organic materials can be layered onto semiconductor substrates without inducing large strains in either material, hence allowing for a wide range of material combinations with a similarly broad range of optoelectronic functions to be realized. As examples, we discuss high bandwidth photodetectors and field effect transistors made using organic/inorganic semiconductor heterojunctions. In addition, we discuss how the optical and electronic properties of the organic films can be modified by irradiation with energetic electron and ion beams.

Introduction

For many years, hope has been extended that organic materials have potential in active electronic device applications. Although this promise has remained largely unfulfilled, it has developed out of the realization that the variety of organic compounds with a wide range of both optical and electronic properties is unlimited, with several thousand compounds being readily available. The source of the disappointing performance of organic materials comes from several sources. Among the most important drawbacks of organic materials are: 1) Many compounds tend to have unstable chemical or electrical properties. This instability can be exacerbated by exposure to adverse environments such as humidity or high temperature. 2) Organic semiconductors are difficult to dope with trace impurities in order to alter their majority carrier type. 3) The mobilities of electrons and holes are generally low at room temperature. Mobilities of ≤ 1 cm^2/V-s are typical of many organic semiconductors. 4) Good crystallinity, or long range crystalline order, are difficult to achieve.

In spite of these difficulties, considerable progress has been made in the last decade in realizing practical, active electronic and optoelectronic devices where an organic material forms an integral part of the device structure. One promising approach employs an organic film that is layered onto the surface of a conventional inorganic semiconductor substrate such as Si, GaAs, or InP to form an insulating or conducting layer that controls the distribution of electric fields and hence the transport of charge within the device. An attractive feature of such devices is that the composition of the organic film can be altered only slightly to effect large changes in its optical and electronic properties. Furthermore, the cohesive forces that bind molecules within these organic semiconductors are due to the relatively weak Van der Waals dipolar attraction. Hence, the materials are somewhat soft (as compared to Si, for example), and can thus be layered without inducing strain onto a variety of semiconductor substrates. Thus, the organic/inorganic semiconductor growth process need not be limited by the constraints of lattice–matching. This restriction imposes extremely narrow boundaries on the particular inorganic semiconductors that can be combined without inducing large strains, and hence lattice defects. However, no such constraint exists for organic-on-inorganic (OI) semiconductor material combinations.

In this paper we describe OI devices employing thin layers of crystalline molecular semiconductors commonly used as richly colored dye pigments. These films form rectifying heterojunction contacts when deposited onto the surfaces of inorganic semiconductor substrates. The resulting devices are extremely simple to fabricate, and their electronic properties are often similar to those obtained using painstakingly processed, ideal inorganic semiconductor devices.

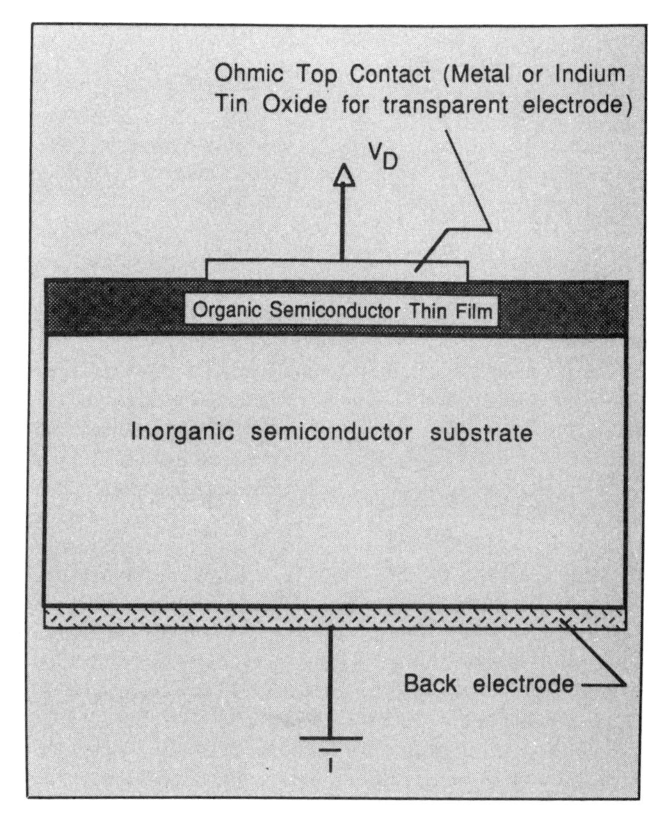

Fig. 1: Schematic cross-sectional view of an organic-on-inorganic semiconductor heterojunction device.

Reprinted from *IEEE Circuits and Devices Magazine*, Vol. 5, No. 3, pp. 33-37, 41, May 1989.

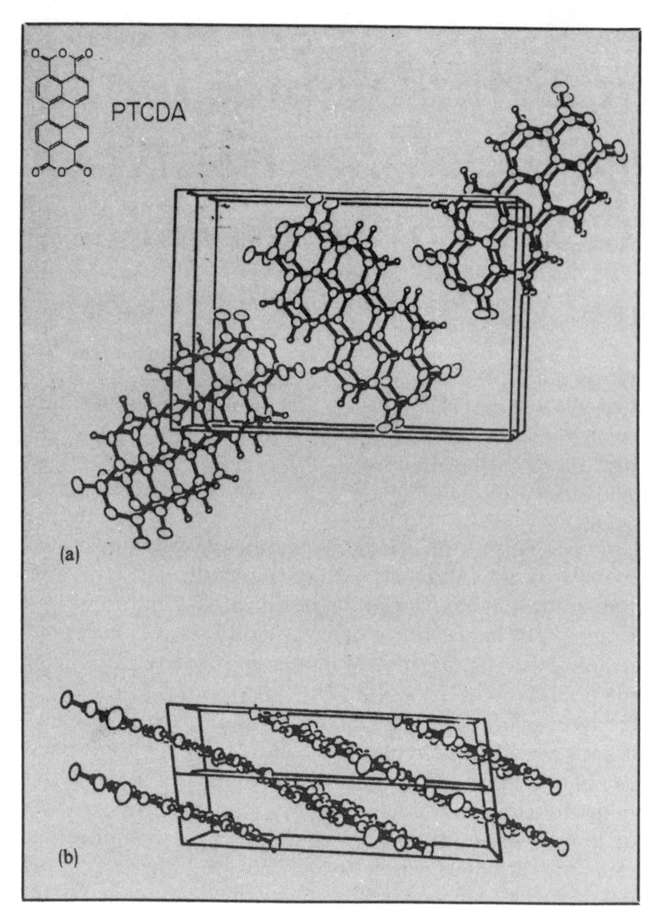

Fig. 2: A unit cell of PTCDA as viewed along two orthogonal directions. Inset: Structural formula for PTCDA.

Indeed, the OI semiconductor heterojunction appears to be an advantageous combination of many of the properties of both organic and inorganic materials, and hence may open many new opportunities for fabricating optoelectronic devices whose applications extend well beyond those accessible to conventional semiconductor devices currently being investigated.

OI Semiconductor Device Operation

Figure 1 shows a schematic cross-sectional view of a typical OI heterojunction rectifier. To fabricate this device, a purified sample of a crystalline molecular semiconductor, such as 3,4,9,10 perylenetetracarboxylic dianhydride (or PTCDA, see inset, Fig. 1), is loaded into a vacuum system equipped for thermal evaporation. Typically, 50Å to 2000Å of the organic material is then deposited by sublimation in vacuum onto the clean surface of a semiconductor substrate. Many different p- or n-type semiconductors such as Si, Ge, GaAs, InP and HgCdTe can be used in OI device structures. Ohmic contact is made to the organic thin film by deposition of the appropriate metals (such as indium or titanium) through a shadow mask that defines the contact pad. Ohmic contact is also made to the entire substrate surface by metal deposition.

To understand how charge is transported in crystalline molecular semiconductors, Fig. 2 illustrates a unit cell of PTCDA as viewed along two crystal axes. It is apparent that the molecules form stacks where the intermolecular spacing in the stacking direction is small (3.2 Å for PTCDA, for example). The valence, hybrid s-p orbitals of such molecules form rings that lie parallel to, and slightly above the molecular plane. Hence, the electronic orbitals from two adjacent molecules in a stack overlap to a considerable degree, resulting in the delocalization of the electrons in the stacking direction. This leads to a relatively high charge mobility in a direction perpendicular to the molecular plane. On the other hand, mobility from stack to stack along the molecular planes is considerably lower since no extensive orbital overlap exists in that direction.

In the case of PTCDA, the stacking direction is approximately perpendicular to the substrate surface plane. Furthermore, under appropriate deposition conditions, a single crystalline sheet of this material can be formed on the substrate surface. However, under most circumstances, polycrystalline films are produced with typical grain sizes on the order of 5000 Å. It has been observed that grain boundaries produce a significant resistance to charge flow, but given the typical grain size, charge propagating perpendicular to the thin film plane does not encounter such grain boundaries. When we combine the limitations on mobility due to the intrinsic stacking habit of the molecules, along with grain boundary effects, it is observed that the conductivity transverse to the film plane can be several orders of magnitude higher than that parallel to the plane. Hence, charge is constrained to flow only directly below the ohmic contact pad that therefore defines the OI diode area without any necessity of further confinement of the structure by subsequent patterning. This feature of the device makes it extremely simple to fabricate as compared with conventional semiconductor device processing.

Deeper understanding of the charge transport mechanisms in OI heterojunction diodes can be derived from the energy band diagram for a metal/PTCDA/p-Si device shown in Fig. 3. Here it is shown that PTCDA is a relatively large band gap (2.2 eV) semiconductor whose "valence-like" band, or highest occupied molecular orbital (HOMO), is offset from the Si valence band maximum by an energy, ΔE_v = 0.5 eV. The "conduction-like" band, or lowest unoccupied molecular orbital (LUMO), is then offset from the Si conduction band minimum by $\Delta E_g - \Delta E_v$ = 0.6 eV, where ΔE_g

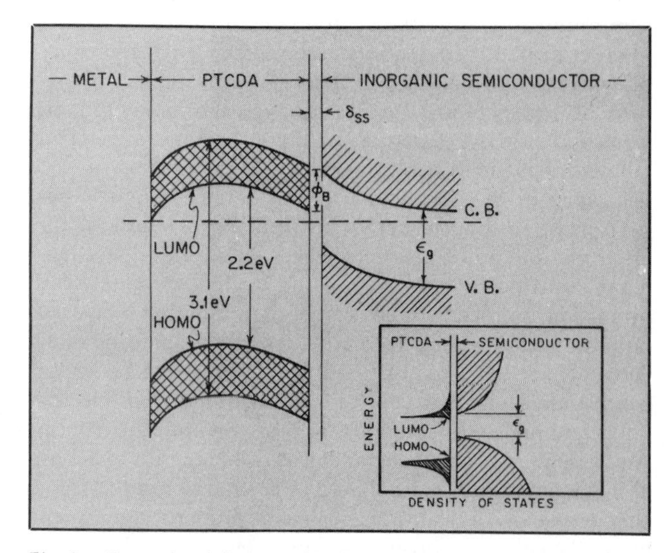

Fig. 3: Energy band diagram of PTCDA showing a thin interfacial layer of thickness, δ_{ss}, and a band offset potential of ϕ_B. Inset: The density of states at an OI heterojunction.

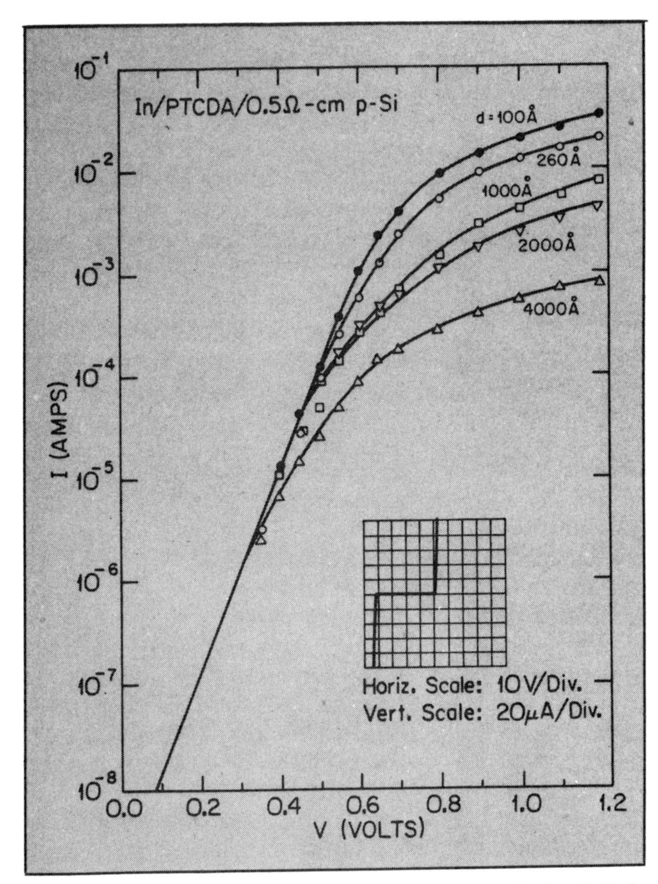

Fig. 4: Forward current-voltage characteristics of a PTCDA/p-Si diode for various PTCDA layer thicknesses, d. Inset: Bipolar current-voltage characteristics of a PTCDA/0.5 Ω-cm p-Si diode.

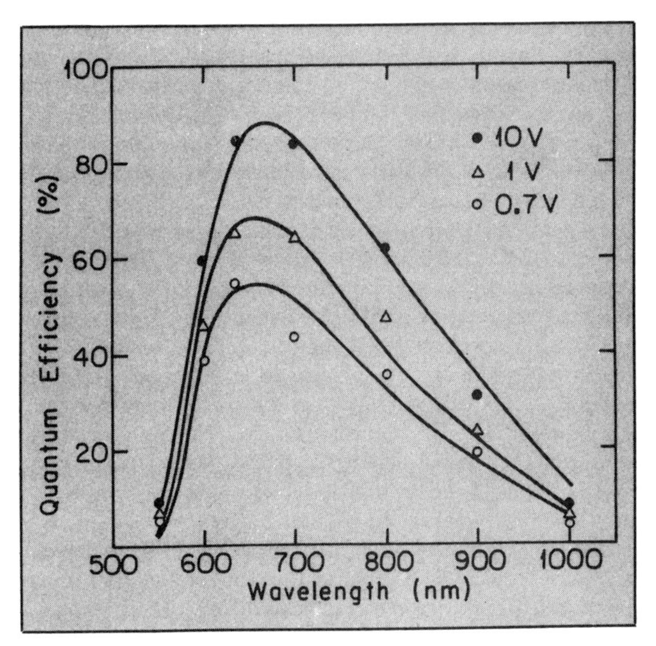

Fig. 5: External quantum efficiency of an ITO/PTCDA/p-Si detector.

= 1.1 eV is the energy band gap difference between the two semiconductors. This energy band diagram implies that charge is injected into the organic film from either the metal contact or the inorganic semiconductor substrate, and then is thermionically emitted over the energy barrier at the heterojunction contact. Transport across the thin film itself is limited either by its internal resistance or by space charge build-up, depending on the current density and the density of free carriers in the unbiased organic film.

Due to the nearly symmetrical barriers at the valence and conduction band edges, and also due to approximately equal mobilities for electrons and holes within the organic films (although for many organic materials such as PTCDA, the hole mobility is somewhat larger), such OI heterojunction contact diodes can be made on both p-type and n-type subtrates with only minor differences in performance.

The bipolar room temperature current-voltage (I-V) characteristics of a typical In/PTCDA/p-Si OI diode are shown in the inset of Fig. 4. The overall characteristic is similar to that of an ideal diffused p-n junction p-Si diode. A detailed examination of Fig. 4 indicates that the maximum current that can be achieved under forward bias is limited by the film thickness (d), with the current at a particular forward voltage scaling as d^{-3}—a behavior characteristic of space charge limited transport. At lower forward voltages, the current rises exponentially due to thermionic emission over the valence band offset barrier at the OI heterojunction. Under reverse bias, the dark current is limited by leakage processes occurring in the Si substrate, and at 240 V, avalanche breakdown occurs in the substrate.

OI Photodetector

Given these rectifying characteristics, several optoelectronic devices can be demonstrated that utilize the many properties made available by the large family of molecular compounds. In Fig. 1 the schematic cross-section of an OI photodetector consisting of a transparent indium tin oxide (ITO) contact pad that was sputter-deposited onto the organic thin film surface is shown. The PTCDA layer for this device was 2000 Å thick and was deposited onto a p-Si substrate. The device I-V characteristics are similar to those in Fig. 4, except that avalanche breakdown accompanied by photocurrent multiplication occurred at 60 V, consistent with the lower resistivity of the particular Si substrate used. In Fig. 5 is shown the quantum efficiency of the detector as a function of wavelength and reverse bias. At wavelengths longer than 1.1 μm, absorption in the Si substrate vanishes due to the band edge cutoff at this energy. At shorter wavelengths, photons are absorbed in the depletion region in the Si bulk beneath the OI contact. Electron-hole pairs created by the absorption process are subsequently separated by the electric field and are collected at the contact terminals. At even shorter wavelengths (<6000 Å), the red PTCDA film becomes absorbing, and hence photons can no longer penetrate to the electric field region in the Si. The voltage dependence of the quantum efficiency is a result of the long absorption lengths of near-band-edge photons in Si. At low voltages, the electric field extends only a short distance from the OI heterojunction into the Si bulk, and hence photons not absorbed in the high field region have a larger probability of generating electron-hole pairs that recombine before they can be collected. As the bias is increased, the region of non-zero electric field extends further into the substrate, thus allowing for a larger fraction of the photogenerated carriers to be collected. A maximum quantum efficiency of 85 percent was observed for this diode, although higher efficiences should be attainable if the thickness of the organic layer and the ITO contact are adjusted to the appropriate thickness for an antireflection coating.

Speed of response to fast 0.8 μm wavelength optical pulses was also measured for this diode, and it was found that diffusion through the organic film provides a fundamental transient response time limit of 5 ns. However, for a diode with an organic film thickness of 50-100 Å, the rise and fall times can be as short as 100 ps, allowing for the use of OI diodes in high bandwidth applications.

The refractive index of PTCDA has been found to be between 2.2 and 2.3 in the near-infrared wavelength region that is of interest in optical communications. Thus, it is possible to use these films as waveguides, provided that bulk and surface scattering losses can be made sufficiently small. Interest in such guides arises largely from the potential to combine optical (waveguides) and optoelectronic (e.g., photodiodes) structures in a single monolithic chip using the OI heterojunction. To reduce scattering losses, the films must have extremely smooth surfaces with features that are small compared to the wavelength of the guided waves. It has been found that very smooth surfaces can be obtained by depositing the organic material onto substrates that have been cooled to below 100°K. Deposition at higher temperatures results in the formation of unacceptably rough surfaces that result in a large outcoupling of guided light.

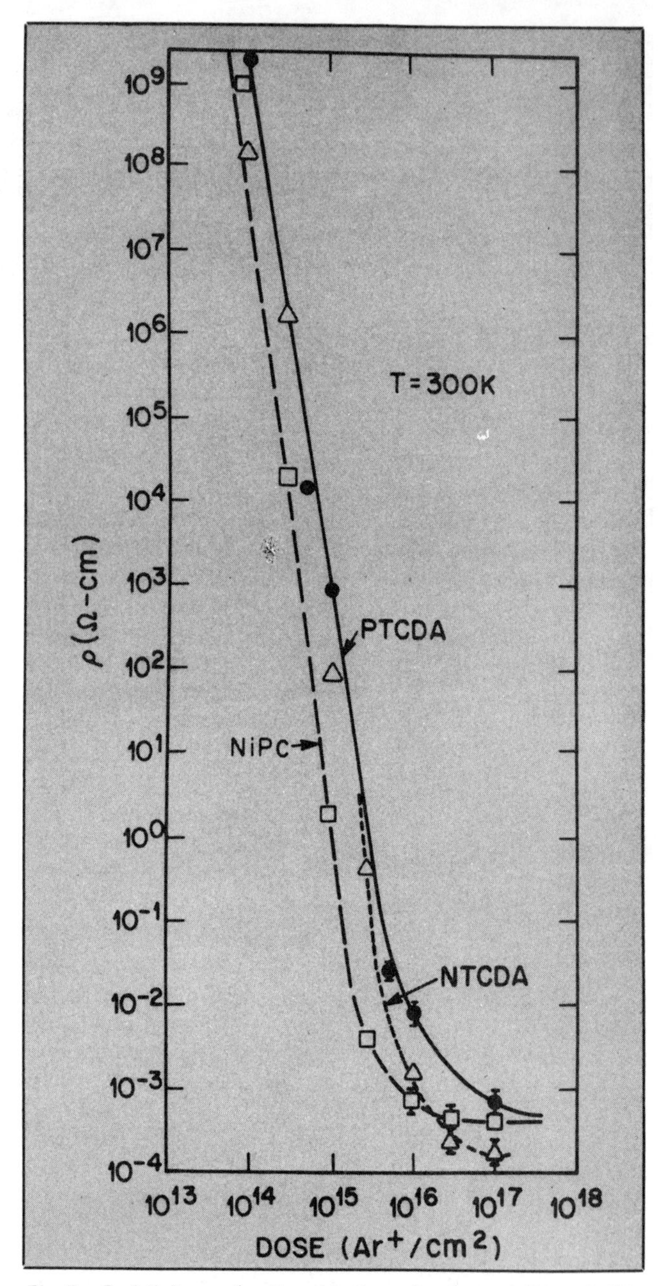

Fig. 7: Resistivity as a function of ion beam dose for several organic thin film semiconductor materials.

OI Field-Effect Transistor

A second interesting device is the "OISTR", or Organic-on-Inorganic Semiconductor Transistor shown schematically in Fig. 6. Here, an organic contact was used to replace the conventional gate contact in a metal-semiconductor gate field-effect transistor (or MESFET). This latter structure has unacceptably high gate leakage currents when fabricated using InP and related compounds for the channel material. For this particular device, after the channel and contact layers were formed by Si implantation into a semi-insulating InP substrate, a SiN_x etch mask was deposited and patterned to form the gate region. Using etchants that expose (111) crystallographic planes, a groove was formed in the substrate as shown in Fig. 6, with the top opening only 1 μm wide. By deposition of the organic material (*N, N'* dimethyl-3,4,9,10 pereylenetetracarboxylic diimide) and the gate metal through this opening, a short (1 μm) gate was formed without the use of wet chemical processes and photolithography that can lead to potential damage of the relatively soft crystalline organic film.

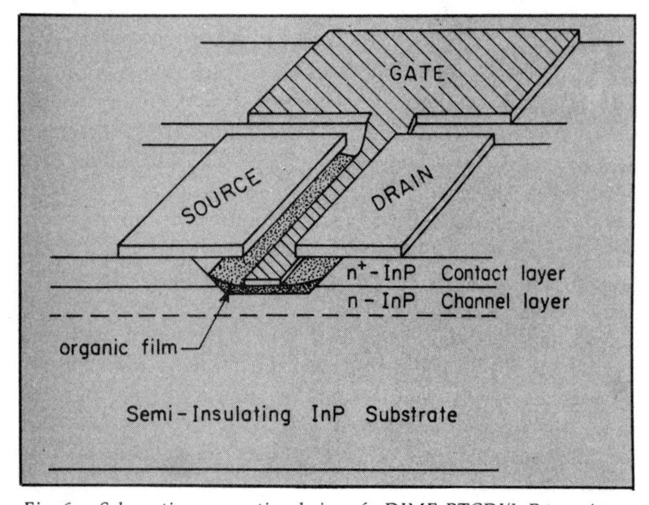

Fig. 6: Schematic cross-sectional view of a DIME-PTCDI/InP transistor, or OISTR.

This device exhibited transconductance values comparable to similar InP FET structures, along with a gate leakage current approximately 20 times less than has been observed for conventional InP MESFETs. More important, however, was the observation that the drain current characteristics of these devices were stable with time, exhibiting none of the slow drift in transconductance and saturation current typical of insulated gate transistors (or MISFETs) made using InP as the channel material.

Irradiation of Organic films

One additional aspect of organic thin films is that their conductivity can be increased by nearly 14 decades when exposed to high energy ion beams. Figure 7 shows the dependence of film resistivity on 2 MeV Ar^{+} ion dose for several different crystalline organic materials. The minimum or "saturation" resistivity is roughly independent of material composition and attains a value of $1-5 \times 10^{-4}$ Ω-cm. This value is approximately that of amorphous metals, and suggests that the irradiated material is carbonized with a high density of microscopic graphite crystallites that enhance the conductivity to well above that of amorphous carbon. Both the chemical and optical properties of the films are changed along with the conductivity. For example, whereas the sublimation temperature of PTCDA is 500°C, the irradiated material is stable to temperatures exceeding 1000°C. Futhermore, at relatively small ion (or electron) doses of 10^{13} cm^{-2}, the material begins to lose its brilliant natural color and begins to darken, suggesting a concomitant change in index of refraction. Thus, irradiation with energetic particle beams provides a means to directly "write," into the organic thin film, gratings and other optical devices, along with very fine electrical interconnects. When we combine this feature along with the waveguiding and active optoelectronic properties of these materials, a considerable array of device functions can potentially be integrated onto a single chip.

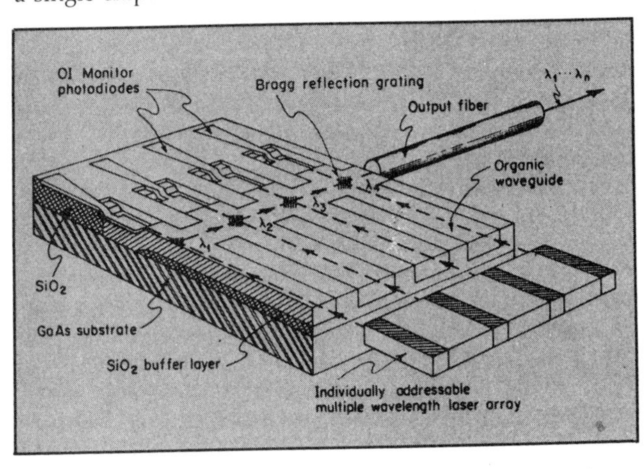

Fig. 8: A conceptual view of an organic-on-inorganic semiconductor wavelength multiplexer array.

Future Prospects

As a "gedanken" device that utilizes many of the diverse features of OI heterojunctions, consider the wavelength multiplexer shown in Fig. 8. Here, light of several different wavelengths is edge-coupled from a multi-wavelength laser array into a matching array of organic waveguides. To form the guides on a semiconductor substrate, the high index of refraction organic material is deposited onto a lower index dielectric buffer layer consisting of, for example, SiO_2 or SiN_x. The array of guides all intersect, at a 90° angle, a similar organic guide. At each waveguide intersection is inscribed, by direct e-beam or ion beam writing, Bragg reflection gratings, each tuned to the wavelength incident from the waveguide array into a given intersection. The effect of these gratings is to bring all of the various wavelength signals into a single coaxial path that is then efficiently coupled into an output transmission fiber. Since the efficiency of each grating depends on its length, we expect there to be some leakage of light from the incident lasers into extensions of the waveguides beyond the grating region. In these extensions, the dielectric buffer layers are removed, hence allowing for the absorption of light into the semiconductor substrate. By placing an electrical contact pad over the absorption region, a highly efficient high bandwidth array of photodetectors can be formed that can be employed to monitor the laser output light in such a way that intensity stabilization of the lasers can be achieved over a broad operating temperature range.

Such a device illustrates only one of the many ways in which organic thin films can be used in devices that perform a variety of optoelectronic functions. The excellent electronic properties of the OI heterojunction, coupled with the ease with which these devices can be fabricated without regard for the constraints which lattice matching between contacting materials places upon conventional inorganic semiconductors, makes the former materials a worthwhile subject of investigation. However, as in the case of all new semiconductor materials, there are still a considerable number of problems to be solved before they can be seriously considered for use in real systems. For example, at present, means for passivating OI diodes so that they are stable under many different environmental conditions have not been established. Furthermore, due to the relative softness of these materials, application of the photoresists needed in wet processing of semiconductors will undoubtedly damage the crystalline films. Hence, dry processing techniques and patterning to only 1 or 2 μm need to be developed. As exemplified by the OISTR, one means of forming small structures is to first form patterns in a semiconductor substrate, then "replicate" these small dimensions in the organic material by directional vacuum deposition of the organic material onto the preformed patterns. Finally, robust and reliable means for contacting the organic films must be established.

In spite of the difficulties that remain, it appears that crystalline organic thin film semiconductors present both a challenge and a promise for a new generation of versatile optoelectronic components to the device engineer.

Acknowledgements

The author thanks F. F. So, M. L. Kaplan and P. H. Schmidt who have contributed enormously to his understanding and appreciation of organic semiconductors. Also, the author is indebted to the Air Force Office of Scientific Research (C. Litton) and Rome Air Development Center (J. Lorenzo) and a grant from the Powell Foundation for their support of this work.

References

[1] For a review on other applications of OI devices, see S. R. Forrest, M. L. Kaplan and P. H. Schmidt, *Ann. Rev. Mat. Sci.*, 17, 189, 1987.

[2] For details on OI detectors, see F. F. So and S. R. Forrest, *IEEE Trans. Electron Dev.*, 36, 66, 1989.

[3] For the effects of beam irradiation, see M. L. Kaplan, S. R. Forrest, P. H. Schmidt and T. Venkatesan, *J. Appl. Phys.*, 55, 732, 1984.

[4] For a review of the details of transport and optical properties of organic materials, see F. Gutman and L. E. Lyons, *Organic Semiconductors*, New York, John Wiley & Sons, 1967.

[5] For a review of optical devices similar to that shown in Fig. 8, see H. Nishihara, M. Haruma, and T. Suhara, *Integrated Circuits*, New York: McGraw–Hill, 1989.

Semiconductor Optical Amplifiers

Gadi Eisenstein

Abstract

Semiconductor optical amplifiers are emerging as important componenets in many optical communication, switching, and signal processing systems. An overview of semiconductor optical amplifier characteristics and their potential applications is presented.

Introduction

Recent demonstrations of high performance semiconductor optical amplifiers, and the successful incorporation of such amplifiers in laboratory-type systems, have advanced the prospects for practical use of these important devices in future optical communication, switching, signal processing, and other optical systems. The research on semiconductor optical amplifiers is currently driven by the needs of optical fiber communications systems. [1-3] Therefore, most reported devices are made of InGaAsP and operate in the 1.3-1.55 μm wavelength range, which is the wavelength window where optical fibers have their most favorable properties: minimum loss at 1.55-μm and zero dispersion at 1.3-μm. There have also been reports of AlGaAs operating at ~0.85 μm.

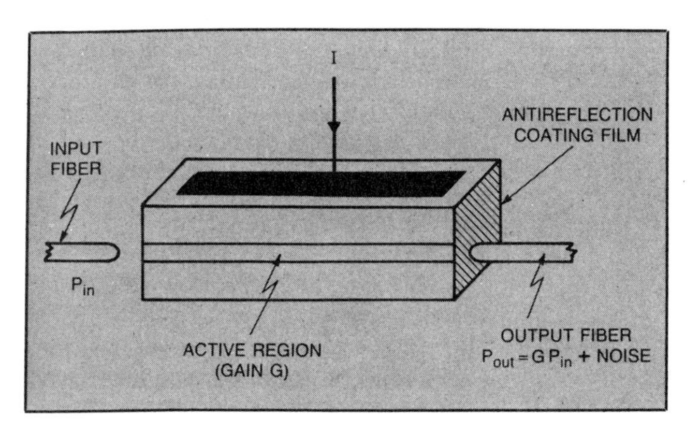

Fig. 1 Schematic representation of a semiconductor optical amplifier.

The optical amplifier shown schematically in Fig. 1 consists of a laser diode with both its facet reflectivities reduced to very low levels. The device contains several epitaxially grown layers. The most important is the active region into which carriers are injected from an external bias source. The injected carriers are confined to the active region because it is surrounded by layers of material with higher energy band gaps. If a sufficient number of carriers are injected, then an optical signal impinging on the active region will induce stimulated emission and consequently be amplified. The active region also serves as the core of a waveguide in which the light, propagating along the device as it is amplified, is confined. In a semiconductor laser, the end facets are cleaved perpendicular to the waveguide. The semiconductor-air interface forms a 35 percent mirror which provides the feedback required for laser oscillations. In an amplifier, the reflectivity (feedback) has to be eliminated so that oscillations are prevented and single-pass gain, sometimes called traveling-wave operation, is obtained.

The reduction of facet reflectivity is most often achieved by depositing a dielectric anti-reflection coating film. [4] Other, less common, methods to reduce the reflectivity are tilting the waveguide direction with respect to the facet [5] or adding a non-guiding region, of a few μm length, between the waveguiding region and the diode facet (window structure). [6] It is the lack of a mature technology to reduce the facet reflectivity that is principally responsible for the lag in the development of practical semiconductor optical amplifiers behind the development of sophisticated laser diodes. Recent advances in dielectric coating technology have led to a proliferation of high quality research devices and consequently, to many important experiments that have confirmed most previously developed theoretical models and predictions regarding the performance of these amplifiers. The availability of practical semiconductor optical amplifiers and their emergence as routine components incorporated in various systems is expected in the next few years.

Characteristics of Semiconductor Optical Amplifiers

The most common way to use a semiconductor optical amplifier is shown in Fig. 1. The device is electrically driven by a dc bias (in some cases it may have an ac drive) and has optical input and output ports consisting of optical fibers coupled to the amplifier waveguide. A weak optical signal P_{in} is injected into the amplifier and emerges at the output as a larger signal with added noise, $P_{out} = GP_{in} + P_{noise}$. The amplifier provides gain G over a large bandwidth. As long as it operates in the linear regime, it is practically transparent to the wavelength, modulation format and frequency, as well as the power of the input signal.

The overall small-signal gain spectrum of an optical amplifier is very broad (approximately 4000 GHz) and is determined by the semiconductor material properties. The broad gain function is modulated by Fabry-Perot modes resulting from the finite residual reflectivities of the amplifer facets. The gain spectrum of an amplifier with single-pass small-signal gain g, length L, and facet reflectivities R_1 and R_2 is [7,8]

$$G(f) = \frac{kg(1-R_1)(1-R_2)}{(1-g\sqrt{R_1R_2})^2 + 4g\sqrt{R_1R_2}\sin^2\left[\frac{2\pi fnL}{c}\right]}$$

with k being the total coupling loss, c the velocity of light in vacuum, f the optical frequency, and n the effective index of refraction in the amplifier waveguide. The modulation of the gain has a period determined by the amplifier length. For typical lengths of 250-500 μm in the wavelength range

Reprinted from *IEEE Circuits and Devices Magazine*, Vol. 5, No. 4, pp. 25-30, July 1989.

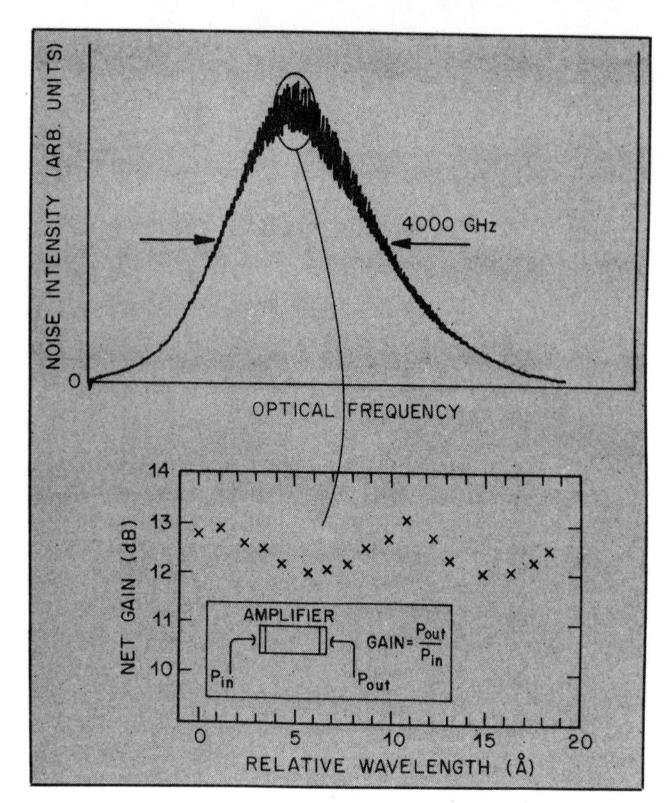

Fig. 2 Measured gain spectrum of a semiconductor optical amplifier.

of 1.3-1.5 μm, the spacing between peaks is 50-150 GHz so that there are approximately 30-100 modes within the overall gain bandwidth of the amplifier. The depth of the gain modulation is determined by the product $g\sqrt{R_1R_2}$. The ideal amplifier has zero reflectivity and no gain modulation. Practical amplifies have some residual facet reflectivity hence a finite gain modulation. An examination of the expression for the gain reveals the requirement for the very low reflectivities. For example, if a net gain of 20 dB is required and the total coupling losses are approximately 10 dB, then, if a 3 dB gain modulation is acceptable, the reflectivity must be 2×10^{-4}. If a 1 dB gain modulation is required, the reflectivity has to be 5×10^{-5}.

The most common method to achieve low facet reflectivities is by high quality anti-reflection coating. The parameters of an anti-reflection coating film matched to a laser waveguide are different from the conventional quarter wavelength film. Several theoretical designs of antireflection coating films for laser facets [9] as well as experimental demonstrations of reflectivities on the order of 10^{-4} have been published [3,10-12] by researchers from AT&T Bell Laboratories, CNET, NTT, and BTRL. Fig. 2 shows a 4000 GHz wide measured noise spectrum (which is proportional to the gain spectrum) of an amplifier operating near 1.3 μm. The figure also shows, in detail, the gain and gain variations (which are ± 1 dB) of one Fabry Perot mode. The corresponding facet reflectivities in this device are 3×10^{-4}. The dielectric coatings are processed with electron-beam or thermal evaporations or by RF sputtering. The most common materials are SiO_x, ZrO_x, Si_xN_y, PbO_x/SiO_y, all of which are non-stoichiometric materials having an index of refraction that can be adjusted using the deposition conditions. The film thickness is usually controlled by sophisticated in-situ monitoring techniques. [4,10] Other methods to obtain low reflectivities are tilted laser waveguides [5] or window structures [6] in combination with simple anti-reflection coatings. Such amplifiers have recently been demonstrated at Bellcore, AT&T Bell Laboratories, Imperial College London, STL Research Laboratories, and NEC.

The single-pass small-signal gain g is a very weak function of optical frequency. [7] Its value is determined by the amplifier material, structure, and driving conditions. In a given optical frequency range, it has the general form [7]

$$g = \exp\{L[\Gamma g_0(N-N_0)-\Gamma\alpha_1]-\alpha_2 L\}$$

in which L is the amplifier length, g_o is a linear gain coefficient which is a device parameter, N_o is the carrier density required for transparency, N is the carrier density in the active region which is related to the drive current, Γ is the mode confinement factor which is determined by the amplifier waveguiding characteristics, and α_1 and α_2 are optical losses in and out of the waveguiding region, respectively. The dependence of the gain on Γ, the waveguide confinement factor, gives rise to a major problem with most semiconductor optical amplifiers, namely, the dependence of the gain on polarization. Since to date most amplifiers have been obtained by anti-reflection coating of conventional semiconductor lasers, and since the confinement factor in injection lasers is significantly larger for light polarized in the junction plane (TE) than for light polarized in the orthogonal polarization (TM), most amplifiers have a polarization-dependent gain. To overcome this adverse property of amplifiers, new gain structures with more symmetric waveguides are now being designed and processed, with very promising results [3] obtained in several laboratories such as CNET and AT&T Bell Laboratories. In amplifiers with low reflectivities, the gain g may be as high as 30-35 dB for a 500 μm long device. The usable gain is reduced by coupling losses that are typically 3-5 dB per facet. Significant advances in permanent coupling of fibers to the input and output of amplifiers have been reported by BTRL.

For large input signals, the amplifier gain is reduced due to nonlinear effects. If the gain is compressed due to a strong input pulse, it takes a finite time to recover to its initial value. If a second pulse arrives before the gain has fully recovered, it will experience a reduced gain and cause errors in a communication system. The gain recovery time is determined by complicated carrier dynamics which have relatively slow time constants. Measurements of gain recovery time in semiconductor optical amplifiers have recently been performed [13] at AT&T Bell Laboratories. The shortest measured times are on the order of 100 ps. The nonlinear gain compression also limits the output power that can be extracted from a semiconductor optical amplifier. The maximum output power can be increased by reducing the gain recovery time.

Semiconductor optical amplifiers generate amplified spontaneous emission noise which is added to the amplified signal. This noise is spread over the bandwidth of the gain spectrum (several thousand GHz). In general, most of this noise can be filtered out using optical filters such as gratings, passive or active Fabry-Perot etalons, electro-optic filters, etc. In coherent systems, the filtering is an inherent part of the detection mechanism. The noise bandwidth after filtering can be as narrow as the bandwidth of the information carried by the modulated optical signal. The impact of the added noise depends on the amplifier application. The noise and amplified signal are detected simultaneously and generate a mixed noise current which may affect the

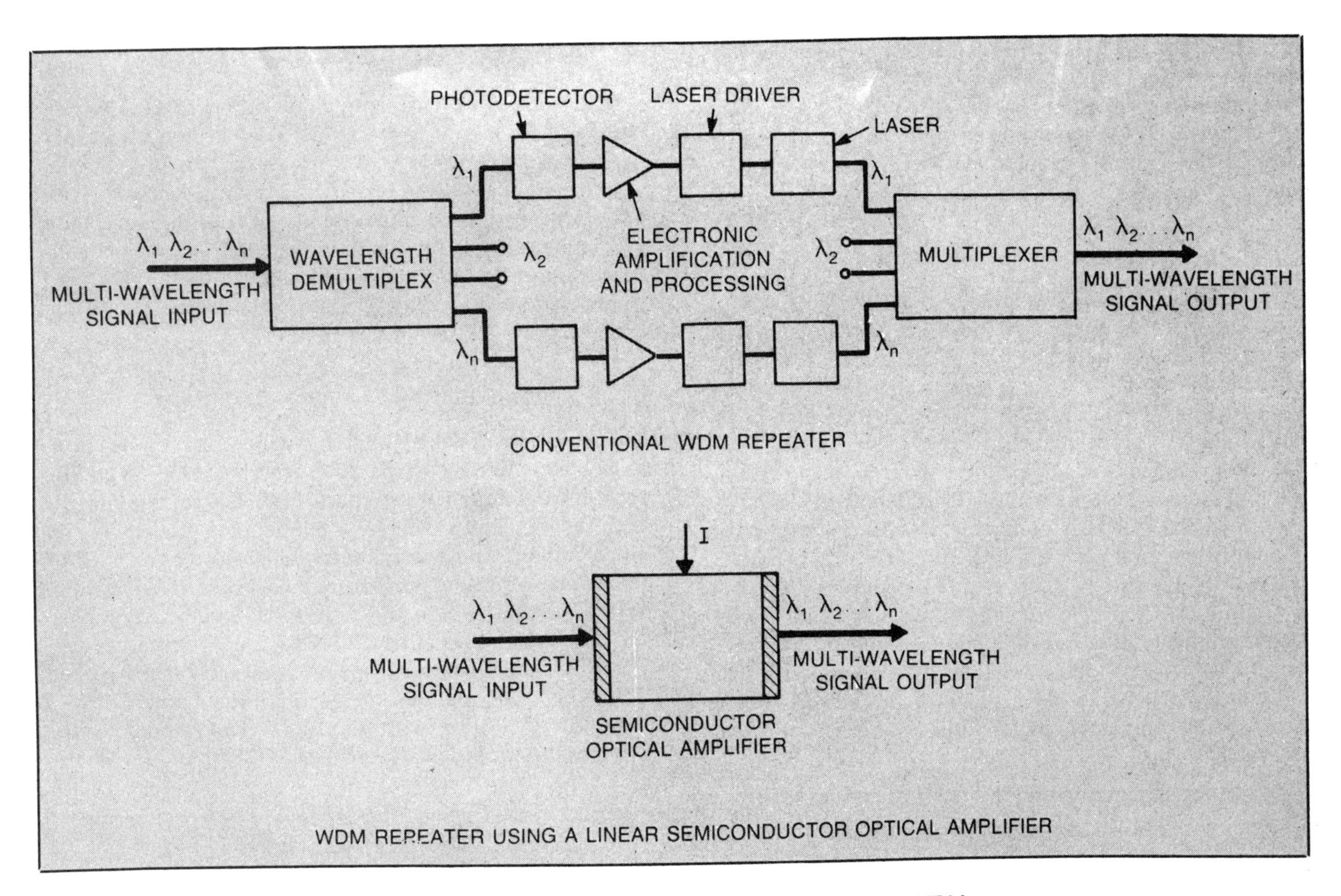

Fig. 3 Comparison between a conventional WDM repeater and a WDM repeater that employs a single linear optical amplifier.

detection sensitivity. There are two major types of mixing products. [7,14] For low amplifier input power, the main mixing product is due to beating between different portions of the noise spectrum and is called spontaneous-spontaneous beat noise. For large amplifier input signals, the dominant noise is due to beating between the signal and the spontaneous emission noise and is the favorable mode of operation. The effects of the amplifier noise on a system are usually characterized by the degradation in signal to noise ratio due to the amplifier. This degradation (usually called the noise figure F) is equal to 3 dB for an ideal amplifier. In any practical amplifier $F>3$ dB. For high quality amplifiers, operating under high gain conditions, $F \sim 2n_{sp}$. The factor n_{sp} is a device parameter called the population inversion parameter which is equal to the ratio of spontaneous to stimulated transition rates in the amplifier. The effective noise figure of a practical amplifier equals F plus the input coupling loss. The lowest noise figure for an amplifier to date was reported by NTT to be (excluding the coupling loss) $F = 5.2$ dB (corresponding to $n_{sp} = 1.65$) under the condition of 20 dB gain.[10]

Applications of Semiconductor Optical Amplifiers

The list of potential applications for semiconductor optical amplifiers is very long. The most commonly suggested and experimentally investigated applications are described below.

Non-Regenerative Repeater

In long-haul optical transmission systems, repeater spacing is determined by either fiber dispersion or fiber loss. In the latter case, a linear optical amplifier can be used to compensate for the loss and thereby avoid the need for complicated regenerative repeaters. Such in-line amplifiers have been demonstrated in systems using either direct or coherent detection schemes. The number of amplifiers that can be cascaded is limited by the accumulation of amplifier noise and by fiber dispersion. Using four optical amplifiers placed approximately every 70 km, AT&T Bell Laboratories researchers have transmitted a 1 Gb/s signal over 310 km in a direct detection system [15] and 400 Mb/s over 370 km in a coherent system.[16] The main advantage of in-line amplifiers is in multi-channel, wavelength division multiplexed (WDM) systems. In such systems, a single amplifier can amplify simultaneously all channels. In contrast, conventional WDM systems require a demultiplexer, a regenerator for each channel and a multiplexer at each repeater. The two types of WDM-system repeater are illustrated in Fig. 3. The advantage of the simple in-line amplifier repeater is quite dramatic, particularly when the number of channels is large. A variety of laboratory-type WDM systems using a single optical amplifier have been demonstrated. Most notable is a ten-channel coherent system experiment operating at $\lambda = 0.85$ μm performed at the Heinrich Hertz Institute [17] and a four channel, densely spaced coherent system operating at $\lambda = 1.3$ μm demonstrated at AT&T Bell Laboratories. [18] Simultaneous amplification of a multi-wavelength signal may cause crosstalk between the channels. [18-20] This crosstalk results from non-linear effects such as four-wave mixing, intermodulation distortion, and gain saturation. It can be minimized by increasing channel spacing, limiting the input power to the amplifier, and by using frequency shift or phase shift keying rather than intensity modulation as the data encoding scheme.

Optical Receiver Pre-Amplifier

Optical receivers usually employ either PIN or APD photodetectors in conjunction with an FET. The largest gain-bandwidth demonstrated for an InGaAs APD is ~ 70 GHz. For bandwidths of up to a few GHz, the most sensitive receivers employ InGaAs APDs. However, for wider bandwidths, the APD gain bandwidth is not sufficient and a fast PIN photodetector in combination with an optical pre-amplifier [1] (which may have up to 20 dB gain with a bandwidth of ~4000 GHz) is preferable. A typical input stage of a receiver employing an optical pre-amplifier is shown in Fig. 4. High efficiency coupling to the amplifier input is of paramount importance in this application, and the receiver requires a filter to eliminate the wideband amplifier noise. Experimental demonstrations of receivers with sensitivities that are higher than the best reported APD based receivers for comparable bandwidths have been reported.

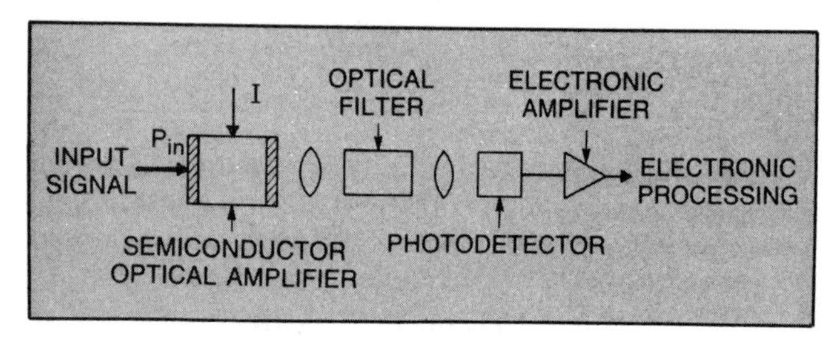

Fig. 4 Schematic of an optical receiver employing a semiconductor optical preamplifier.

Examples include a 10 GHz receiver reported by researchers from BTRL [1] as well as 4 Gb/s and 8 Gb/s receivers demonstrated at AT&T Bell Laboratories. [11,21]

Power Amplifier and Amplifier/Modulator

Power amplifiers and amplifier/modulators are placed in optical transmitters. Power amplifiers may be used to compensate for losses due to optical components such as external $LiNbO_3$ modulators, wavelength or time multiplexers, power splitters, etc. Power amplifiers have been demonstrated in conjunction with a coherent transmission system by BTRL [22] and in an 8-Gb/s optical time division multiplexed experiment performed at AT&T Bell Laboratories. [23] Researchers from Bellcore have used a power amplifier to compensate for signal splitting losses in a subcarrier multiplexed video distribution system. [24] The limiting factor in power amplifiers is usually gain and power saturation. Amplifier/modulators speeds are determined by the modulation bandwidth of the amplifiers, which are limited by carrier dynamics to several GHz. The main attraction of amplifier/modulators lies in the potential to monolithically integrate them with the laser source. Amplifier/modulators operating in conjunction with short pulse lasers have been demonstrated at AT&T Bell Laboratories.

General Optical Gain Block

Semiconductor optical amplifiers can be inserted in various systems to compensate for losses of optical components and for splitting losses in networks. For example, researchers at Ericsson Telecommunication Research Laboratory have demonstrated the use of an optical amplifier to compensate for the losses of a $LiNbO_3$ optical circuit. Semiconductor optical amplifiers are also used as the gain element in linear or ring configuration external cavity lasers.

Narrow Band Tunable Amplifiers

Advances in tunable single-frequency laser diodes such as DFB and DBR lasers (which incorporate a built-in grating section for frequency selection) have led to experiments with devices which combine the functions of filtering and amplification. These amplifiers, which also generate less noise than wideband amplifiers, have been incorporated in WDM systems where they serve as active demultiplexers.[25] Researchers at CNET, NEC, and Bellcore have reported a variety of experiments with such narrow band amplifiers. The operation of a narrow band amplifier as a tunable active filter (with gain) for WDM demultiplexing applications is shown schematically in Fig. 5.

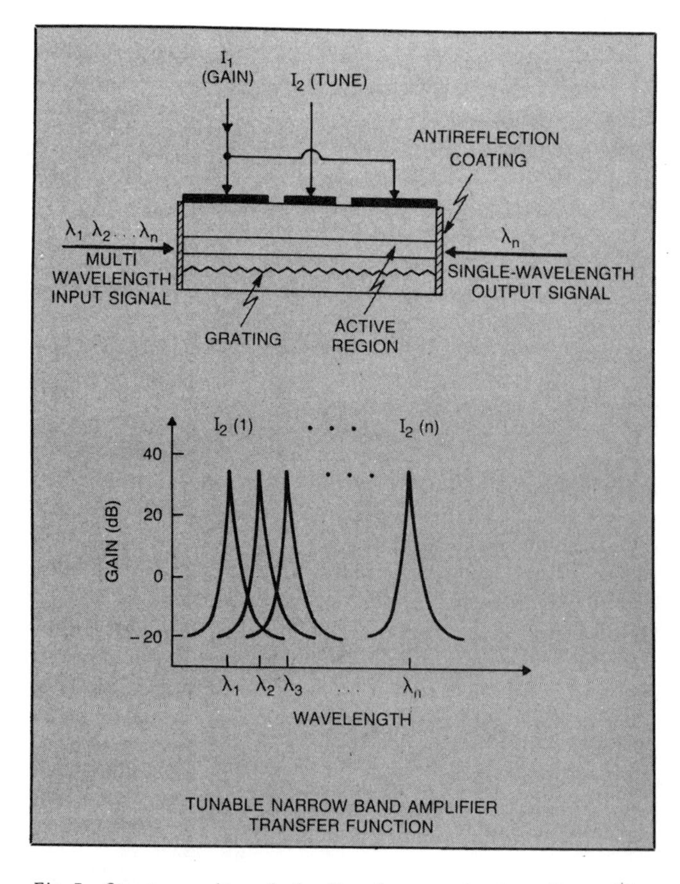

Fig. 5 Structure and transfer function of a narrow band tunable amplifier operating as a WDM demultiplexer.

Bi-Stable Amplifiers

Under some conditions, the amplifier non-linear characteristics may be used to perform useful functions. Optical amplifiers which exhibit bi-stability have been used to dem-

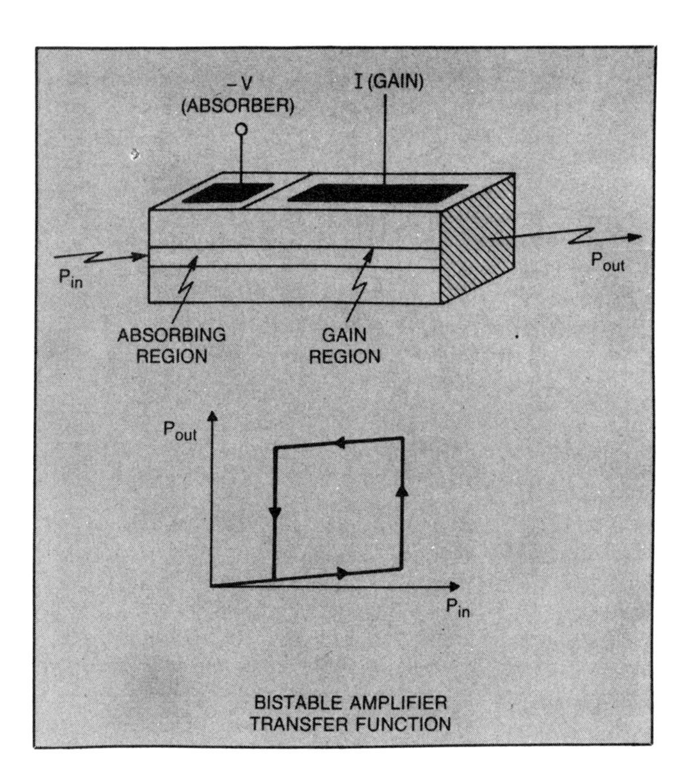

Fig. 6 Structure and transfer function of a two-section bi-stable optical amplifier.

onstrate some basic logic functions required for optical processing, switching, and all-optical fully regenerative repeaters. Bi-stability has been demonstrated in two types of amplifiers. The first is a simple uncoated laser diode operating below threshold (sometimes called a Fabry-Perot amplifier). Such an amplifier has a highly resonant gain transfer function which is very sensitive to the intensity of the input signal hence is very non-linear and exhibits bistability. These simple bi-stable amplifiers have been demonstrated at BTRL, GTE, and NTT. A more advanced bistable amplifier, shown schematically in Fig. 6, employs two integrated sections. [26] The first performs a nonlinear thresholding function by saturable absorption, and the second provides gain. This type of bi-stable amplifier has been used for switching and pulse shaping by BTRL researchers. The major limitations of bi-stable amplifiers are relatively low speed [27], difficulties in obtaining large extinction ratios, and stability.

Future Trends

Semiconductor optical amplifiers are likely to become the next opto-electronic component to have a significant impact on optical communication, switching, and signal processing systems. It is expected that the designers of future optical systems will be able to incorporate optical amplifiers into their systems at will, in a manner similar to the way amplifiers are incorporated in present day electronic systems.

Several technologies have to be perfected before practical semiconductor optical amplifiers become abundant. The first is a batch process to obtain low reflectivity facets. To date, all reported high quality amplifiers were processed as carefully controlled individual devices. Recent advances in high reflectivity coatings of mirrors, which produced commercially available Fabry-Perot etalons having a low loss mirror with reflectivities of 99.999 percent, have proven that dielectric coatings can be controlled to very tight tolerances. Similar control methods may be employed in a batch process for anti-reflection coatings on laser diode facets. Other possibilities include further developments of tilted facet and window structures. The second problem which needs to be addressed is the design and fabrication of symmetric waveguide gain structures that will overcome the unacceptable polarization dependence of the gain. The issue of amplifier packaging is already quite well developed with fiber pig-tailed amplifiers whose packages also includes isolators being made available for commercial use. Finally, semiconductor optical amplifiers will very likely play a major role in integrated photonic circuits, where they may be used without the need to anti-reflection coat them and where coupling in and out of them will be accomplished by well matched, monolithically integrated waveguides. [28] Moreover, all the necessary optical filters as well as the electronic drive and monitoring circuits will probably be integrated on the same chip.

References

[1] M. J. O'Mahony, "Semiconductor Laser Optical Amplifiers for Use in Future Fiber Systems," *J. Lightwave Technology*, vol. 6, no. 4, p. 531, 1988.

[2] T. Saitoh and T. Mukai, "1.5 μm GaInAsP Traveling-Wave Semiconductor Laser Amplifier," *IEEE J. of Quantum Electronics*, vol. 23, no. 6, p. 1010, 1987.

[3] J. C. Simon, "InGaAsP Semiconductor Laser Amplifiers for Single-mode Fiber Communications," *J. of Lightwave Technology*, vol. 5, no. 9, p. 1286, 1987.

[4] G. Eisenstein, G. Raybon, and L. W. Stulz, "Deposition and Measurement of Electron Beam Evaporated SiO_x Antireflection Coatings on InGaAsP Laser Facets," *J. Lightwave Technology*, vol. 6, no. 1, p. 12, 1988.

[5] C. E. Zah, C. Caneau, F. K. Shokoohi, S. G. Menocal, F. Favire, L. A. Reith, and T. P. Lee, "1.3 μm Near-Traveling-Wave Laser Amplifiers Made by Combination of Angled Facets and Antireflection Coatings," *Electron. Lett.*, vol. 24, no. 20, p. 1275, 1988.

[6] I. Cha, M. Kitamura, and I. Mito, "1.5 μm Band Traveling-Wave Semiconductor Optical Amplifiers with Window Facet Structure," *Electron. Lett.*, vol. 25, no. 3, p. 242, 1989.

[7] T. Mukai, Y. Yamamoto, and T. Kimura, "Optical Amplification by Semiconductor Lasers," in *Semiconductors and Semimetals*, vol. 22, part E, p. 265, 1985.

[8] G. Eisenstein and R. M. Jopson, "Measurements of the Gain Spectrum of Near Traveling-wave and Fabry-Perot Semiconductor Optical Amplifiers at 1.5 μm," *International J. Electronics*, vol. 60, no. 1, p. 113, 1986.

[9] G. Eisenstein, "Theoretical Design of Single-Layer Antireflection Coatings on Laser Facets," *AT&T Bell Laboratories Technical J.*, vol. 63, no. 2, p. 357, 1984.

[10] T. Saitoh and T. Mukai, "Recent Progress in Semiconductor Laser Amplifiers," *J. Lightwave Technology*, vol. 6, no. 11, p. 1656, 1988.

[11] N. A. Olsson, M. G. Oberg, L. D. Tzeng, and T. Cella, "Ultra-low Reflectivity 1.5 μm Semiconductor Laser Preamplifier," *Electron. Lett.*, vol. 24, no. 9, p. 569, 1988.

[12] G. Eisenstein, B. C. Johnson, G. Raybon, "Traveling-wave Optical Amplifier at 1.3-μm," *Electron. Lett.*, vol. 23, no. 19, p. 1020, 1987.

[13] G. Eisenstein, R.S. Tucker, J. M. Wiesenfeld, P. B. Hansen, G. Raybon, B. C. Johnson, T. J. Bridges, F. G. Storz, and C. A. Burrus, "Gain Recovery Time of Traveling-wave Optical Semiconductor Optical Amplifiers," *Appl. Physics Lett.*, vol. 54, no. 5, p. 454, 1989.

[14] Y. Yamamoto, "Noise and Error Rate Performance of Semiconductor Laser Amplifiers in PCM-IM Optical Transmission Systems" *IEEE J. Quantum Electron.* vol. 16, no. 10, p. 1073, 1980.

[15] M. G. Oberg, N. A. Olsson, L. A. Koszi, and G. Przybylek, "313 km Transmission Experiment At 1 Gb/s Using Optical Amplifiers and a Low Chirp Laser," *Electron. Lett.*, vol. 24, no. 1, p. 38, 1988.

[16] N. A. Olsson, M. G. Oberg, L. K. Koszi, and G. Przybylek, "400 Mb/s 372 km Coherent Transmission Experiment Using In-line Optical Amplifiers," *Electron. Lett.*, vol. 24, no. 1, p. 36, 1988.

[17] R. P. Braun, R. Ludwig, and R. Molt, "Ten-Channel Coherent Fiber Transmission Using an Optical Traveling-Wave Amplifier," Post Deadline Paper, *European Conf. on Optical Comm.*, Barcelona, 1986.

[18] B. Glance, G. Eisenstein, P. J. Fitzgerald, K. J. Pollock, and G. Raybon, "Optical Amplification in a Multichannel FSK Coherent System," *Electron. Lett.*, vol. 24, no. 18, p. 1157, 1988.

[19] K. Inoue, T. Mukai, and T. Saitoh, "Nearly Degenerate Four-Wave Mixing in a Traveling-Wave Semiconductor Laser Amplifier," *Appl. Physics Lett.*, vol. 51, no. 14, p. 1051, 1987.

[20] T. E. Darcie and R. M. Jopson, "Nonlinear Interactions in Optical Amplifiers for Multifrequency Lightwave Systems," *Electron. Lett.*, vol. 24, no. 10, p. 638, 1988.

[21] R. M. Jopson, A. H. Gnauck, B. L. Kasper, R. E. Tench, N. A. Olsson, C. A. Burrus, and A. R. Chraplyvy, "8 Gb/s 1.3 μm Receiver Using Optical Preamplifier," *Electron. Lett.*, vol. 25, no. 3, p. 233, 1989.

[22] R. C. Steele, M. J. Creaner, G. R. Walker, N. G. Walker, J. Mellis, S. A. Al-Chalabi, J. Davidson, I. Sturgess, M. Rutherford, and M. Brain, "Field Trial of 565 Mb/s DPSK Heterodyne System Over 108 km," Post Deadline Paper, *European Conf. on Optical Comm.*, Brighton, 1988.

[23] G. Eisenstein, R. S. Tucker, G. Raybon, and S. K. Korotky, "8-Gb/s Transmission over 40 km in a Two-Channel Single-Laser Optical Time Division Multiplexed System Experiment," *Optical Fiber Comm. Conf.*, Houston, 1989.

[24] W. I. Way, C. E. Zah, G. Menocal, N. K. Cheung, and T. P. Lee, "Carrier to Noise Ratio Performance of a Ninety-Channel FM Video Optical System Employing Subcarrier Multiplexing and Two Cascaded Traveling-Wave Amplifiers," *Optical Fiber Comm. Conf.*, Houston, 1989.

[25] M. Nishio, T. Numai, S. Suzuki, M. Fujiwara, M. Itoh, and S. Murata, "Eight-Channel Wavelength-Division Switching Experiment Using Wide-Tuning-Range DFB LD Filters," Post Deadline Paper, *European Conf. on Optical Comm.*, Brighton, 1988.

[26] I. W. Marshall, M. J. O'Mahony, D. M. Cooper, P. J. Fiddyment, J. C. Regnault, and W. J. Devlin, "Gain Characteristics of a 1.5 μm Nonlinear Split Contact Laser Amplifier," *Appl. Physics Lett.*, vol. 53, no. 17, p. 1577, 1988.

[27] M. J. Adams, "Time Dependent Analysis of Optical Bistability in Semiconductors," *Proc. IEE*, Part J, vol. 132, no. 6, p. 343, 1985.

[28] U. Koren, T. L. Koch, B. I. Miller, G. Eisenstein, and R. H. Bosworth, "Wavelength Division Multiplexing Light Source with Integrated Quantum Well Tunable Lasers and Optical Amplifiers," *Appl. Physics Lett.*, vol. 54, no. 17, p. 2056, 1989.

Lasers Primer for Fiber-Optics Users

A bewildering variety of semiconductor lasers are now available. What properties are important for real-world applications?

by Joanne LaCourse

The volume of information concerning semiconductor lasers can be overwhelming. More than 20 manufacturers sell a wide variety of semiconductor-lasers in all price ranges, and the device handbooks grow larger each year. At every conference we hear of new advances in laser performance. In all this "optical noise," it's all too easy for a user to lose sight of the essentials. And the essential question is straightforward: Which semiconductor laser properties are important for your application? That's the question we will explore in this article, and it leaves us with a lot of ground to cover. To narrow the focus, we will concentrate on lasers with wavelengths between 1.3 and 1.55 μm, the range most applicable to fiber-optic communications. In this range, light experiences lower loss in silica fiber; in addition, 1.3 μm is the wavelength of zero dispersion, where the refraction index and, consequently, the speed of propagation are independent of wavelength.

But why semiconductor lasers? Simply, they are ideally suited to fiber-optic systems. Their small size makes it easy to couple their laser light into the small (~8 μm) core of a singlemode fiber. They have low input-power requirements (typically, 50 to 100 mA at 1 volt), and an enviable power conversion efficiency of a few tens percent. They are rugged, long-lived, compact, high-speed, and available at the appropriate wavelengths.

Let's look at some basics. The semiconductor laser is a diode consisting of multiple epitaxial layers grown on an indium phosphide (InP) substrate (Fig. 1). An "active layer" (typically indium gallium arsenide phosphide, InGaAsP) is sandwiched between n- and p-type cladding layers of InP. A bias current applied to metal contacts provides the active layer with a supply of electrons and holes, which are confined to the active layer by the difference in energy gaps between the active and cladding layers. When the electrons and holes recombine, light is generated at a wavelength determined by the active layer's energy gap. The relatively high refractive index of the active layer confines the light to the vicinity of that layer in much the same way as light is confined in the higher-index core of an optical fiber.

The active layer is typically 200 to 500 μm long, with a small cross-section, usually about 0.2 μm thick and 1 or 2 μm wide. In "quantum-well" lasers, the active layer is a stack of thin (~10 nm) low-bandgap layers sandwiched between higher-bandgap barrier layers. These lasers exhibit some improved properties over conventional lasers, including lower threshold currents, higher output powers, and narrower linewidths, but they are not yet commercially available at suitable wavelengths.

Different lasers have different characteristics, which may or may not be suitable for a specific application (see *Meeting Application Requirements*). Let's examine some of the key properties of semiconductor lasers. These include power-current

Reprinted from *IEEE Circuits and Devices Magazine*, Vol. 8, No. 2, pp. 27-32, March 1992.

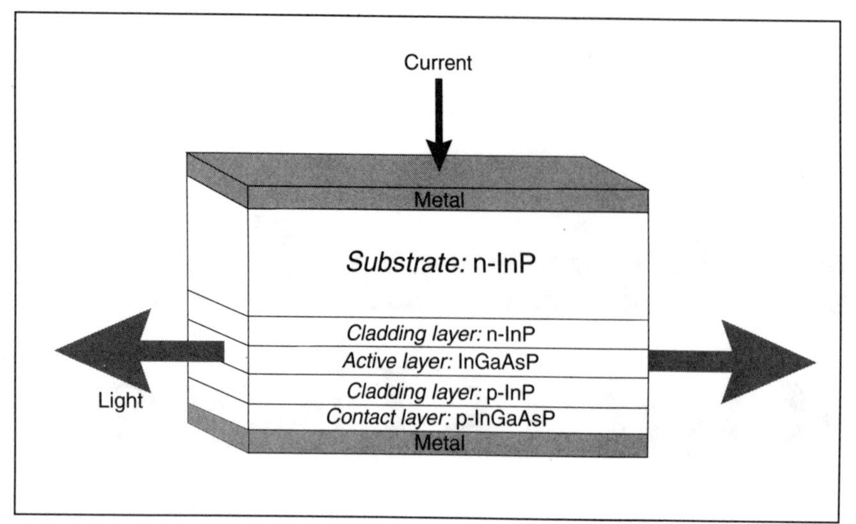

1. The semiconductor laser is a diode consisting of multiple epitaxial layers grown on an indium phosphide (InP) substrate. Light is generated in the active layer when current is applied to the metal contacts.

characteristics, beam shape, spectra, modulation, noise characteristics, and reliability.

Power-Current Characteristics

At low bias currents, the laser acts essentially as a light-emitting diode, with little output power. As the current increases, the optical gain in the active layer also increases until it equals the optical loss. The bias current at this point is known as the lasing threshold current. The device functions as a laser only at currents above threshold. Here, the light output increases more-or-less linearly with bias current (Fig. 2). The slope of the power-vs.-current curve depends on the laser's length and facet coating, and can vary from 0.1 to 0.8 mW/mA. Facet coatings are often used to ensure most of the light is emitted from one facet. For a fiber-coupled laser, the slope is smaller in proportion to the fiber-coupling efficiency.

The optical power is modulated by varying the laser's drive current (Fig. 2). The optimum modulation condition (bias current and modulation depth) are determined by the power-current curve, and the requirements of the particular application. For example, analog systems are sensitive to the distortion that results if the modulation current reaches below threshold, or into the nonlinear regions at high current.

Threshold Current

The threshold current of most commercial lasers is in the range of 5 to 20 mA. Approximately one milliampere has been achieved in quantum-well lasers with high-reflectivity facet coatings, although the total output power is less than 1 mW [1]. The threshold current is not critical for most applications, but very low threshold currents offer the bonus of a simpler drive because no bias current is needed. More important is the operating current needed for the desired amount of optical power, especially for applications with high packing densities, such as optical interconnects, where heat dissipation is a concern.

Output Power

Recommended operating powers of semiconductor lasers are usually several milliwatts. The maximum power is limited by leakage currents associated with the active layer, and by internal heating created by the input current. Unlike shorter-wavelength lasers, these lasers are not subject to optical facet degradation, even at high power. As much as 312 mW has been obtained from specially designed lasers with antireflective coatings on the front facet and reflective coatings on the rear facet [2].

Such high output power is usually not necessary— with one important exception. An erbium-doped, fiber-optic amplifier requires as its power source 0.98µm or 1.48-µm wavelength lasers delivering as much as tens of milliwatts of fiber-coupled power. Pump lasers at these wavelengths are now available commercially.

Temperature sensitivity

One unfortunate but unavoidable characteristic of semiconductor lasers is a decrease in output power at high temperature. This condition arises because the threshold current rises exponentially with temperature, i.e, $\exp(T/T_0)$, where T_0 is typically 65°C for 1.3-µm lasers, and 45°C for 1.55-µm lasers. Thus, in cases where ambient conditions are subject to change, temperature sensors and controllers should be incorporated in the laser module to keep it's temperature constant.

Beam Shape

Unlike the pencil-thin beams associated with gas lasers, semiconductor lasers have highly divergent beams, with full angles of 5-25° parallel to the active layer, and 25-50° perpendicular to it. The divergence is due to diffraction arising from the small size of the light-emitting area. The small size ensures a single spatial lobe, which is essential for good fiber-coupling efficiency, linear power-current curve, and minimum noise. Unfortunately, the asymmetry and divergence of the beam make coupling the laser beam to a fiber more difficult, especially for singlemode fiber. Coupling efficiencies to an as-cleaved singlemode fiber are about 10 percent, so most packaging schemes use a lens or lensed fiber to achieve coupling efficiencies ranging from 25 to 40 percent. Non-fiber-coupled lasers

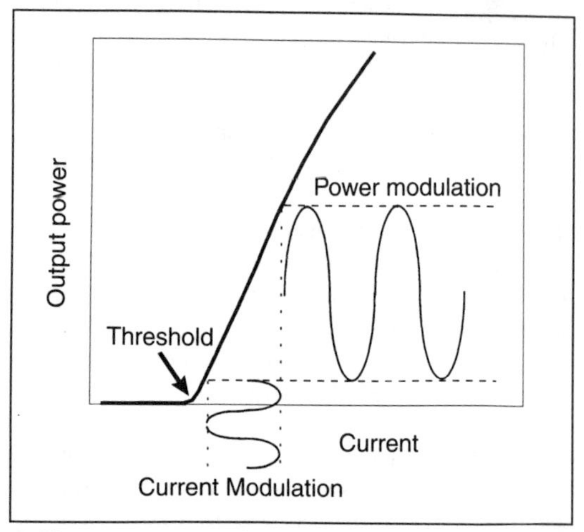

2. The output power of a semiconductor laser is modulated by varying the input current.

are less costly than fiber-coupled types, but they are far more difficult to use because the user is responsible for performing the optical coupling, as well as for providing the protection against physical damage.

Optical Spectrum

The simplest laser structure is the Fabry-Perot cavity, which has cleaved facets at both ends (Fig. 3, top). Fabry-Perot lasers emit at resonant wavelengths called "longitudinal modes" (Fig. 4, top). The total spectral width is about 3 nm, and broadens still further during current modulation.

Single-longitudinal-mode spectra are achieved by using more complicated structures. The distributed feedback (DFB) laser has an internal grating (a layer with periodically varying thickness) that preferentially selects one wavelength (Fig. 3, bottom). In a typical DFB laser's spectrum the peak longitudinal mode (determined by the grating's pitch) is about 30 dB larger than the other modes (Fig. 4, bottom). DFB lasers with 1.3-μm and 1.55-μm outputs are widely available, although at prices significantly higher than those for Fabry-Perot lasers.

There is more to a laser's spectrum than the wavelength of its dominant mode. The broad spectral width of a Fabry-Perot laser produces temporal pulse broadening with increasing distance because of chromatic dispersion, each wavelength traveling at a slightly different speed. This characteristic can be a severe limitation in 1.5-μm, high-bit-rate, long-haul systems. Solutions include the use of DFB lasers or fiber having its zero-dispersion point at 1.5 μm (instead of the usual 1.3 μm). DFB lasers therefore permit a larger bit rate-transmission distance product. DFB lasers also offer the advantage of lower noise, which is critical for AM video systems where the need for a high signal-to-noise ratio sets a severe limit on allowable noise. DFB lasers are also essential to experimental applications where multiple longitudinal modes can't be tolerated. Among these are coherent systems, where the signal is recovered by beating the received power against a local oscillator; and wavelength-division multiplexed (WDM) systems, where a single longitudinal mode is needed to prevent crosstalk between channels. DFB lasers are also critical for experimental wavelength-division-multiplexed (WDM) systems to prevent crosstalk between channels. Channel spacings in such WDM systems range from a few nanometers to a few *hundredths* of a nanometer— a channel spacing so small it is usually described in terms of optical frequency (a few GHz). When channel spacing is small, these frequency-division-multiplexed (FDM) systems require extremely tight frequency stabilization.

Linewidth

A coherent system needs more than just a single longitudinal mode. That mode must have a narrow spectral linewidth— less than 1 percent of the bit rate— for phase-shift keying (PSK), a coding scheme that embodies the signal as a change in optical phase. Similarly, a linewidth of less than 10 percent of the bit rate is required for frequency-shift keying (FSK). The linewidth requirements can be relaxed by using phase-cancellation techniques. Linewidths are broadened by random fluctuations in the electron population, which affect the wavelength through the gain and refractive index. Linewidths of 10 to 100 MHz are typical for commercial DFB lasers, but experimental DFB laser structures have produced linewidths as narrow as 170 kHz

Meeting Application Requirements

Long-haul, digital systems: Commonly used by telephone companies, these systems are optimized by using single-wavelength (DFB) lasers to minimize dispersion penalties. If the laser is directly modulated, low chirp may be desirable (depending on the transmission distance and fiber dispersion).

AM video systems: The commercial systems installed by cable television companies must have SNR > 50 dB, which requires low-noise, high-linearity DFB lasers. RIN must be less than -150 dB/Hz, necessitating optical isolators to suppress reflections. The CSO and CTB distortion must be less than about -60 dBc. Fortunately, the cost of these relatively expensive lasers is shared among many subscribers. (A rival technology— externally modulated, high-power Nd:YAG lasers— is also available for these systems.)

Digital and FM SCM systems: These commercial and field-trialed systems are much more forgiving than AM video systems because they require a SNR of only 16 dB or so. The RIN requirement can be relaxed to less than -135 dB/Hz. Broadband SCM systems need high-speed lasers with a high resonance frequency that is well above the signal band in order to keep RIN and distortion down.

Subscriber systems: Cost and input-power requirements are the major concerns. Lasers must be inexpensive, operate reliably under extreme weather conditions, and consume little input power. Ruggedized commercial laser modules are now available that operate from -40° to +85°C without requiring a thermoelectric cooler (thus reducing power consumption), but the cost is still a few hundred dollars. Since this is potentially a very large market, researchers are examining new ways to simplify the packaging and reduce the cost to a few tens of dollars.

WDM systems: These laboratory-demonstrated systems have channel spacings as small as 1 nm, and currently rely on experimental tunable lasers or preselected commercial DFB lasers for different channels with temperature tuning for fine control. Simple dual-wavelength (1.3 μm and 1.5 μm) WDM systems have relatively loose wavelength requirements.

FDM systems: These highly experimental systems can efficiently utilize the huge potential bandwidth of optical fiber. With channel spacings on the order of 5 GHz (< 0.04 nm), many channels can be transmitted simultaneously. Unfortunately, these complex systems place extraordinary demands on lasers. They require laser tunability and stringent optical frequency stabilization, which has been achieved only in the laboratory.

Pump lasers for erbium-doped fiber-optic amplifiers: These devices have already had a tremendous impact on laboratory research and on systems planned for installation, and they will play an increasing role in the future for applications at 1.5 μm. The required output power from a 0.98 μm or 1.48 μm pump laser depends on the amplifier's use as an in-line, power, or pre-amplifier. Commercial lasers are available, although they do not presently offer adequate reliability at very high power.

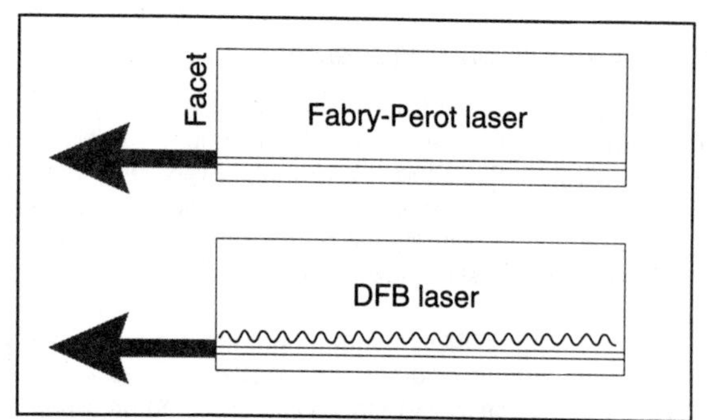

3. The profile of a Fabry-Perot laser reveals uniform layers, unlike the internal grating of a DFB laser.

[3]. The record-breaking results were obtained by using special grating structures, increasing the laser length to over one millimeter, and minimizing the electron-induced change in refractive index relative to gain, which are achieved by using quantum-well lasers or by "detuning"— adjusting the grating pitch of a DFB laser such that lasing occurs on the short-wavelength side of the gain peak. Obtaining narrow linewidths may also require optical isolators, which transmit light in only one direction, to prevent reflections from the fiber back to the laser from exceeding -50 dB.

Wavelength Tunability

WDM and FDM systems require tunable lasers so the transmitter wavelength can be placed in the appropriate channels. Lasers can be tuned over a few tenths of a nanometer by changing the laser temperature, but tuning over 10 nm would be preferred. One approach uses feedback from an external grating. As the grating is moved, the lasing wavelength shifts as much as 80 nm (or more than 200 nm for special quantum- well lasers) but not continuously— some wavelengths inside the tuning range cannot be obtained [4]. This technique has the advantages of commercial availability and narrow linewidths, and the disadvantages of high cost and instability under current modulation. Excellent results have also been obtained with experimental multiple-section internal-grating lasers. The current to one section provides gain while current to a separate "tuning" section determines the wavelength. Researchers have reported tuning ranges of about 2 nm for continuous wavelength tuning and about 10 nm for discontinuous tuning (including the jumps from one longitudinal mode to the next) [4].

Chirp

The linewidth broadens during modulation because of fluctuations in the electron population. This broadening, known as "chirp," increases with modulation frequency and peak-to-peak power swing. Typical values are on the order of 1 nm (~10 GHz) at a few Gb/s, although chirp has been substantially reduced in experimental, detuned DFB quantum-well lasers.

For high-bit-rate (Gb/s), long-haul systems operating at 1.5 μm, temporal pulse broadening due to chirp may limit the bit-rate-transmission-distance product. On the other hand, chirp makes frequency modulation of lasers more efficient; consequently, it can benefit high-speed coherent systems.

Modulation Behavior

Intensity modulation is the most common type of modulation in today's optical communication systems. One example is digital pulse-code-modulated (PCM) baseband systems, in which the laser is directly driven with the baseband signal. In subcarrier-multiplexed (SCM) systems— another example of an intensity-modulated system— the laser is driven by an RF or microwave subcarrier modulated by the message signal. The subcarrier modulation format can be AM, FM, or digital.

The important laser property for these systems is the modulation response (Fig. 5). A flat, high-efficiency, wide-bandwidth response is preferred. The bandwidth of most lasers is limited to a few GHz by electrical parasitics, which include diode resistance, capacitance, and bond-wire inductance. For low-parasitic lasers, the 3-dB bandwidth is normally about 1.5 times the laser's resonance frequency— the frequency at which energy oscillates between the electron and photon populations. The resonance frequency, and consequently the bandwidth, increases with the square root of the output power. The highest 3-dB electrical bandwidths achieved thus far are 24 GHz for a 1.3-μm Fabry-Perot laser, and 17 GHz for a 1.55-μm DFB laser [5,6].

The modulation response tends to peak at the resonance frequency, although the peak may be flattened by rolloff produced by electrical parasitics, or by an intrinsic damping effect arising from interaction between the photon density and the optical gain. In the ideal laser, i.e., one having no electrical parasitics and unlimited power, the ultimate bandwidth is limited by critical damping at high power. Recent reports of very lightly damped quantum-well lasers suggest that the ultimate bandwidth could be as high as 70 GHz [7]. There is plenty of room for improving the 4-to-18-GHz bandwidths available in today's commercial devices.

Surprisingly, most broadband optical systems benefit from using lasers with a much wider bandwidth than the system bandwidth. For modulation frequencies well below resonance, the response is flatter and more efficient; and the relative intensity noise, distortion, and phase deviation are reduced.

Frequency modulation— a more exotic format that includes coherent FSK systems— has been demonstrated only in the laboratory. Modulation of the optical frequency is achieved by the modulation current's effect on the gain and refractive index. A flat, high-efficiency, wide-bandwidth response is desired, but "structure"

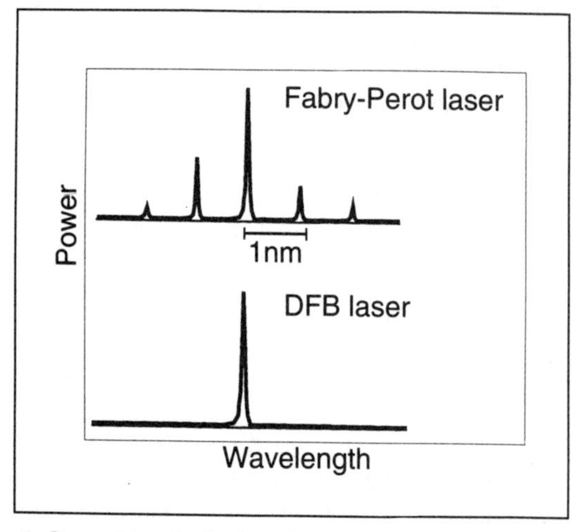

4. Several longitudinal modes appear in the spectra of Fabry-Perot lasers, while the dominant mode in DFB lasers is typically 30 dB larger than side modes.

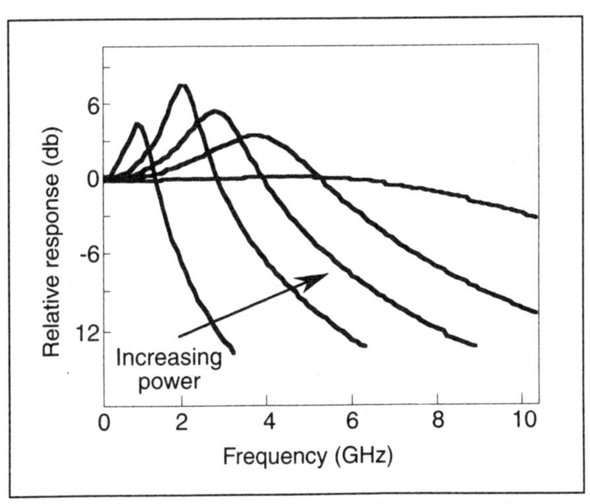

5. The laser's modulation bandwidth increases with output power.

may appear in the response curve at low frequencies (<1 MHz) because of thermal effects and at high frequencies (GHz) due to resonance effects. Experimental lasers used for frequency modulation include DFB lasers (with bandwidths up to 15 GHz and FM response of about 0.1 GHz/mA); and two-section DFB lasers, with a larger response (1 GHz/mA) but a smaller bandwidth (~1 GHz) [4].

Noise

Noise is undesirable because it reduces the signal-to-noise ratio (SNR). For some applications, the laser noise is unimportant compared to thermal noise in the receiver. For applications which must maintain a high SNR, the laser noise is critical. For multi-channel AM video systems, which need a SNR>50 dB, the laser noise must be maintained at less than -150 dB/Hz by using commercial, low-noise DFB lasers with optical isolators.

Relative intensity noise (RIN), or fluctuations in the output power, arises from fluctuations in the electron density and spontaneous emission. RIN peaks at the resonance frequency (a few GHz) and generally decreases at higher output power (Fig. 6). The sharpness of the resonance peak is reduced at high powers by the same damping effect that flattens the modulation response.

Fabry-Perot lasers exhibit greater RIN at low frequencies (<1 GHz) compared to DFB lasers. In a system with dispersive fiber, mode partition noise, i.e., fluctuations in power between different modes, causes additional low-frequency noise in the total transmitted power. This noise can produce problems even in higher-frequency systems, such as SCM systems, by beating with the signal, and translating into the signal band. Low-frequency noise can be minimized by operating at the zero dispersion wavelength and using a DFB laser with a side mode suppression ratio greater than 30 dB.

Even small amounts of optical feedback caused by reflections into the laser from the fiber end or other components can seriously degrade a laser's noise and distortion characteristics. Feedback of as little as -40 dB can significantly increase RIN for DFB lasers, which are especially sensitive to reflection. The cure for this problem is simple but expensive: include optical isolators to suppress reflections.

Phase noise (fluctuations in the optical phase) arises from fluctuations in the electron population, like RIN, and peaks at the resonance frequency. It is not a serious problem for most applications, although it may be converted to intensity noise by phase-sensitive components such as connectors.

Distortion

Distortion is interference arising from a laser's nonlinear response to current modulation. Like noise, it reduces the SNR. In digital systems it is not usually troublesome, but in analog systems it can be a serious problem, especially for multi-channel AM video systems that require a high SNR. Sources of distortion include nonlinear power-current curves, operating too close to the resonance frequency, and clipping (which results when the modulation current drives the laser below threshold).

Distortion is quantified by describing the Composite Second Order (CSO) and Composite Triple Beat (CTB), which are measures of the worst-case carrier-to-interference ratio generated by all the second-order and third-order intermodulation products, respectively. In general, a CSO and CTB less than -60 dBc are necessary for a 40-channel AM video system with a SNR of 50 dB. These conditions are met by commercial, high-linearity DFB lasers. Such systems may also require an optical isolator, because reflections can lead to enhanced distortion.

Reliability

No discussion of laser properties is complete without mentioning reliability. This is a particularly difficult area to pin down, because determining reliability is a time-consuming task and the results can vary widely, depending upon the laser's structure, package, and operating conditions. Nevertheless, it is an area that has shown steady improvement.

The laser's intrinsic reliability is determined by the migration of defects to the active or current-confining layers. This migration produces increased optical loss and reduced radiative efficiency, or increased leakage currents around the active layer. Most manufacturers use careful laser design coupled with screening tests at high temperatures and current to eliminate lasers that might fail prematurely.

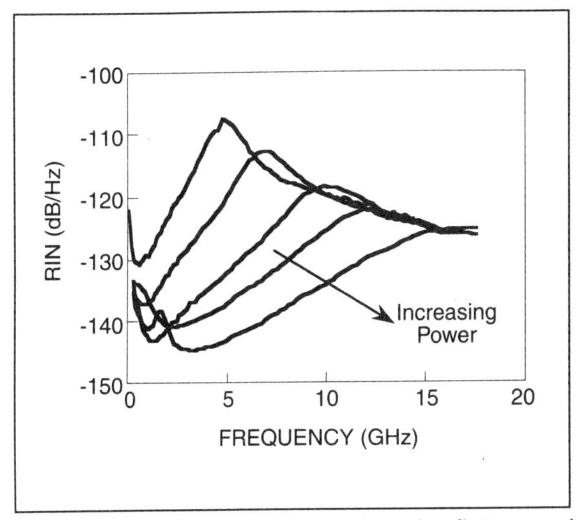

6. The peak of the relative intensity noise flattens and moves to higher frequencies with increasing output power.

A generally recognized limitation to reliability is the laser's package. Even a slight fiber shift during aging or temperature extremes may cause the output power to drop significantly because the coupling efficiency is sensitive to sub-micron changes in position. Many laser manufacturers now include thermal cycling in their screening procedures, and offer ruggedized packages that can operate from -40° to +85°C.

How much reliability is needed? A lifetime of 25 years is desired, but it easier to achieve in undersea systems operating at a nearly constant 10°C than in subscriber loop systems that occasionally experience elevated temperatures. Today's commercial laser modules have sufficient reliability for undersea systems, but more demanding applications such as AM video systems may need more extended reliability assurance with regard to critical laser properties.

Things to Come

Which of the competing technologies will eventually "win out" for tomorrow's optical networks? Not surprisingly, the answer depends partly on the performance, cost, and availability of the semiconductor components, especially the laser. Many of the remaining challenges are those related to engineering issues; more practical packaging, for example. Others challenges require ingenious solutions relating to new materials or structures. Strained quantum well material, for example, has recently shown improved characteristics— lower thresholds, higher power, narrower linewidths—compared to lattice-matched materials. Surprisingly, the reliability appears to be comparable, at least from preliminary reports.

One novel structure now being investigated is a vertical-cavity laser. The resonant cavity is perpendicular to the active layer instead of parallel to it. This structure offers the advantages of two-dimensional arrays, high packing densities, high fiber-coupling efficiency, and wafer-scale testing. Possible applications include optical interconnects, spatial light modulators, and signal processing. Such experimental lasers have been highly successful at shorter wavelengths (0.8-1.0 µm), but have not been demonstrated in room-temperature dc operation at longer wavelengths.

Putting lasers in arrays and integrating them with other devices such as modulators, power combiners and optical amplifiers, may also play a key role in enabling the manufacture of inexpensive, reliable, high-performance laser transmitters. ***C&D***

Acknowledgment

I thank the many friends who provided helpful suggestions for this paper, especially Dr. Robert B. Lauer, Dr. Robert Olshansky, and Mr. Paul Hill.

Joanne LaCourse [SM] is a Principal Member of Technical Staff with the Optoelectronics Devices and Materials Department at GTE Laboratories Inc., Waltham, Mass.

References

1. C. Zah *et al.*, "Submilliampere threshold 1.5 µm continuously graded index separate confinement strained quantum well lasers, "IEEE Photon. Technol. Lett. **2**, p. 852, 1990.

2. H. Karnei, et al., "Ultrahigh output power of 1.48-µm GaInAsP/GaInAsP strained-layer MOW laser diodes," paper TuH6, Optical Fiber Communication Conference, Sa Jose, CA, 1992.

3. M. Okai *et al.*, "Corrugation-pitch-modulated MQW-DFB laser with narrow spectral linewidth (170 kHz)," IEEE Photon. Tech. Lett. **2**, p. 529, 1990.

4. T. Koch and U. Koren, "Semiconductor Lasers for Coherent Optical Fiber Communications," J. Lightwave Tech. **8**, p. 274, 1990.

5. E. Meland, et. al., "Extremely high-frequency (24 GHz) InGaAsP diode lasers with excellent modulation efficiency," Electron. Lett. **26**, p. 1827, 1990.

6. K. Uomi, et al., "Ultrahigh-speed 1.55 µm 1/4-shifted DFB PIQ-BH lasers with bandwidth of 17 GHz," Electron. Lett. **25**, p. 668, 1989.

7. Y. Hirayama, et al., "Extremely reduced non-linear K-factor in high-speed strained layer multi-quantum well DFB lasers," Electron. Lett. **27**, p. 875, 1991.

For more information, see G. P. Agrawal and N. K. Dutta, *Long Wavelength Semiconductor Lasers*, Van Nostrand Reinhold Company, New York, 1986; and *Optical Fiber Communications II*, S. E. Miller and I. P. Kaminow, editors, Academic Press, 1988.

CHAPTER 5

OPTICAL FIBER

Optical Fiber – The Expanding Medium

Suzanne R. Nagel

Introduction

Lightwave communication systems using hair-thin silica glass optical fibers have rapidly become the preferred method to transmit huge quantities of information from one place to another. Such systems are typically digital in nature: information is encoded into a binary format of zero's and one's for transmission. In a lightwave system, a one corresponds to the transmission of a "pulse" of photons of light, while a zero corresponds to the absence of such a pulse. A basic system consists of a transmitter with a laser that is modulated or turned on and off millions of times per second, a transmission medium that is a glass fiber carrying injected laser light, and a receiver having a photodiode that detects arriving photons or pulses of light after they have travelled some distance through the fiber. For transmission over long distances, repeaters are used to reamplify these light signals in order to maintain the quality of light transmission.

A key element in the dramatic implementation of lightwave communication technology has been the tremendous advances in the fabrication and performance of silica-based optical fiber lightguides. The realization of very low signal attenuation and low dispersion or distortion of the light pulses in fibers allow very high information rates (bits/sec) to be transmitted over long distances between repeaters. High tensile strengths and excellent long term mechanical performance of long lengths of fiber allow practical handling, cabling and installation of glass fibers. Tight dimensional control resulting in low attenuation splicing has also been achieved. Overall, fabrication technology to achieve economical large scale manufacture with optimized properties are now routine. A number of reviews have reported on the details of fiber technology. [1-3]

This paper will examine the basic principles and properties of optical fibers and their fabrication, with particular focus on one of the more stringent applications for fibers – their use in the undersea environment. Historically, the demand for transoceanic communication has been increasing exponentially over the past 30 years, with an average growth rate in the North Atlantic of about 24 percent per year, and steady growth in the Pacific, Mediterranean and Caribbean as well. [4,5] A guiding force in the design of such systems has been high reliability with lowest possible cost/circuit mile, which directly is addressed by reducing the numbers of repeaters in a system. Thus, the choice of lightwave communications systems using high reliability, high bandwidth, low loss lightguides was an extremely attractive option, leading to the proposal for the first optical Submarine Lightguide (SL) System. [6] By 1988 the first such transoceanic system, known as TAT-8, will be installed [4] and next generation systems with increased capacity and repeater spans are already being planned. [5] Further into the future, new fiber materials with even greater theoretical transparency than silica fibers have been proposed that might allow even more expanded repeater spacings to be realized. [7]

LightGuide Materials Considerations

In order to make a practical fiber transmission medium for an optical communications system, a number of importance materials and design factors consistent with manufacturability concerns must simultaneously be realized. Fiber lightguides based on high silica glasses rapidly became the preferred technological choice, after first being proposed in 1966. [8] High silica glasses could be made into a variety of fiber designs with the requisite propagation characterisics, and theory rapidly evolved to describe and predict their performance. Such glasses had intrinsically low optical loss and high bandwidth and thus allowed high bit rate systems with improved repeater spacing relative to copper-based coaxial systems. These fibers also had excellent chemical durability and thermal stability – important reliability issues for long lifetime systems. Other reliability criteria related to radiation damage and static fatigue have also been realized. The intrinsic tensile strength of silica surpassed that of steel by a factor of three. Silica is one of the most abundant materials in nature and raw materials to fabricate high silica glasses proved to be quite inexpensive. Small amounts of dopants such as germania, fluorine and phosphorus could be stably incorporated into the glass to make the various fiber designs discussed in the next section. Most important, a variety of innovative processing techniques emerged to allow fabrication of low-cost, long length fibers with excellent and reproducible optical and dimensional properties. [3] In addition, by coating the fiber with protective plastic, practical long lengths of high strength could be realized to allow subsequent cabling and installation. Thus, all performance considerations have been met in high silica fiber technology.

A variety of other glass systems are now under investigation for potential use in future systems. In particular, glasses based on light and heavy metal halides are of interest because of the potential for lower optical attenuation. [7,9] Considerable work, however, must be directed at examining and technologically realizing appropriate glass materials, designs, physical properties and processing technology resulting in fibers with the requisite reliability. In general, their hydroscopic nature and tendency to crystallize present imposing technological challenges.

Light Guidance, Fiber Types, and Dispersion

To realize the high bandwidth potential of systems operating at optical frequencies, fiber designs that guide light over distances with minimal distortion of the light pulses

Reprinted from *IEEE Circuits and Devices Magazine*, Vol. 5, No. 2, pp. 36-45, March 1989.

were required. To understand light guidance, we must first define some basic aspects of the interaction of light with materials. The refractive index of a material is a measure of the speed of light in a perfect vacuum, relative to its speed through the material. The higher the refractive index, the more the light is retarded or the slower it travels. Secondly, when light travels from one medium to another, it is refracted or "bent" by some angle proportional to the relative refractive indices of the two media: this is known as "Snell's Law of Refraction." For light travelling in a higher refractive index medium, if it impinges on the interface of a lower refractive index medium at an angle which that some critical angle, θ, it will be reflected rather than refracted. This principal, known as "total internal reflection," was first discovered by Tyndall in 1870 for light travelling in water relative to air, and is the basic principle used for making a lightguide structure.

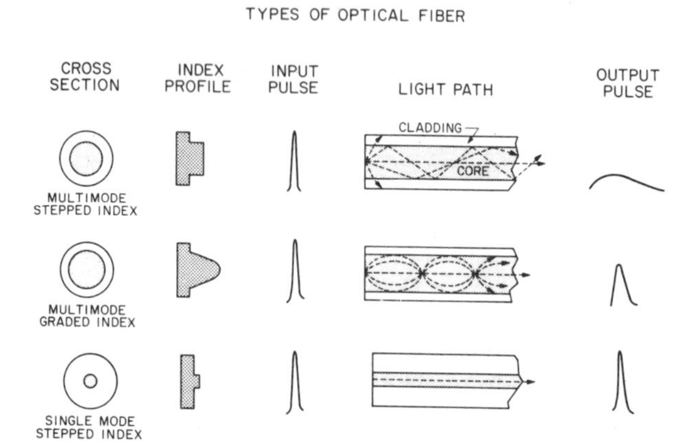

Fig. 1. Characteristics of the three basic types of optical fibers.

The simplest type of lightguide, known as a step index multimode fiber, is illustrated in the top portion of Fig. 1. It consists of a circular core of material of constant refractive index n_2 surrounded by a cladding material of lower refractive index n_1. When a pulse of light is injected into the core of this type of fiber, certain modes of light at particular angles in the core will meet the critical angle criteria for total internal reflection and will be guided in the core; others will be refracted and therefore lost. A basic parameter of lightguides is known as the numerical aperature (*na*), which defines the maximum angle, θ_0, of incident light that can be totally internally reflected:

$$NA = \sin\theta_0 = \sqrt{n_2^2 - n_1^2} \approx n_2\sqrt{2\Delta} \quad (1)$$

where

$$\Delta = (n_2 - n_1)/n_2 \quad (2)$$

The numerical aperature is thus a measure of the light collecting capability of the fiber. The number of modes of light that can propagate in a fiber are governed by Maxwell's electromagnetic field equations, and are related to a dimensionless quality V:

$$V = \frac{2\pi a}{\lambda}(NA) \approx \frac{2\pi a n_2}{\lambda}(2\Delta)^{1/2} \quad (3)$$

where λ is the wavelength of light and α is the core radius. When V is less than 2.405 only a single mode of light can be propagated; all other modes are cut-off.

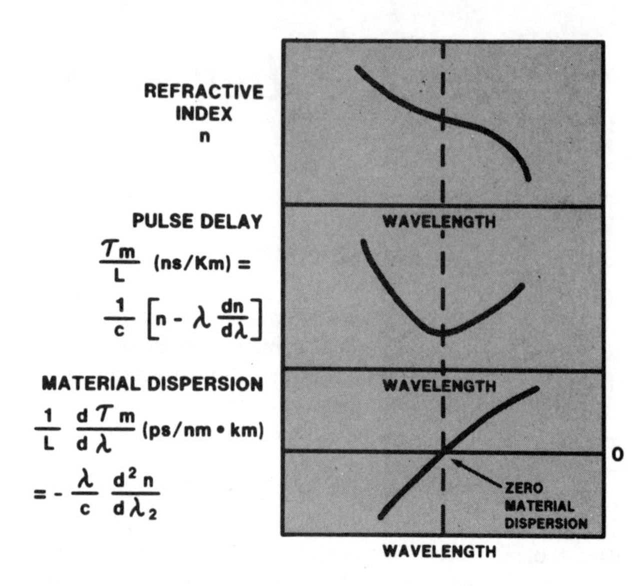

Fig. 2. Relationship of refractive index versus wavelength behavior of glasses to the resultant material dispersion. At any wavelength, a pulse will be delayed by a time τ_m per unit length L according to the equation shown, where c is the speed of light. The material dispersion is a measure of the delay with respect to wavelength, and is minimized at the zero material dispersion wavelength.

While the ray diagram does not represent a rigorous description of the detailed characteristics of light propagation in fibers, it is a useful way to understand the origin of pulse broadening of light as it travels through the lightguide, as well as the basic types of fibers used for long distance communication. [10] When a temporally narrow pulse of light is injected into the fiber depicted in the top portion of Fig. 2, each guided mode or angle of light has a different path length to travel through the fiber. Thus, while every mode was injected into the fiber at approximately the same time, they arrive at the far end after different times, leading to broadening of the initial pulse of light. This important phenomena, known as intermodal dispersion, limits the bandwidth, or rate at which individual pulses can be injected into the transmission medium and still individually detected without overlapping after traveling some length through the fiber. Bandwidths of 10-20 MHz-Km are typical for step index multimode fibers.

To overcome the intermodal dispersion in these lightguides, cores with graded refractive index profiles are used, such as depicted in the middle portion of Fig. 2, and hence are called graded index multimode fibers. In this type of structure, different angles of light injected into the fiber core are refracted rather than reflected, resulting in gentle periodic paths of propagation. Typically, the refractive index at any radial position r, is given by:

$$n(r) = n_2[1 - \Delta(r/a)^{\alpha}] \quad (4)$$

where α is defined as the profile parameter. While the detailed propagation of each mode is complex, in general higher order modes travel over longer path lengths than lower order modes, but travel faster in the vicinity of lower index. Thus, by choice of the proper α for a given material composition, the transit times for each mode can be roughly equalized. Near parabolic profiles ($\alpha = 2$) are optimal for minimizing intermodal dispersion, but the optimal values depends on wavelength and composition. Use of graded index profiles can reduce the pulse broadening by over three orders of magnitude relative to step index fibers.

Another source of pulse spreading is due to material dispersion, which results from the fact that the index of refraction of glass varies with wavelength as shown in Fig. 2. For high silica glass compositions, such effects are minimum near 1.3 μm, while for glasses such as the heavy metal halides, this zero material dispersion occurs at longer wavelengths. Thus, there will be some pulse spreading associated with the spectral width, $\Delta\lambda$, of the light source. Material dispersion effects on bandwidth of multimode fibers are typically small relative to intermodal dispersion effects, but increase as the spectral width increases. However, it is a major factor in the dispersion of single mode, the third type of fiber to be described.

The very highest bandwidth is achieved in a single mode fiber, as depicted in the bottom of Fig. 1; it represents a lightguide that only supports a single fundamental mode of light. This mode, as it propagates through the fiber, is actually characterized by a near-Gaussian power distribution with some fraction of the power travelling in the cladding. To achieve single mode transmission in an idealized step index fiber, V must be less than 2.405 as described in equation 3. The wavelength λ_c at which the fiber becomes single mode is termed the cutoff wavelength and is determined by the core radius, profile and Δ. Pulse spreading in single mode fibers is only due to chromatic dispersion that consists of two terms, material and waveguide dispersion. At short wavelengths material dispersion is dominant, but near 1.3 μm for silica based fibers the magnitude of such effects are similar, as illustrated in Fig. 3. The zero dispersion wavelength, λ_0, is the wavelength at which these two effects exactly counterbalance each other, and represents the maximum bandwidth wavelength for such a fiber. The values of Δ, α, and profile can be chosen to tailor λ_0 to a desired operating wavelength.

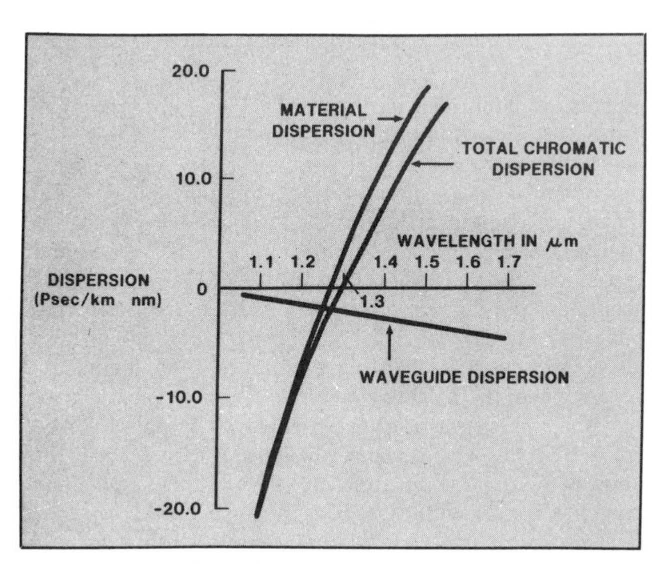

Fig. 3. Dispersion in single mode lightguides. The fiber depicted is typical of a 1.3 μm optimized design.

Thus the dispersion of a given fiber is determined by the specific fiber design parameters as well as the wavelength of operation. In general, single mode fibers have several orders of magnitude lower dispersion than multimode fibers, but in each case the resultant bandwidth will be determined by the spectral width of the source. The small core size of the single mode fiber in general favors its use with a laser in high bandwidth applications while multimode fibers are used with both LED's and lasers for low to intermediate bandwidth systems.

For high silica lightguides, a number of optimized fiber designs have emerged. Single mode fibers typically have core diameters of 8-10 μm, 125 μm outer diameters and Δ = 0.3-0.5%. For multimode fibers, core diameters of 50, 62.5 and 85 μm, outer diameter of 125 μm and Δ = 1-2% are most common, although core diameters of 100-200 μm are also in use.

Optical Transmission Loss

The next important concept to consider in lightguides is how the intensity of the light signal is affected as it travels through the fiber structure. The fiber is not a perfect conductor of light and important intrinsic material properties as well as extrinsic effects can cause the signal to be attenuated. Such effects are discussed in detail in a variety of references. [1,2,11,12] This phenomena is referred to as optical loss and defined in dB/km, by

$$\text{Loss} = 10\log(P_i/P_0) \quad (5)$$

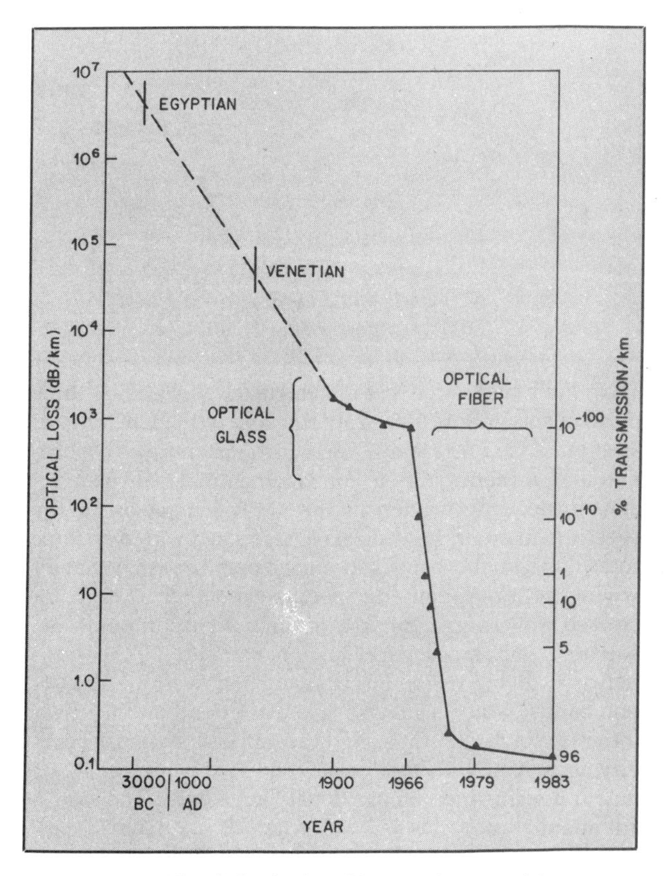

Fig. 4. Historical reduction of loss as a function of time.

where P_i and P_0 are the input and output power intensity at any given wavelength. The dramatic improvement in optical loss achieved as a function of time in silica-based optical fibers relative to conventional glasses is shown in Fig. 4. This extraordinary transparency has allowed transmission over distances greater than 200 km without the need for reamplification.

TABLE 1. SOURCES OF TRANSMISSION LOSS IN LIGHTGUIDES

Mechanism	Cause
1. Absorption	
• INTRINSIC	- UV ELECTRONIC TRANSITIONS
	- IR MOLECULAR VIBRATIONS
• IMPURITY	- TRANSITION METALS
	- RARE EARTHS
	- INTERSTITIALS
	- MATRIX IMPURITIES
	- OH VIBRATION
	- H_2 VIBRATION
• DEFECTS	- VACANCIES
	- RADIATION INDUCED
	- THERMALLY INDUCED
	- H_2 INDUCED
2. Scattering	
• RAYLEIGH	- MINUTE DENSITY AND CONCENTRATION FLUCTUATIONS
• BULK IMPERFECTIONS	- BUBBLES, INHOMOGENEITIES, CRACKS
• WAVEGUIDE IMPERFECTIONS	- CORE, CLAD INTERFACIAL IRREGULARITIES
• BRILLOUIN, RAMAN	- SPONTANEOUS
3. Waveguide	
• MACROBENDING	- CURVATURE INDUCED
• MICROBENDING	- PERTURBATION INDUCED
• DESIGN	- RADIATIVE
• STIMULATED RAMAN, BRILLOUIN	- DEPENDS ON POWER DENSITY

Optical transmission loss in lightguides is caused by three mechanisms: absorption, scattering and waveguide effects, as summarized in Table 1. Each glass has intrinsic absorption and scattering due to fundamental material phenomena. Intrinsic absorption in the ultraviolet region of the spectrum in amorphous media is associated with electronic transitions in the band gap and shows Urbach behavior where the absorption edge decays exponentially with increased wavelength. Intrinsic far infrared absorption is associated with molecular vibrational modes in the glass network, giving rise to fundamental and overtone absorption bands, which also have tails that extend into the near infrared. Rayleigh scattering is an intrinsic material property associated with small scale refractive index variations due to density and compositional fluctuations. The resultant attenuation varies as $A\lambda^{-4}$, where A is a material constant that depends both on composition and on the glass preparation methods. Fluctuations can be "frozen" in depending on the quench rate of the fiber during processing. In addition, spontaneous non-linear Raman and Brillouin scattering can occur. The magnitude of such effects in high silica is typically very small if the optical power density is below a threshold level. For new lightguide materials, only limited data are available, and the magnitude of such affects must be assessed for new potential low compositions.

Superimposed on these intrinsic absorption and scattering effects, a variety of extrinsic process related loss mechanisms can limit the achievable optical loss. Impurities can cause very broad, strong absorptions in glass. Control of 3d transition metals such as iron and copper to the parts per billion level are necessary in high silica glasses. For longer wavelength materials (2-12 μm), rare earth cations such as cerium and neodymium, as well as oxygen and carbon oxides can cause strong wavelength dependent absorptions, and thus must also be controlled to the parts per billion level. OH control is extremely important in both silica and non-silica fibers since its strong fundamental absorption is in the 2.7-3 μm region, with overtone and combination bands occurring at shorter wavelengths. Dissolved H_2 is IR active and must be avoided; if dissolved in the glass network it can further react to cause new absorption bands to appear with time. In addition, a variety of defect related absorptions can occur due to structural imperfections, and are closely related to the specific processing parameters. A variety of extrinsic scattering mechanisms due to particles, bubbles, inhomogeneities, strain and waveguide irregularities must be avoided. Lastly, waveguide designs that eliminate or minimize radiative, macrobending and microbending losses must be chosen. Thus, the achievement of low losses in lightguide structures requires not only proper choice of materials and designs with low intrinsic loss, but highly controlled processing to avoid extrinsic effects at a given wavelength of operation.

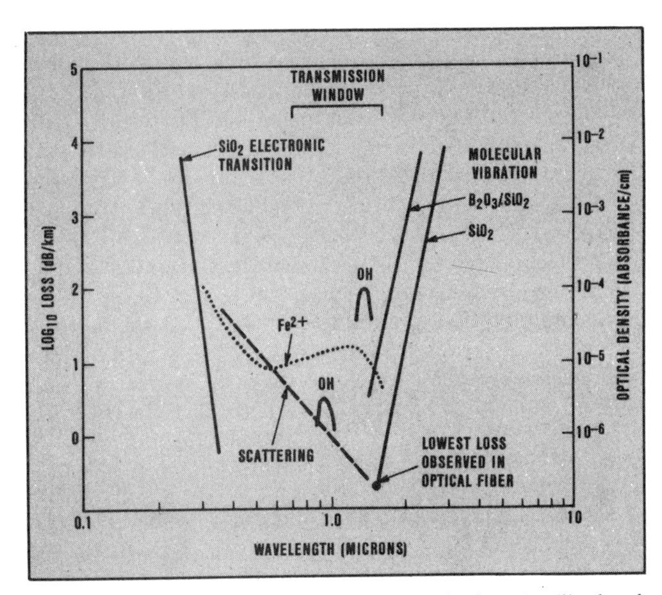

Fig. 5. Schematic representation of loss in mechanisms in silica based lightguides as a function of wavelength.

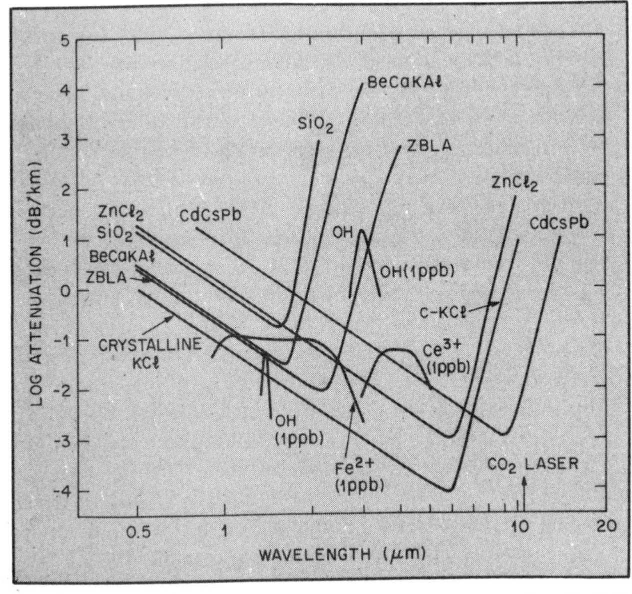

Fig. 6. Theoretical losses of lightguide material versus wavelength. [12] BeCaKA1 are fluoride glasses made with those cations; ZBLA represents a zirconium barium lanthanum aluminum fluoride glass: CdCsPb is a chloride glass made from those cations.

A schematic of transmission loss versus wavelength for high silica lightguides is illustrated in Fig. 5. Note that an intrinsic transmission window occurs in the 0.6-1.6 μm region with the very lowest losses occurring in the 1.5-1.6 μm wavelength region. The exact values of the intrinsic Rayleigh scattering and UV and IR edges will depend on the specific composition. Also note the strong absorptions associated with impurities. Losses as low as 0.16 dB/km have been achieved in silica at 1.57 μm, representing the only fibers to date showing near-intrinsic loss behavior. [13] The projected theoretical losses for a variety of low loss materials is depicted in Fig. 6. [14, 15] The future challenge lies in realizing such losses in practice. Much recent work has focused on the ZBLA glass family, zirconium barium lanthanum aluminum fluorides, where losses as low as 0.7-0.9 dB/km at 2.3 μm have been reported. [16, 17] In general, glasses based on fluoride, chlorides and other halides have intrinsic vibrational absorptions that occur at longer wavelengths, thus transmit further into the infrared if impurity absorptions can be avoided. Some of these glass systems have very low scattering coefficients, A, if crystallization and extrinsic scattering mechanisms can be avoided in processing.

Mechanical Properties

Glasses used as optical fibers behave as elastic bodies up to their breaking strength, with the theoretical strength determined by the cohesive bond strength. [1] For silica fiber, intrinsic strength is estimated to be about 2 million (13.8 GPa), and strengths of this order have been measured at liquid N_2 temperatures and in high vacuums where moisture is absent. In room atmospheres, reduced strengths of 800 ksi are typical for short lengths. Much less is known about the theoretical strength of other glasses, but values as high as 500 ksi have been estimated for the heavy metal halides. [18]

However, most glasses do not exhibit these theoretical strengths, since they are brittle materials that are weakened by the presence of inhomogeneities and flaws that act as stress concentrators, leading to local failure initiated at these flaws. The variability of strength is typically described by the Griffith model of brittle fracture, [19] where the stress causing failure at a given flaw is inversely proportional to the flaw size, as shown in Fig. 6 for high silica fibers. [20] Thus, the resultant strength of fibers is statistical in nature related to the distribution of flaws. Typically, plastic coatings are applied on-line to protect the pristine surface of the fiber. However, as the length of a fiber is increased, the probability of encountering increasingly larger flaws increases. This statistical length versus strength dependence of silica is depicted by a Weibull distribution such as shown in Fig. 7. [20] For a given fiber length, the proba-

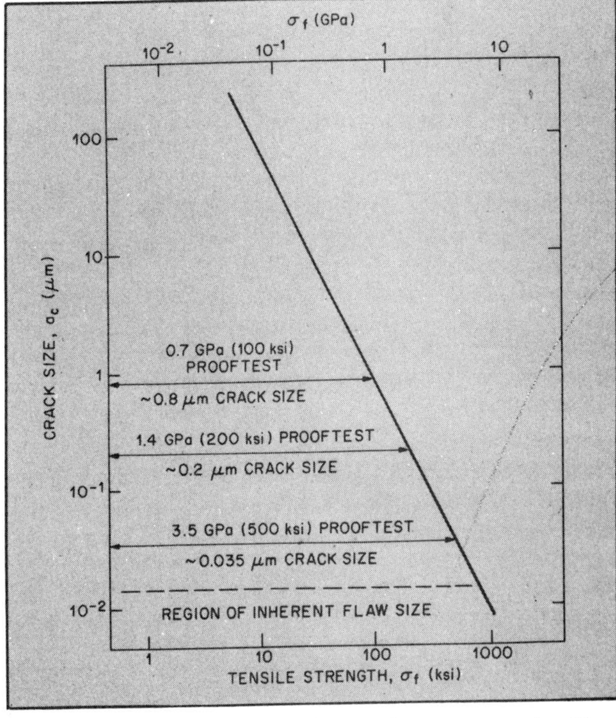

Fig. 7. Crack size associated with different fracture stresses for silica-based fibers. [18]

bility of failure at a given tensile stress can be estimated. In short lengths, very high unimodal distributions of strength are obtained, reflecting flaw free fiber. At length greater than 1 km, a bimodal dependence is shown, with the lower strength mode reflecting isolated surface flaws introduced during processing.

Since it is difficult to detect and eliminate these flaws, tensile proof testing is used, where all of the fiber is subjected to some prooftest stress in order to truncate the flaw distribution and guarantee some minimum strength level. Note in Fig. 6 that proof test levels ranging from 100-500 ksi correspond to elimination of flaws greater than 0.8 to 0.035 μm in diameter. In order to make long lengths of fiber with such strength levels, very controlled processing is required and will be discussed in the next section. For new materials, achievement of high strengths in long lengths has yet to be demonstrated. For heavy metal halide glasses, lower strength is attributed to both flaws and surface crystallization.

Another important mechanical property of fibers is delayed failure, or static fatigue due to stress enhanced slow crack growth in the presence of moisture. In the presence of an applied load, this can lead to failure after some time at a stress lower than that which would cause instantaneous failure. The static fatigue characteristics can depend on the local environment surrounding the crack, the temperature and the type of fiber coating material used. Thus it is very important to take such effects into account when using a fiber in a given service environment.

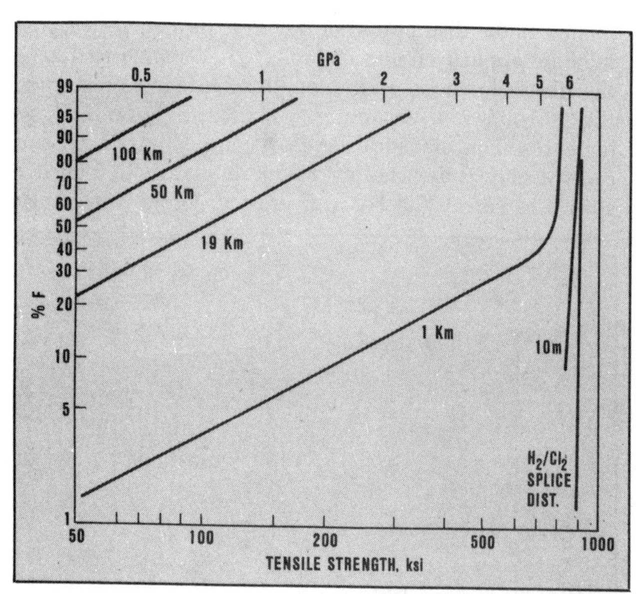

Fig. 8. Failure probability versus tensile stress for high strength silica fiber. [18]

Fiber Fabrication

Optical fiber fabrication technology for communications requires realization of fiber designs in multikilometer lengths using specific glass compositions that have the requisite optical, mechanical and other physical properties important for controlled manufacture. Reproducibility, low cost, and high reliability of the resultant fiber are critical. A number of fabrication techniques have evolved for large scale manufacture of high silica fibers, [3] while a variety of approaches are being explored in research laboratories for the longer wavelength halide-based glasses. [7,9]

Silica-based fibers are manufactured in two steps: a preform or boule of glass containing the core-cladding structure is first made, followed by fiber drawing and coating. Three major manufacturing techniques for making the precursor glass are depicted in Fig. 8. They are all based on vapor deposition techniques in which semiconductor grade liquid halide reactants such as $SiCl_4$ and $GeCl_4$ are entrained in a carrier gas and then reacted to form high silica glass. By using vapor transport, very low transition metal impurities are achieved since they are nonvolatile compared to the glass dopants. Most high silica glasses consist of a SiO_2, GeO_2-SiO_2, or GeO_2-P_2O_5-SiO_2 cores, clad with SiO_2 or fluorosilicate to achieve the desired refractive index structure.

In the outside vapor phase deposition (OVD) technique, invented by Corning Glass Works, a burner is used to generate particles by hydrolysis. The burner flame is directed at a rotating mandrel, resulting in the deposition of a porous layer. The burner or mandrel can be translated back and forth, resulting in layer-by-layer build-up of material. First, the core, then the cladding composition is deposited. After deposition, the mandrel is removed, and this porous soot boule is transferred into a consolidation furnace where it is sintered to yield a solid preform that contains the core-clad structure. By controlling the chemical atmosphere during consolidation, OH contamination can be reduced to the part per billion level.

The vapor axial deposition (VAD) technique, developed by many Japanese manufacturers, is based on the same principle of flame hydrolysis. However, in this case porous material is deposited on the end of a rotating bait rod, and by use of specialized torch designs and control of many process flows and variables, the chemical composition across the soot boule can be controlled to form both graded and step index structures. This growing boule can be simultaneously sintered by placing a consolidation furnace in-line. Cladding material are added by either using secondary torches to deposit from the side, and/or by shrinking a silica tube around the consolidated deposited material. In this process, OH is also removed during consolidation.

The last process is the Modified Chemical Vapor Deposition (MCVD), invented by AT&T Bell Laboratories, which is based upon the high temperature oxidation of the halide reactants. Chemicals are injected into a rotating silica tube that is heated by an external traversing oxyhydrogen torch.

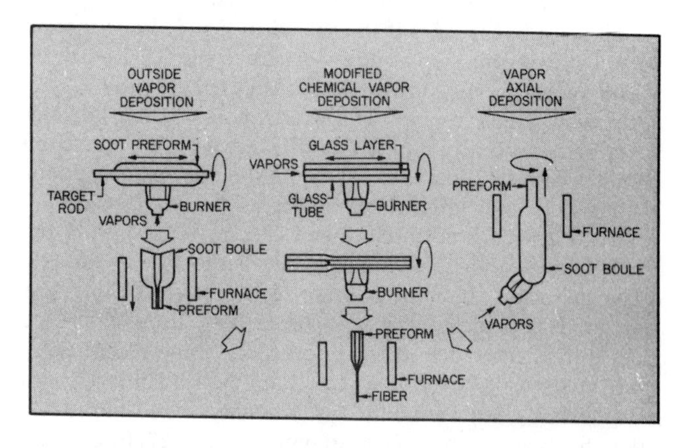

Fig. 9. Schematic representation of three major manufacturing techniques for making high silica fibers.

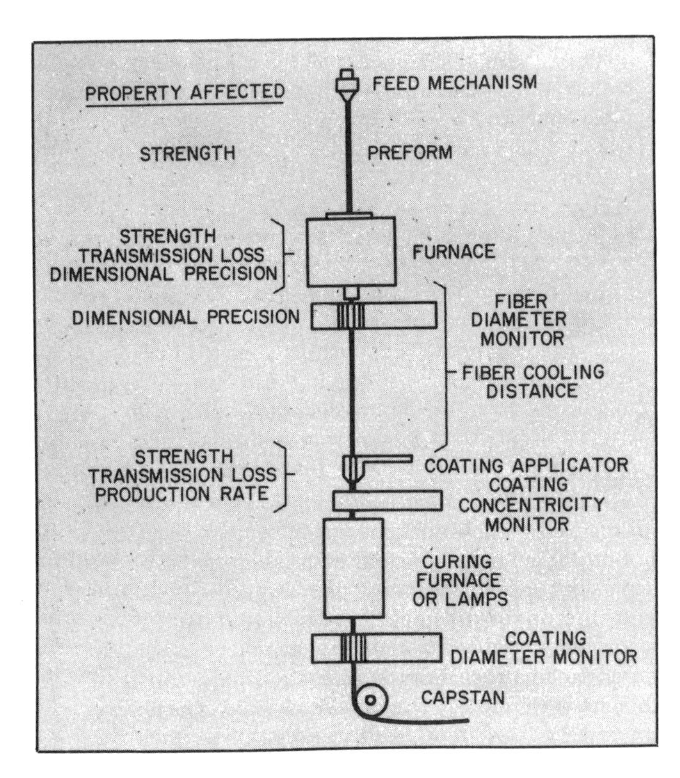

Fig. 10. Schematic of silica fiber drawing apparatus.

This tube provides the outer-most cladding in the resultant fiber. When the chemicals reach the hot zone they are oxidized to form particles that are deposited and are fused to form a thin glass layer. First cladding, then core material are deposited. The tube is then further heated, causing it to collapse down upon itself to form a solid core-clad preform rod. Thus deposition and consolidation occur simultaneously, and low OH is achieved by low hydrogen containing chemicals and controlled chemistry during deposition and collapse.

In all cases, the preform or boule of glass is then transferred to a fiber drawing apparatus, schematically depicted in Fig. 10. The glass is lowered into a high temperature furnace operating at 2100-2300 °C, where the glass softens enough to be drawn into fiber. A capstan provides the pulling force, and the preform feed rate and capstan speed determine the draw-down ratio. Preforms typically range from 1-6 cm in diameter and are drawn down to 125 μm fiber. Just below the furnace, a contactless diameter monitoring devices is used to detect the fiber diameter and feedback to the capstan for fast response, precise diameter control. In line coatings are applied in a contactless manner by use of a coating cup containing a liquid coating material. This material can be a thermal set material that is cured by a furnace or a UV curable polymer that is cross-linked by exposure to UV lamps. In this manner, protective coatings are applied to the pristine fiber for protection before it is touched by the drawing mechanism. In addition, the coating provides a lower modulus buffer against external perturbations such as encountered in cabling, and its characteristics and concentricity are key features in the resultant strength and optical performance. A coating concentricity monitoring system can be used to both characterize the coating thickness and control its concentric application.

Drawing and coating speeds of 1-10 m/sec are typical and preforms that yield 15-250 km are drawn. A minimum fiber cooling distance must be maintained so that the fiber is sufficiently cool when it enters the coating cup so that it does not thermally degrade the organic coating materials. An additional consideration is to maintain a very clean, particle-free environment during drawing so that high strength is achieved. Furnaces that are particle free have been developed and the fiber drawing path is purged with clean air to avoid air-borne particles. Clean, particle free-coating materials are used, and any contact with the fiber is avoided. For very high strength applications, the preform is fire-polished before drawing to eliminate any surface flaws that might not heal during the drawing process. In this manner, very long-length, high-strength fibers can be achieved routinely. All fibers are prooftested to assure some minimum tensile strength level. This can be done in-line or off-line on a separate prooftester.

Fabrication of new low loss glass preforms are in their earliest stages. Volatile chemicals for vapor phase processing are typically not available, and the vapor pressures of some contaminants are similar to those of materials from which candidates glasses are composed. Typically, bulk melting techniques are used where core and cladding glasses derived from high purity materials are made in dry box atmospheres, then rotationally cast or poured into molds to make a core-clad structure. Controlled melting and cooling are required to avoid contamination and crystallization.

Drawing and coating of new glass materials present a variety of challenges. These materials are typically much lower melting (300-500 °C) and their viscosity changes rapidly with temperature, requiring precise and narrow hot zones. In addition, many of these glasses are prone to crystallization, and crystals can readily nucleate in the hot zone during drawing. Because many of these materials are also hydroscopic, application of hermetic coatings may be required and very dry fiber drawing environments must be maintained. Coatings that are compatible with these materials must also be developed.

Glass Fibers For Undersea Applications

High capacity undersea cable systems based on optical transmission offered tremendous economic advantages over coaxial cable technology due to their combined high bit rate – long repeater spacing potential. The first SL system pointed to a 274 Mbit/sec transmission at 1.3 μm using single mode fibers and injection laster light sources, based on technological feasibility considerations. [6] When this system was formally proposed in 1980, very stringent requirements were placed on the transmission medium; a low loss, dispersion optimized design was required to allow repeater spacings up to 54 km. The fiber had to be insensitive to microbending losses on cabling, deployment and service, and able to be spliced with low loss and high strength. In addition, optical and mechanical reliability had to be suitable for 25 year system lifetimes. Long continuous lengths of fiber capable of being cabled, deployed, and should cable repair be necessary, withstanding recovery strains of up to 1 percent for 5 h, were deemed essential. This consideration dictated minimum initial strengths on the order of 200 ksi, and underscored the need for both long-length, high strength fiber drawing as well as a high strength splicing technique. Large scale manufacture of multimode fiber for terrestrial applications had only begun

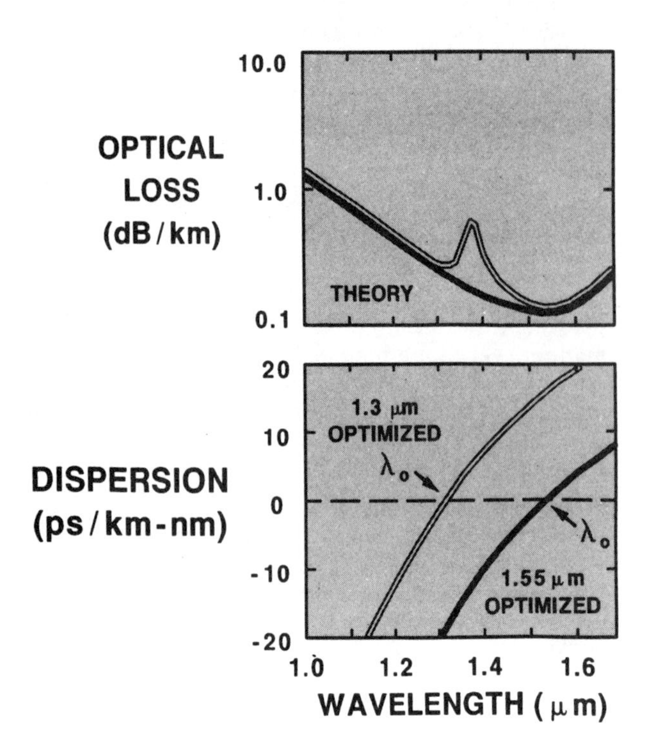

Fig. 11. Loss and dispersion characteristics of 1.3 and 1.55 μ single mode fiber designs.

in 1979, and the challenges of advancing the technology to meet such objectives looked formidable. Tremendous advances in the technology occurred, as recently reviewed, and installation of a number of undersea systems worldwide are planned in the 1988-1993 time frame. [21]

Single mode technology based on two specific 1.3 μm designs have emerged as the dominant lightguide technology for both terrestrial and undersea applications. One is a matched index cladding design, referring to the fact that the entire cladding has a single value of refractive index. The second design is a depressed cladding lightguide in which the index of the inner cladding next to the core is lower than that of the outermost cladding. Both of these fiber designs have minimum dispersion at 1.3 μm, with low losses of 0.35 dB/km at 1.3 μm and 0.20 dB/km at 1.55 μm, such as shown in Fig. 11. Such fibers are planned for use at 1.7 Gbit/s rates with 30 km-40 km repeater spacings, and further bit rate increases are possible. The depressed cladding design was initially specifically developed for SL undersea systems in order to simultaneously optimize the dispersion, loss, cut off wavelength, and microbending sensitivity of the lightguide, and this fiber typically demonstrates much improved bending performance relative to the matched cladding design at both 1.3 and 1.55 μm. [21] SL performance criteria for 274 mbit/s operation required dispersion less than 2.3 ps/nm·km between 1.29-1.33 μm that are readily achieved.

Fibers for SL are subjected to a 200 ksi prooftest stress, based on a conservative static fatigue design diagram analysis, such as shown in Fig. 12. [18] Each curve is constructed based on time-to-failure data for fibers in 100% water at temperatures ranging from 3-90 °C when subjected to an applied stress. The lowest curve is the derived design curve for 200 ksi prooftested fiber operating at ocean temperatures of 3 °C. As long as the time duration of applied stress is below the design curve, failure due to static fatigue will not occur. It shows that all the calculated stresses that would be imposed upon the fiber during cable installation, recovering and ocean bottom lifetime are well below the curve, thus assuring mechanical reliability over the system lifetime.

High strength fusion splicing has also been developed for SL, resulting in joining techniques with strengths much higher than the prooftest level. A modified oxyhydrogen or chlorine oxyhydron torch is used that produces splice strengths in excess of 500 ksi, and has achieved unimodal high strength of 800 ksi, such as depicted in Fig. 8.[18] A splice overcoating technique has also been developed that allows the assembly of long lengths of fiber with strengths of 200 ksi. [21] Static fatigue characterization of these splices indicate similar behavior to that of the fiber. Other reliability concerns of the performance of the transmission media have also been recently reported. [18] The specific development of the depressed cladding fiber for use in SL represents the first time that very high strength, high reliability and high performance have been simultaneously realized for lightguide technology.

Current focus is on the development of 1.55 μm fiber designs for the next generation of undersea systems. [5] At this wavelength, the intrinsic loss of the fiber is lower and thus repeater spans of 100 km or greater might be realized. Dispersion shifting to achieve zero dispersion at 1.55 μm has been demonstrated in a number of fiber designs, with losses that are as similar to those achieved in 1.3 μm designs, as shown in Fig. 11. [22–25] However, these fibers are typically more difficult to process and much attention is being focused on developing designs with the requisite reliability, high strength, dispersion characteristics, and bending induced-loss insensitivity, as well as assessing large scale manufacturability issues. If such designs can be realized, relaxed requirements on 1.55 μm lasers as well as high bit rate long repeater span operation is possible.

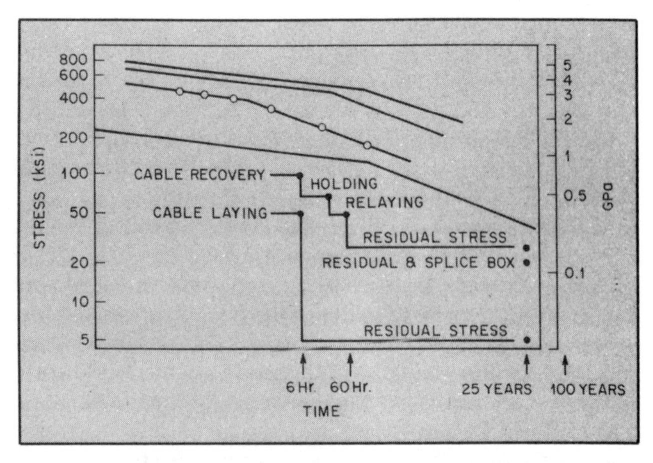

Fig. 12. Static figure of silica fiber in water at various temperatures and design curve for SL. [18]

Future Directions and Summary

Many people have expressed the dream that some day repeaterless transoceanic lightwave systems may be possible using glass fibers whose transmission losses are projected to be in the range of 0.002-0.01 dB/km at wavelengths in the 2-6 μm region. This review has tried to address the

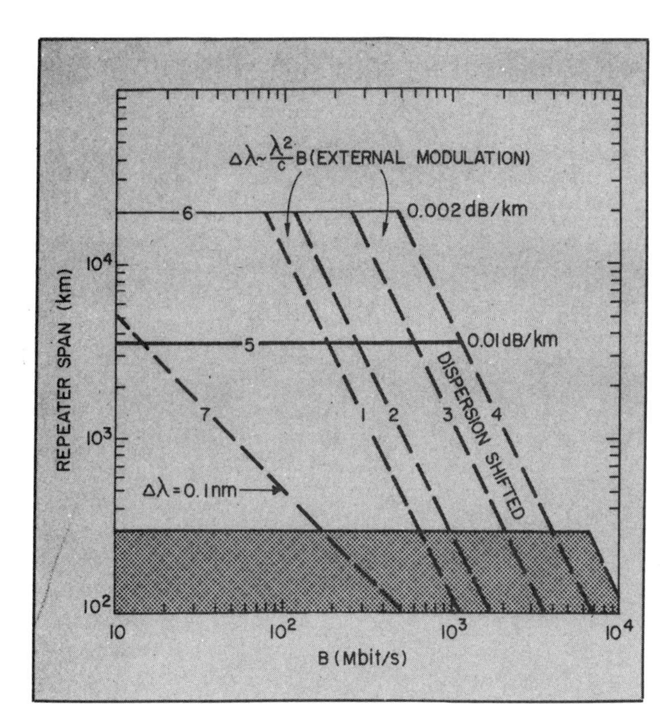

Fig. 13. Maximum repeater span length versus bit rate for various possible lightguide systems. The shaded region is representative of high silica technology where losses as low as 0.16 dB/km have been achieved. Curves 2, 4, 5 and 7 are projected for ultra low loss fluoride glasses while curves 1, 3 and 6 apply to possible low loss chloride glasses.

important optical and mechanical properties, as well as processing technology that are critical for the realization of long length of low loss, low dispersion optical fibers. The limits of system performance if fibers and the other requisite components were available, and in that sense represents the ultimate in bit rate-distance performance that might be achieved is shown in Fig. 11.[26] In this treatment, 500 km repeater spans might be achieved in silica-based fibers up to bit rates of about 10 6km 6 bits/sec, where dispersion limited performance sets in to decrease the achievable span length. Curve 7 is indicative of the performance of a very low loss (0.02 dB/km) halide fiber operating at 2.5 μm with a laser with $\Delta\lambda = 0.1$ nm and no dispersion optimization (20 psec/nm·km), and points at the need for having excellent lasers as well as dispersion control in such fibers, if improved distances are to be realized at bit rates greater than about 200 Mbit/sec. Curves 1 and 3 represent dispersion limited operation for fibers with 20 psec/nm·km dispersion operating at 6 and 2.5 μm, respectively with laser line width limited bit rates and coherent detection. Curves 2 and 4 represent dispersion limited operation for fibers with 1 psec/nm·km dispersion (dispersion optimized) operating at 6 and 2.5 μm, respectively. Up until bit rates where the dispersion limits set in, these results indicate that repeater spacings, in the range of 1200-3600 km are theoretically possible if such low losses can be achieved. Another important feature of this analysis is that at bit rates greater than 246 bits/sec, the advantages of halides relative to silica disappear, despite their low loss, since inter symbol interference sensitivity increases with wavelength of operation.

Thus, new glasses have the potential for achieving very long repeater spans at moderate bit-rates if the many technological obstacles to their fabrication can be overcome. Increasing effort is being directed at seeing if the spectacular advances in high silica fiber technology can be repeated for the halide glasses.

References

[1] *Optical Fiber Telecommunications*, ed: S.E. Miller and A.G. Chynoweth, Academic Press, Inc., New York, p. 705 (1979).

[2] J. E. Midwinter, *Optical Fibers for Transmission*, John Wiley and Son, New York, p. 410 (1979).

[3] *Optical Fiber Communications*, Vol. 1, Fiber Fabrication, ed. T. Li, Academic Press, Inc., New York, p. 363, 1985.

[4] P. K. Runge and P. R. Trischetta, "The SL Undersea Lightguide System," *IEEE J. Select. Areas Commun.*, Vol. SAC-1, pp. 459–466, Apr. 1983.

[5] R. E. Wagner, "Future 1.55 μm Undersea Lightwave Systems," *IEEE J. Select. Areas Commun.*, Vol. SAC-2, pp. 1047–1055, Nov. 1984.

[6] C. D. Anderson, R. F. Gleason, P. T. Hutchison, and P. K. Runge, "An Undersea Communication System Using Fiberguide Cables," *Proc. IEEE*, Vol. 68, pp. 1299–1303, Oct. 1980.

[7] For review, see T. Miyashita and T. Manabe, "Infrared Optical Fibers," and *IEEE Trans. Micr. Theory Tech.*, Vol. MII-30, pp. 1420–1438, 1982.

[8] K. C. Kao and G. A. Hockham, "Dielectric-fiber surface waveguides for optical frequencies," *Proc. Inst. Elec. Eng.*, Vol. 113, pp. 1151–1158, July 1966.

[9] D. C. Tran, G. H. Sigel, and B. Bendow, "Heavy Metal Fluoride Glasses and Fibers: A Review," *J Lightwave Tech.*, Vol. LT-2, pp. 566–586, Oct. 1984.

[10] For a discussion of detailed propagation theory see: A. W. Snyder and J. D. Love, *Optical Waveguide Theory*, Chapman and Hall, London, p. 734, 1983; M.J. Adams, *An Introduction to Optical Waveguides*, John Wiley and Sons, New York, p. 401, 1981.

[11] *Optical Properties of Highly Transparent Solids*, ed. S. S. Mitran and B. Bender, Plenum Press, New York, p. 538, 1975.

[12] I. Fanderlik, *Optical Properties of Glass, Glass Science and Technology*, 5, Elsevier, Amsterdam, p. 320, 1983.

[13] R. Csenscits, P.J. Lemaire, W. A. Reed, D. S. Shenk and K. L. Walker, "Fabrication of low-loss single mode fibers," Tech. Dig., Opt. Fib. Comm. Conf., TV13, New Orleans, LA, Feb., 1984.

[14] M. M. Broer, AT&T Bell Laboratories, private communication.

[15] M. E. Lines, "Scattering Losses in Optic Fiber Materials. 11. Numerical Estimates, J. Appl. Phys., Vol 55, pp. 4058–4063, June, 1984.

[16] D. C. Tran, "Low Optical Loss Fluoride Glass Waverguides," Tech. Dig., Opt. Fib. Commun. Conf. TuAl, p. 20–21, Atlanta, GA, Feb., 1986.

[17] S. Yoshida, "Progress in Fiber Preparation in Japan," Nato Advanced Research Workshop, Halids Glasses for Infrared Fiberoptics, Vilamoura, Portugal, Mar. 31-Apri. 4, 1986.

[18] J. J. Mecholsky, J. Lau, J. D. Mackenzie, D. Tran and B. Bendow," Fracture Analysis of Fluoride Glass Fibers," Paper 32, Proc., Second Intern. Symp. Halide Glasses, Troy, NY, Aug. 1983.

[19] A. A. Griffith, "The Phenomena of Rupture and Flow in Solids," Phil. Trans. Roy. Soc., London, Ser. A, 22, 163, 1920.

[20] J. T. Krause, A. A. Meade, S. Shapiro. "Assuring Mechanical Reliability of High Strength Fiber Cable for SL," Proc., Suboptic Intern. Conf. Opt. Fib. Sub. Telecommun. Syst., Paris, France, pp. 117–122, Jan. 1986.

[21] S. R. Nagel, "Review of the Depressed Cladding Single-Mode Fiber Design and Performance for the SL Undersea System Application," J. Lightwave Tech., Vol. LT2, pp. 792–801, Dec. 1984.

[22] B. J. Ainslie, K. J. Beales, D. M. Cooper and C. R. Day, "Monomode Optical Fibres with Graded-Index Cores for Low Dispersion at 1.55 μm," *Br. Telecom. Technol J.*, Vol. 2, pp. 25–34, April 1984.

[23] A. D. Pearson, L. G. Cohen, W. A. Reed, J. T. Krause, E. A. Sigety, E. V. DiMarcello and A. G. Richardson, "Optical transmission in Dispersion Shifted Single Mode Spliced Fibers and Cables, *Lightwave Tech.*, Vol. LT-2, pp. 346–349, Aug. 1984.

[24] T. D. Croft, J. E. Ritter, and V. A. Bhagavatula, "Low Loss Dispersion Shifted Single Mode Fiber Manufactured by the OVD Process," *J. Lightwave Tech.*, Vol. LT-3, pp. 391–934, 1985.

[25] R. Yamauchi, M. Miyamoto, T. Abiru, K. Nishide, T. Ohaski, O. Fukuda, and K. Inada, "Design and Performance of Gaussian Profile Dispersion Shifted Fibers Manufactured by VAD Process," *J. Lightwave Tech.*, Vol. LT-4 pp. 997–1004, Aug. 1986.

[26] M. M. Broer, and L. G. Cohen, "Heavy Metal Halide Fiber Lightwave Systems," *J. Lightwave Tech.*, LT4, pp. 1509–1513, Nov. 1986.

VIEWPOINT

Richard K. Snelling

Bringing Fiber to the Home

Fiber to the home will soon be a reality. It is no longer a question of whether it is economically feasible. The remaining questions concern when installation will begin on a large scale, and when fiber will attain cost parity with copper for various deployment strategies.

The reasons fiber will replace copper all the way to the subscriber's home are clear. Today's distribution plant is aging. It provides limited bandwidth focused on voice; it is based on a mature technology, and maintaining it is labor intensive. Tomorrow's distribution plant will be composed of a cost-effective, mass-deployed architecture and technology that supports existing services as well as new revenue opportunities. The technology for the future distribution plant is being chosen today, and the choice is overwhelmingly fiber. Moreover, that fiber is being driven deeper and deeper into the network toward the customer.

At Southern Bell, we shipped approximately 132,000 miles of fiber last year (Fig. 1). Approximately 90% of our central offices have at least one fiber-optic trunk linking it to the rest of the network. More than half of the 2,584 feeder routes in our local exchanges contain some fiber. Essentially all of the new Digital Loop Carrier (DLC) deployed in Southern Bell was implemented with fiber feeder facilities in 1990. We also replaced a portion of the embedded copper base. As 1991 begins, Southern Bell has at least 50,000 fewer copper pairs on the mainframe of the Central Offices than existed one year ago.

Approximately 500 miles of fiber are built each day in our four-state area, 353 miles of which are in the loop. We estimate an average loop length of 2.5 miles, and thus project that approximately 160 fiber feeds are added per day. The ultimate SONET (Synchronous Optical Network) 2.4 Gb capacity of these feeds is approximately 4,500,000 voice channels. Southern Bell presently serves over 10 million access lines, so we are placing the current requirements for all voice channels in the loop every two and a half days. And fiber is cheaper than copper so it meets all technology displacement criteria.

Deploying single-mode fiber in the trunk and local-loop feeder networks has produced dramatic advantages: reduced maintenance, immunity to electrolysis and power interference, reduced splicing frequency, and extraordinary bandwidth. These favorable experiences have prompted operating companies to investigate the practicality of extending the fiber optics from the DLC remote-terminal (RT) site to the curb and

Reprinted from *IEEE Circuits and Devices Magazine,* Vol. 7, No. 1, pp. 22-25, January 1991.

to the customer's premise, which would provide an all-fiber transmission system.

Southern Bell is deploying fiber to the curb today (Fig. 2), not for future services but as the most economical method of providing POTS; i.e., Plain Old Telephone Service, for certain high-density and upscale communities. The rising costs of copper, coupled with decreasing costs of fiber optics, should make fiber the economical choice (on a life cycle basis) by the end of 1991 for the average growth project. Fiber should be the economical choice for all growth projects by 1992 and for some rehabilitation projects by 1993. Once fiber attains cost parity with copper for rehabilitation, we will see escalating use of fiber optics and true volume deployment. By 1994, fiber should be the universal choice for all applications.

By 1995 fiber will be used routinely for all network growth

Fiber's Future

Over the long term, the loop (or distribution) plant will be composed of fiber all the way to the home. It will be based on SONET, which will also minimize the use of intermediate electronics, provide bandwidth allocation on demand, use standard customer interfaces, and accommodate completely mechanized operations (for provisioning or maintenance) on a real time- basis.

But how will the "fibering" of the network be accomplished? Several unresolved issues still remain. One of the toughest is the matter of power. Traditionally, power has been (and is) provided by the copper-based network (Fig. 3). Since the inception of the crank generator, and later, the central battery, this highly reliable power network has been continuously available even during the loss of commercial power. But fiber deprives us of the metallic path for powering the equipment, so we must find an alternative method to economically deliver power to the end electronics. This method must power a subscriber's equipment reliably, especially during fault conditions, without compromising lifeline service, safety, or cost.

There are at least three ways to provide power in a fiber system. The first is to extend copper along with the optical fiber, either within the same sheath as the fiber or in an external sheath. This system allows power to be provided from a remote terminal site or central office, with batteries and charger located in the remote terminal housing.

The second method terminates the fiber at the curb, or just short of the living unit, and back-feeds power from the subscriber's residence to the pedestal in order to power the optical-network interface and maintain the backup power system. The third architecture extends fiber all the way to the living unit where local power is required for the subscriber's equipment.

The regulatory perspective is that the only acceptable approach is for the operating companies to continue their historically high quality power. That means we can't rely on commercial power, and there must be adequate backup for high priority and lifeline services.

The time has come for us to settle some of the issues surrounding the powering of fiber systems in the local loop and to call for better development of alternatives. These might include solar cells or chemical fuel cells, powering through the strength members and cladding of the fiber cable itself, and an alternate gas-fired co-generation system at the remote terminal. A more elegant approach would be laser power through the transmission fiber for telecommunications services utilizing low-power electronic sets.

Once powering and other issues are resolved, traditional residential telephone services and revenues must be able to support fiber deployment on the basis of life-cycle cost. This is essential because political, regulatory, and legal issues limit the near-term opportunities for telephone companies to transport broadband services. But fiber-to-the-user technology must allow a graceful upgrade to capture future opportunities for broadband revenue, opportunities that are enhanced by starting now to build an embedded base of broadband-ready loops.

What are these opportunities? Broadband services include high-speed data transmission, such as the high-speed facsimile and wide-area-networks interconnection that Switched Multimegabit Data Service (SMDS) will offer beginning in 1991. Broadband Services also include full-motion video services such as video teleconferencing, video education, movies on demand, video phone, targeted advertising, and high-definition TV.

Starting today, the telecommunications industry must prepare itself to offer broadband's vast array of voice, data, and video services. We must reach agreements on broadband standards as soon as possible so standard offerings will not trail market demand. Among the needed standards are those to characterize a next-generation switch: a modular, broadband, dynamic-load- balancing switch that will build on the embedded base of digital switches. We also need to embed distributed intelligence within the next generation of network elements. Customers, telephone companies, carriers, equipment vendors, and governmental bodies must continue to work together if we are to attain the compatibility that will enable broadband to become a timely reality.

By 1995, fiber will be used routinely for all network growth, whether in the distribution or feeder network. This loop network of the future will support both narrowband and broadband services through the local central office. Residential applications of narrowband ISDN will be common in this time frame, but the network can also support ser-

1. At Southern Bell, fiber deployment is rapidly increasing as copper deployment decreases.

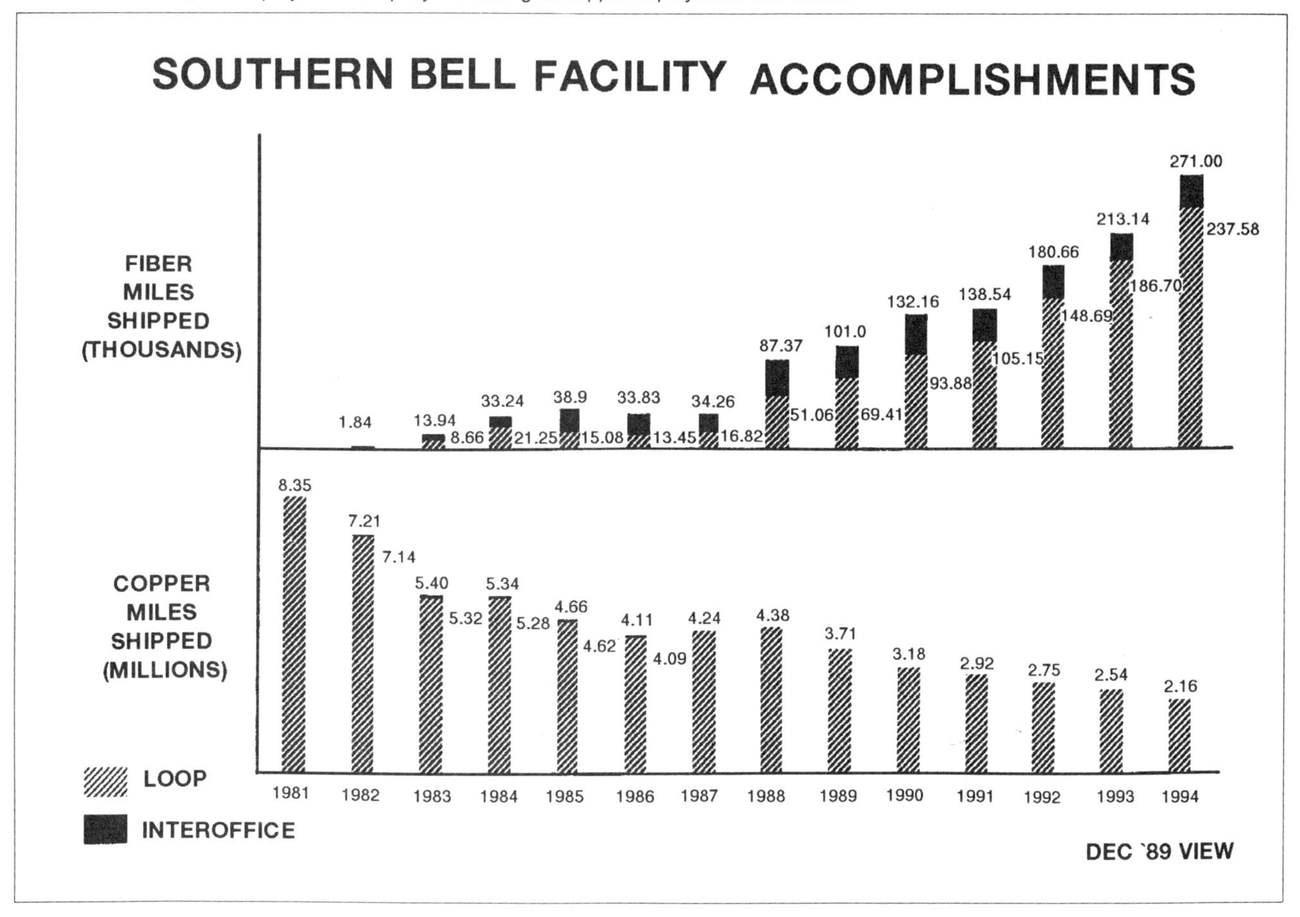

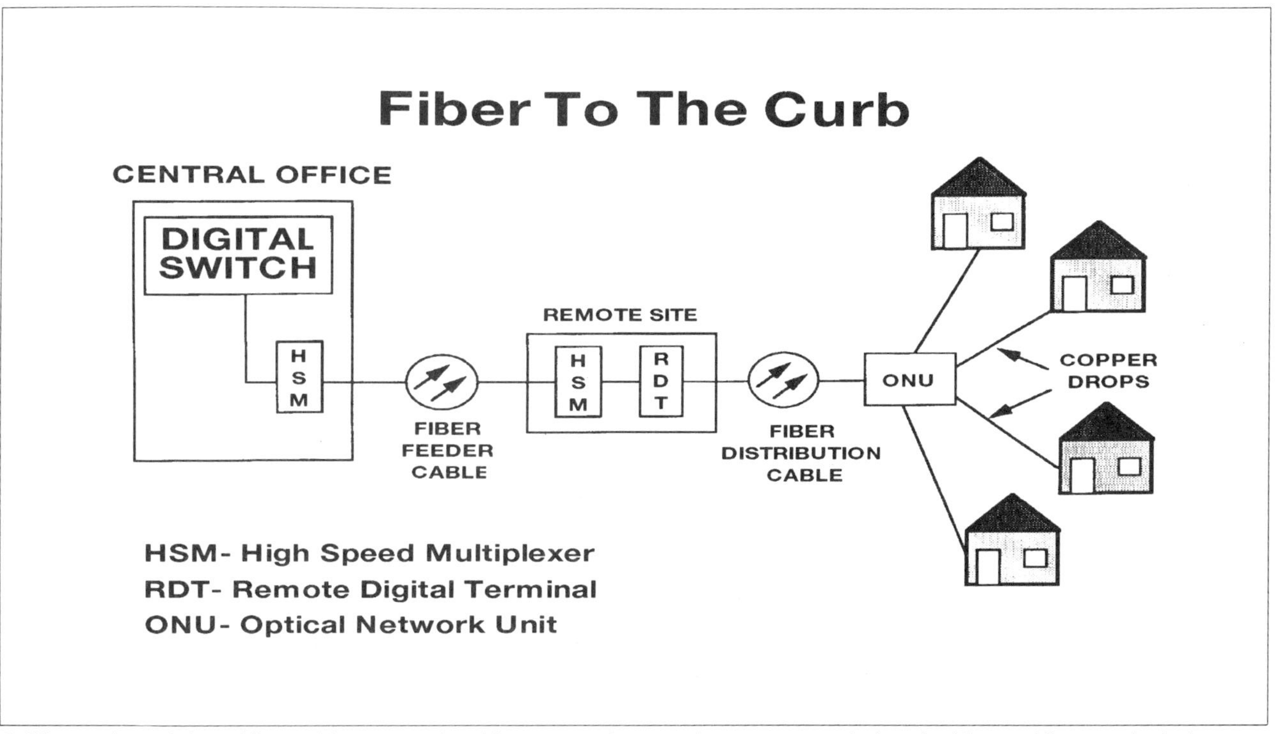

2. Fiber to the curb is rapidly reaching cost parity with copper to become the most economical method for providing standard telephone service.

3. Since the inception of the crank generator, telephone power has been provided by the copper based network; i.e., power that has been continuously available even during the loss of commercial power. But fiber deprives us of the metallic path for powering the equipment, so we must find an alternative method.

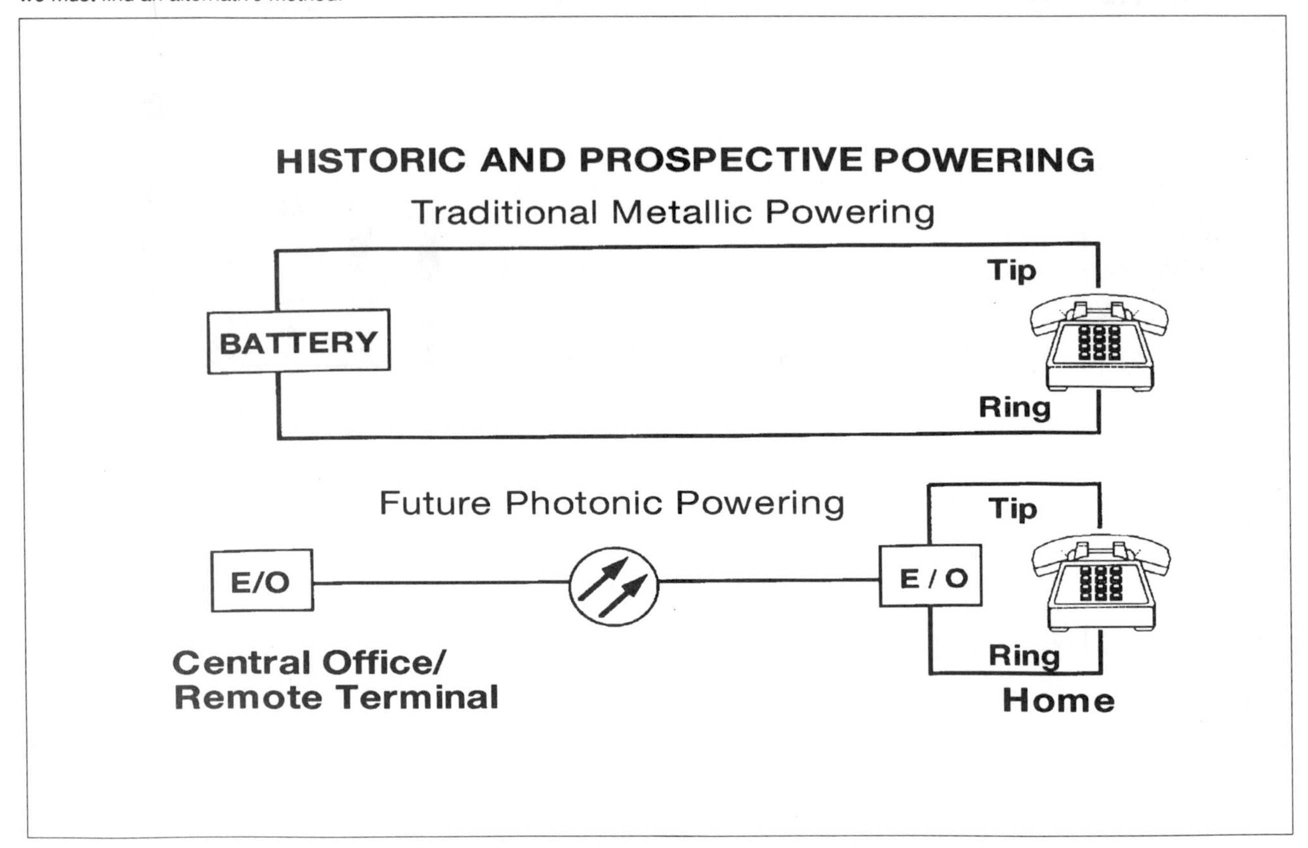

vices based on broadband ISDN. In a broadband ISDN network, the desired bandwidth can be allocated to the end customer in a dynamic, customized manner. This network will be controlled remotely by a network resource manager. It will be fully automatic and self-healing, and be able to isolate and correct problems prior to customers experiencing a service outage. Permanent repairs will be performed on a programmed basis, rather than reactively when customers lose service. This network of the future will be rich with services, but it will require far less labor to operate than the copper analog networks of today. As a result, its operation will be far less costly. For example, current indications are that the cost of maintaining a fiber cable is less than ten per cent the cost (per sheath mile) of maintaining an equivalent copper sheath.

We must settle on broadband standards as soon as possible

Fiber to the user may seem revolutionary, but it is actually the logical continuation of long-term trends in fiber-optic deployment. Optical fiber was first deployed in 1979 in the interoffice trunk network. With increasing experience and a maturing technology, the operating companies began placing fiber optics in feeder routes. By mid 1983, Southern Bell was placing fiber-optic cables into high-rise office buildings and other large customer locations. This policy has been pursued in earnest because it makes the most sense from a first cost and operations point of view. Now, hundreds of office buildings around the world are being served with fiber.

We are now laying fiber to the residences of individual customers, as we previously did to the office building, because it is the most economical way to provide, administer, and maintain the services that are offered today. Fiber to the user is the next step in the deployment of a technology that is replacing the metallic network, a network that has undergone comparatively minor changes over the last 100 years.

by Scott F. Midkiff

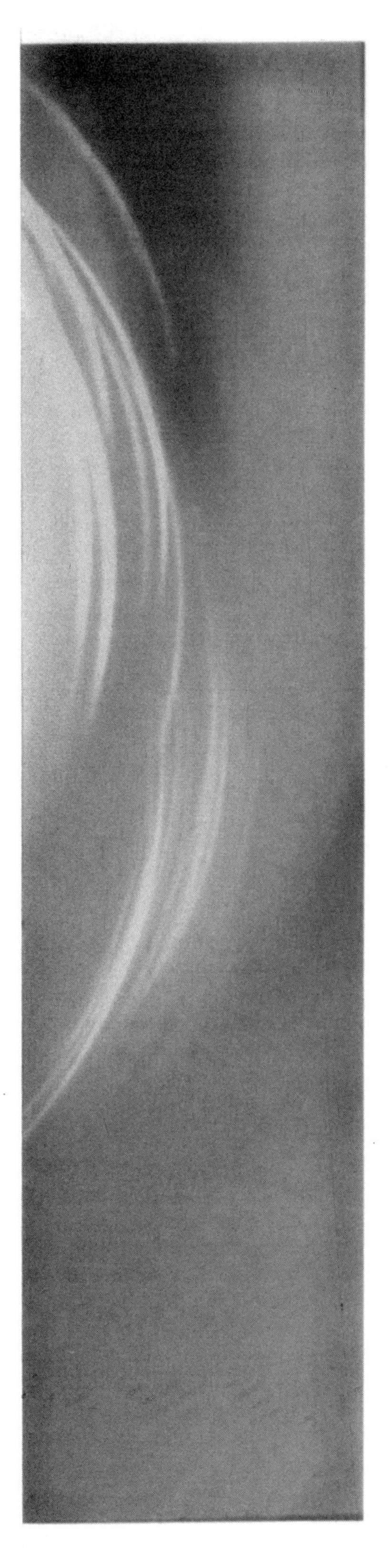

Fiber Optic Backbone Boosts Local-Area Networks

The evolving Fiber Distributed Data Interface (FDDI) for fiber optic networks dramatically increases system capacity and flexibility

Computers of all types are commonly connected by networks that permit the sharing of data, peripherals, and other resources. While steady advances have been made in the capacity of wide area networks that connect geographically distant computers, the capacity of LANs— local area networks— which connect computers within a building or a campus, has not increased significantly for about 15 years. During this time, standard LANs included Ethernet (IEEE 802.3) networks capable of carrying a maximum of 10 megabits/s (Mb/s), and Token Ring (IEEE 802.5) networks with a capacity of 4 or 16 Mb/s. Both IEEE 802.3 and 802.5 networks use coaxial or twisted-pair cable to connect computers.

A local area network standard based on fiber optics and a timed token-ring access method, now supported by suitable products, places users one step closer to more affordable, high-speed networking. With a information-handling capacity of 100 Mb/s, this LAN, the Fiber Distributed Data Interface (FDDI), is much faster than its predecessors.

FDDI, largely developed during the late 80s, employs fiber optics and integrated-circuit technology to provide a reliable LAN with moderate performance at a moderate cost. Currently, its strength lies in connecting multiple LANs in so-called backbone networks. The evolving FDDI standard is certain, however, to greatly strengthen the role of FDDI in a variety of both old and new applications.

Reprinted from *IEEE Circuits and Devices Magazine*, Vol. 8, No. 1, pp. 17-21, January 1992.

The FDDI Standard

The FDDI standard, developed under the auspices of the American National Standards Institute (ANSI) Accredited Standards Committee X3T9, chartered in 1982, takes advantage of fiber optic and VLSI technology to provide a 100-Mb/s LAN. FDDI uses a redundant ring structure along with fiber optic's inherent immunity to electromagnetic interference to achieve high reliability, and the lack of electromagnetic emissions eases security problems. In addition, FDDI exploits the low attenuation properties of fiber optic cables. The initial FDDI standard provides for LANs up to 100 km long with as many as 500 stations. Adjacent stations may be separated by as much as 2 km.

The OSI Reference Model

Computer networks are normally discussed in the context of the ISO's Open Systems Interconnect (OSI) reference model [1]. The OSI model divides a network's functionality into a seven-layer protocol stack (Fig. 1). Each layer provides specific, well-defined services to layers above it in the stack, and uses the services provided by layers below it. Each layer performs its functions in cooperation with peer layers, i.e, layers of the same type, executing on one or more other stations in the network. Any of the seven layers may be divided into sublayers employing similar up-down and peer-to-peer interactions.

The initial FDDI standard defines four protocols: Physical Layer Medium Dependent (PMD), Physical (PHY), Media Access Control (MAC), and Station Management (SMT). PMD and PHY are sublayers that provide the Physical Layer of the OSI model. MAC together with the IEEE 802.2 Logical Link Control (LLC) protocol serves as the Data Link Layer. SMT spans several layers of the idealized OSI model. PMD, PHY, and MAC standards currently exist, and the SMT standard is nearing completion [2-5].

The layered-protocol approach to networks offers a number of advantages, including the flexibility to support various higher-layer protocols, use different sublayer implementations, and to build on different physical media and data communication services. Existing higher-layer protocols, such as DECnet and the TCP/IP suite, can be built on top of LLC and can use the FDDI services. In addition, different sublayers can be employed in FDDI without affecting other parts of the standard (Fig. 2). Various Physical Layer Medium Dependent sublayers, for example, permit the use of different media or allow data on an FDDI network to be transferred over longer distances using SONET, the synchronous optical network.

Physical Layer Medium Dependent

The PMD is concerned with the physical aspects of the optical links and electro-optic interfaces. It defines the types of connectors, the types of cables, and the operation of the data driver and receiver. The initial standard uses an optical source operating at 1300 nm, multimode fiber, an LED transmitter, and a pin-diode receiver.

Alternative PMD standards have been developed or are being developed to improve FDDI's performance, lower its cost, and increase its range of application, but not all these objectives can be simultaneously achieved. One new PMD standard, SMF-PMD, extends the distance allowed between stations from 2 km to 40 km. Such an extension is useful for large FDDI LANs, such as campus-wide backbones. SMF-PMD uses single-mode optical fiber and lasers rather than multimode fiber and less expensive LEDs. A low-cost optical-fiber option, LCF-PMD, has also been proposed. TP-PMD is a PMD standard that is based on shielded copper twisted pair rather

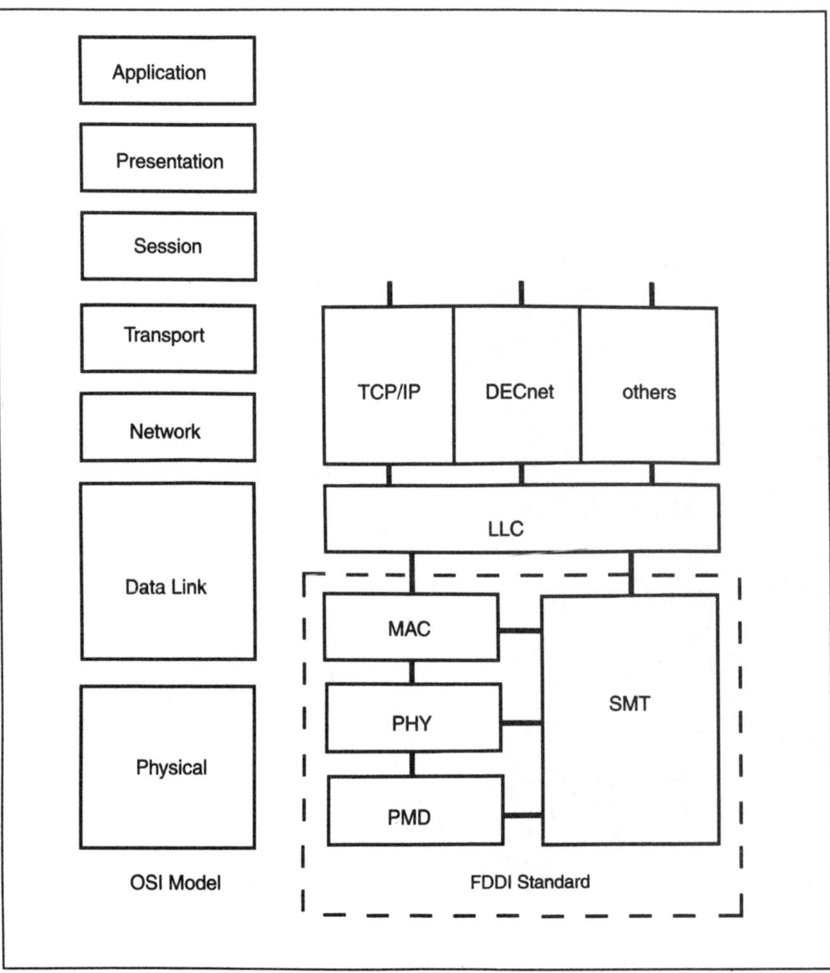

1. The fiber distributed data interface (FDDI) is a set of protocols that fit within the OSI reference model.

than on optical fiber. TP-PMD lowers cost, but decreases the allowable distance between stations. Thus far, it's been shown that 100 Mb/s can be transferred reliably at distances up to 100 m using TP-PMD.

Physical Layer

The PHY standard defines how information is encoded and exchanged between physically adjacent stations. Data is transferred over each link at 100 Mb/s using a 4B/5B non-return to zero invert (NRZI) code. For each 4 bits of data, this code transmits 5 bits, which encode one of the 16 possible data values as well as control symbols such as delimiters and idle codes. The coding scheme limits the maximum time between transitions, which eases the derivation of timing information from the received signal at each station. With this code, one signal transition is required per bit transmitted, so 100 megabits of data are transmitted each second using a 125 megabaud signal.

FDDI uses dual counter-rotating rings and allows configuration variations (Fig. 3). One ring is the primary ring; the other is the secondary or backup ring. Data moves in opposite directions on the two rings. In the event of a link or station failure, the network uses the secondary ring to avoid the failed component. PHY permits two types of physical connections for stations: single-attached and dual-attached. Dual-attached stations connect to both the primary and secondary rings, and single-attached stations connect to only the primary ring. Network adapters for dual-attached stations require two sets of PHY and PMD functions, which makes such adapters more expensive than adapters for single-attached stations. As a result, dual attachment is normally reserved for network equipment (such as routers, bridges, and concentrators) that must be very reliable. User stations such as workstations are often configured as single-attached stations, and may be connected to the FDDI ring by a concentrator.

Media Access Control

The MAC layer sends frames— packets of information— to stations on the ring on a "best-effort" basis. (There is a small but non-zero probability that a frame will not be delivered to its intended destination.) A timed token protocol implemented by the MAC layer prevents multiple stations from sending data on the ring at the same time.

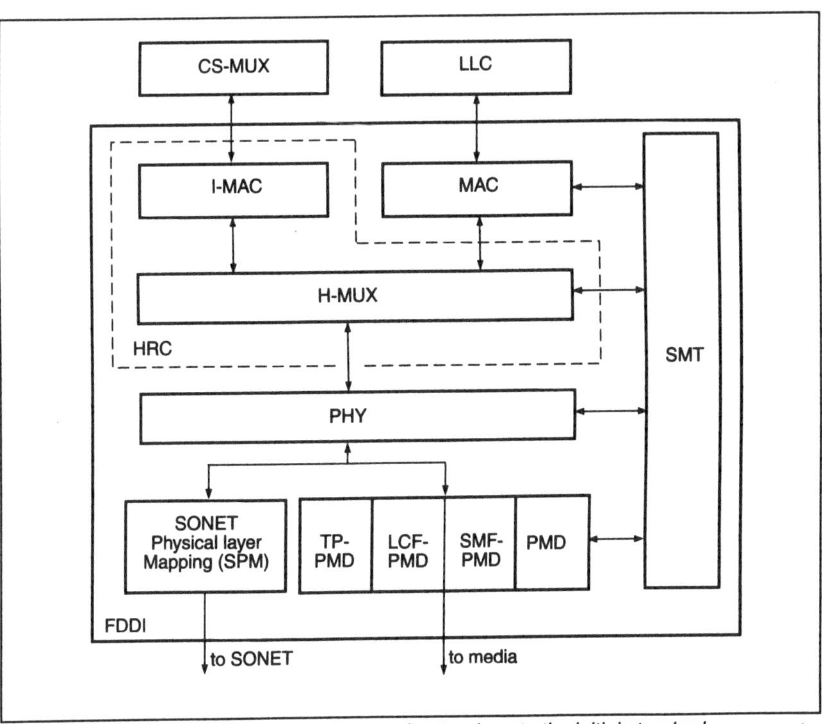

2. FDDI components include alternatives and extensions to the initial standard.

A station may use the ring to transmit data only if it is in possession of the token, which is a special fixed-length frame that circulates around the network. When a station with no data to transmit receives the token, it passes the token to the next station on the ring. When a station is ready to transmit data, it waits until it receives the token and then transmits until the data is sent, or a predetermined time limit (the token rotation time) is reached. Each frame can be no longer than 4500 bytes, but a station may transmit more than one frame while it holds the token. Since each station holds the token for a limited amount of time, it is possible to guarantee a maximum latency between token arrivals at a given station.

When data frames sent by a source station are passed around the ring, each station compares the destination address to its own. If the station is the destination and also has adequate buffer space, it copies the frame. Eventually, the frame is received by the station that sent it. The station checks for success or failure, removes the data frame from the ring, and gives up the held token by sending it to the next station.

The MAC layer is also responsible for initializing the network and detecting frame errors. When the network is initialized, all of the MAC entities in the network execute a distributed algorithm to establish a token rotation time, reconfigure the network if a fault is present, and determine which station will generate the token.

Station Management

SMT provides station-level control of FDDI operations by interacting with local PMD, PHY, and MAC layers, and with SMT entities on other stations. SMT includes a connection-management (CMT) protocol that provides for the addition of stations to a ring (and for their removal), initialization of stations, and changes in configuration. SMT uses a link error management (LEM) protocol to detect network errors and isolate faulty stations. Other functions performed by SMT include scheduling, statistics collection, and address administration.

Unlike the IEEE 802.3 and IEEE 802.5 standards, which codified standards for existing products, the FDDI standard has been developed in advance of FDDI products. The actual integrated circuits and network equipment needed for FDDI are a recent development.

Optical Technology

FDDI's PMD requirements are stringent, so optical components for FDDI tend to be relatively expensive. The desired range and

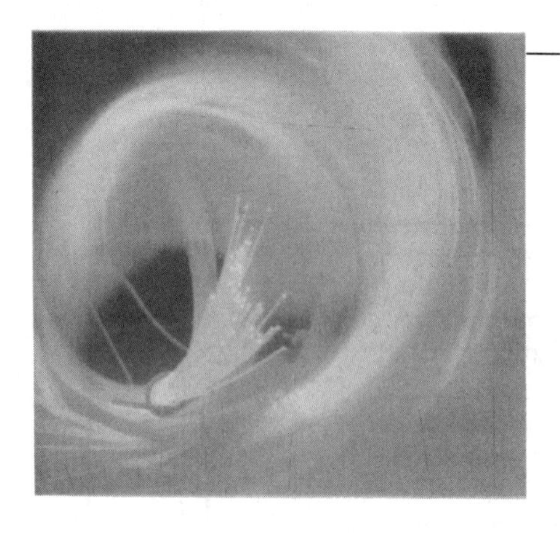

data-rate require a 1300-nm LED and amplitude modulation. The physical layer also requires a bit error rate (BER) of less than 2.5 x 10^{-10}; that is, less than four errors per 10^9 bits. Intersymbol interference is prevented by allocating adequate optical bandwidth, and data-sampling errors are minimized by budgeting jitter accumulation. An 11-dB optical power budget accommodates the several bypass switches encountered in a maximum-length run of optical fiber between stations.

The promise of reduced costs with increasing volume has not prevented the consideration of lower-cost PMD standards. One less stringent option that is suitable for networks within a building would use an 850-nm wavelength to achieve a 500-m range and a 7-dB power budget. A 125-megabaud signal would still be used.

Integrated Circuits

The speed and complexity of FDDI demands the use of application-specific integrated circuits (ASICs). Clearly, the lower FDDI layers (such as PHY) must be implemented in hardware. At the upper layers, however, there is a trade-off between implementing protocols in silicon for performance, and reducing chip functionality to reduce cost. Chip sets that implement FDDI must apply to a variety of products; a high degree of integration could severely limit the market for a chip.

The first commercially available FDDI chip set was the Supernet chip set from Advanced Micro Devices (AMD). Second-generation chip sets are now available from AMD, AT&T, Digital Equipment Corporation (DEC), Motorola, and National Semiconductor, among others. These chip sets typically implement PMD, PHY, MAC, and some SMT functions in silicon. The host processor or a dedicated embedded processor is still called upon for many SMT functions, as well as Network, Transport, and other upper-level protocols.

DEC's implementation of the Physical layer is a good illustration of chip sets for that layer [6]. DEC uses three chips, in addition to a fiber-optic transmitter and a fiber-optic receiver to implement PHY and PMD. The three chips are the ELM chip, which implements the PHY protocol, the clock and data-conversion transmitter (CDCT), and the clock and data-conversion receiver (CDCR). The CDCT and CDCR chips provide interfaces between a 5-bit parallel data path operating at 25 MHz, and a serial data path operating at 125 MHz. They are implemented in custom ECL technology to accommodate the required operating speeds. The ELM chip operates on only the 25-MHz parallel data path, and is implemented as a CMOS gate array to reduce power and cost, and to simplify design changes.

DEC has also implemented a three-chip set for MAC functions consisting of the ring memory controller (RMC), the media access controller (MAC), and a content-addressable memory (CAM)[7]. The RMC provides a direct-memory access interface between the MAC functions and the frame buffer in system memory. It is implemented in full-custom and standard-cell CMOS technology and contains approximately 87,000 transistors. The MAC chip, which performs the MAC protocol functions, is implemented as a CMOS gate array containing about 49,000 transistors. The CAM chip matches destination addresses with 64 48-bit entries that are stored on the chip, and then passes the results to the MAC chip. The chip decides if the frame is to be received or discarded, and which bits are to be set in a forwarded frame. The CAM is a custom CMOS chip containing about 44,000 transistors. Roughly 34,000 of the transistors make up the core-memory array.

Applications

Although FDDI's high data rate makes it an attractive LAN, the high cost limits its application. Currently, FDDI is used primarily in the aforementioned backbone network, i.e., one that interconnects multiple LANs, each of which may be an IEEE 802.3 or 802.5 network rather than a FDDI network. Many organizations with large and heavily used LANs have a critical need for network-to-network connections that cannot be provided by relatively slow IEEE 802.3 or 802.5 LANs.

Networking equipment for FDDI backbone networks, specifically routers, bridges, and concentrators, is available from a number of vendors. A LAN, which may comprise many stations, requires only a single router or bridge with an FDDI interface to connect to the backbone network. Concentrators reduce cost since each concentrator connects a number of stations to the backbone.

Routers typically connect two or more networks by operating at the Network layer of the OSI reference model. In addition to connecting LANs, routers can be used to connect to wide area networks. For example, a router could be used to move packets from an FDDI LAN to a remote LAN using a T1 link at 1.544 Mb/s, a T3 link at 45 Mb/s, or SONET, which has a scalable data rate starting at 155 Mb/s.

A bridge connects a LAN to an FDDI backbone at the MAC sublayer of the Data Link layer. Bridges don't concern themselves with Network or higher-layer protocols, but FDDI bridges must be designed to account for the speed differences between the 100-Mb/s FDDI and slower networks. Other MAC-level differences, including frame formats, bit ordering, padding fields, and maximum frame lengths, must also be considered.

Two widely used approaches to dealing with these differences are translating bridges and encapsulating bridges. Translating bridges convert frame formats from one LAN to another. An IEEE 802.3 frame, for example, might be reformatted by a bridge to form an FDDI frame that is sent over an FDDI backbone to another bridge, where it is translated into a valid IEEE 802.5 frame. Encapsulating bridges are often cheaper, but they are less flexible. An encapsulating bridge does not translate a received frame. Instead, it places the frame into the data field of an FDDI frame and sends it to another encapsulating bridge that must know how to unpack the original frame.

FDDI as a LAN

"Desktop" applications using FDDI to directly connect workstations and personal computers are the key to a large market for FDDI equipment. Applications that often require bandwidths of 20 Mb/s or more, such as image processing, scientific visualization, voice annotation of documents, and centralized file-server backup, could benefit from the high throughput and low latency of FDDI.

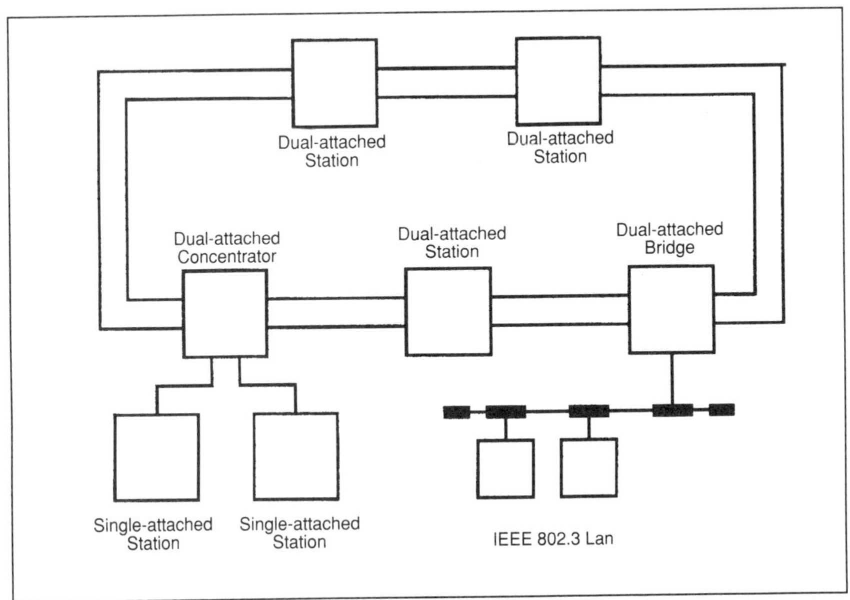

3. Network configurations can exploit FDDI's flexibility for attaching stations and subsidiary LANs.

The current high cost of FDDI limits its use as a general LAN, although wide acceptance would probably lower the cost. Another problem with using FDDI as a LAN is that many workstations and personal computers cannot execute Network and Transport layer protocols fast enough to keep up with a 100 Mb/s data stream. Potential improvements include tuning existing protocols to make them run faster, developing new and faster protocols, and using a dedicated co-processor to execute the protocols.

What's Coming?

FDDI is a fully developed standard, but its evolution will continue in response to pressures for lower costs, improved performance, and support of new applications and services. Within the framework of FDDI itself, alternatives to the initial PMD standard have already been developed to lower cost and increase distance limits. FDDI-II and FFOL (FDDI Follow-On LAN) have been proposed as larger evolutionary steps.

FDDI-II would be compatible with the current packet-switching capabilities of FDDI and would add circuit-switched service, which provides a continuous connection between two stations [8]. With FDDI-II, two stations could establish a circuit-switched connection with a data rate that can be any multiple of 8 kb/s up to 6.144 Mb/s. If necessary, multiple connections could be established to provide higher data rates. Because it is able to carry both packet-switched and circuit-switched traffic, FDDI-II would be well-suited both for backbone applications and for LANs serving emerging applications that require voice, video, or sensor-data transfer.

FDDI-II would include two Media Access Control sublayers, a MAC sublayer for packet-switched data, and an Isochronous Media Access Control (I-MAC) sublayer for circuit-switched data. One or more Circuit-Switching Multiplexers (CS-MUX) could use the services of the I-MAC sublayer. The hybrid functions of the ring, for circuit-switched and packet-switched data, would be controlled by a Hybrid Ring Control (HRC) function. A Hybrid Multiplexer (H-MUX) sublayer would multiplex traffic between the MAC and I-MAC sublayers and the Physical layer (Fig.2).

FDDI-II faces significant problems because it would require complex station-management functions for Hybrid Ring Control. SMT is the most complex part of the current FDDI standard; FDDI-II would significantly increase SMT's complexity.

FFOL is a more likely "next-generation" FDDI [9]. ANSI Accredited Standards Committee task group X3T9.5 has been authorized to start defining six FFOL standards. FFOL PMD would support both multimode and single-mode fibers operating at data rates of 600 Mb/s to 1 Gb/s. The FFOL PHY sublayer would use either PMD or SONET services at scalable data rates, and a Service Multiplexer (SMUX) sublayer would be inserted between PHY and MAC. The SMUX would be responsible for multiplexing asynchronous packet-switched data and isochronous circuit-switched data. Two MAC layers are being defined: IMAC for an isochronous service provided to one or more circuit-switched multiplexers (CS-MUX), and AMAC for asynchronous service provided to the LLC. A FFOL SMT standard is also being defined. The SMT standard would provide layer and configuration control at a station, as required by the current FDDI SMT, and would also provide an interface to a network-management agent. ***CD***

Scott Midkiff is Assistant Professor at the Bradley Department of Electrical Engineering, Virginia Polytechnic Institute and State University, Blacksburg.

References

1. W. Stallings, Data and Computer Communications (2nd ed., Macmillan, 1988).

2. International Standards Organization, Token Ring Physical Layer, Medium Dependent (PMD), ISO 9314-3, 1990.

3. International Standards Organization, Token Ring Physical Layer Protocol, ISO 9314-1, 1989.

4. International Standards Organization, Token Ring Media Access Control (MAC), ISO 9314-2, 1989.

5. American National Standards Institute, FDDI Station Management (SMT) Preliminary Draft Proposed American National Standard, ANSI X3T9/90-X3T9.5/84-49, Rev. 6.2, May 1990.

6. J. D. Hutchison, C. Baldwin, and B. W. Thompson, "Development of the FDDI physical layer," Digital Tech. J., 3, no. 2, pp. 19-30, Spring 1991.

7. H. S. Yang, B. A. Spinney, and S. Towning, "FDDI data link development," Digital Tech. J., 3, no. 2, pp. 31-41, Spring 1991.

8. F. E. Ross, "An overview of FDDI: The fiber distributed data interface," IEEE J. Selected Areas Commun., 7, no. 7, pp. 1043-51, Sept. 1989.

9. R. L. Fink and F. E. Ross, "FFOL - An FDDI Follow-On LAN," Computer Commun. Review, 21, no. 2, pp. 15-16, April 1991.

CHAPTER 6

OPTICAL COMMUNICATION AND SWITCHING

Optoelectronic Integration: A Technology for Future Telecommunication Systems

R. F. Leheny

A principal goal of optoelectronic device research for more than a decade has been increased bit-rates, transmitted over ever increasing fiber distances. Success has made fiber the transmission medium of choice for long haul telecommunications systems. In turn, the expanded information carrying capacity of optical networks has created worldwide expectation of a new "information age," an age in which fiber optic-based networks will provide links of nearly unlimited bandwidth between information providers and information consumers. Optoelectronic device technologies for local loop applications, at a cost compatible with these broad-band network markets, remains one key challenge to realizing this goal. The required device technologies are very different from those developed to meet the "bit rate-distance product at any cost" challenge of the past, but these low cost devices will find significantly larger markets than present long haul systems – markets made even larger when non-telecommunications specific applications, such as dedicated LANs and optical interconnects for high speed signal processors are included.

The integration of multiple optical and electronic functions on a single chip to achieve Opto-Electronic Integrated Circuits, or OEICs, offers one approach to low cost, high performance components to meet these needs. OEICs, such as the proposed integrated coherent receiver illustrated in Fig. 1a, represent the integration of light source with drive electronics, detector with receiver electronics and signal distribution waveguides with photonic and electronic components. The level of integration illustrated is beyond today's capabilities, but research to realize the various components is encouraging.

OEIC chips, integrating very different device technologies, differ significantly from silicon integrated circuits where the commonality of device function facilitates high yields in manufacture. Today the material and device fabrication technologies required to achieve OEICs represent a research challenge, but as was the case for silicon ICs in the past, mastery of fabrication complexities can be expected to reduce costs for subsystems, and in turn generate new applications as the cost of converting between photon and electron becomes insignificant. Reflecting the principal function of these chips, interfacing photons and electrons, the acronym OEIC should be taken to mean Opto-Electronic *Interface* Circuit. In this article we review the current status of OEIC research and look to possible future developments.

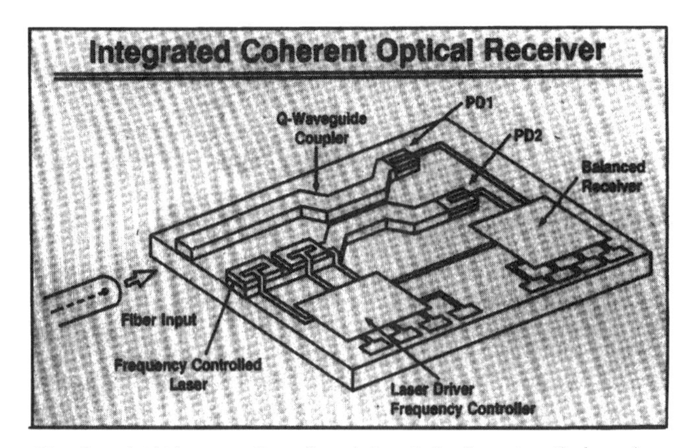

Fig. 1a Artist's conception of an integrated coherent optical receiver OEIC chip incorporating a local oscillator laser, laser frequency control element, waveguide distribution of local oscillator and signal, dual detectors and receiver preamp. At present no report of such a complex OEIC chip has been made, but various combinations of the components illustrated have been investigated.

Beginning in the late 1970s with the pioneering work of Yariv's group at Cal Tech [1], initial OEIC investigations built on the expanding base of GaAs IC technologies being researched for high speed circuits. One of the milestones achieved by the Cal Tech group [2], the realization of a complete transceiver – detector, amplifier, and transmitter – on a single chip is illustrated in Fig. 1b. It also illustrates most of the key problems facing OEIC technologies. The three devices have very different material and structural requirements. The laser requires limited chip width, typically 150–300 μm, corresponding to the laser cavity length, and fairly heavily doped layers of p-and n-type material on either side of the active region. The limited dimensions of the chip restrict the layout of FET amplifier circuits, while FET doping requirements call for tight control over the channel layer doping-thickness product, doping requirements that are incompatible with the laser structure. Finally, the p–i–n photodetector incorporates three distinct doped regions, with the constraint that the lightly doped "intrinsic" i-region, in an optimum design, be thick enough to insure good quantum efficiency with low enough doping to insure full depletion at modest applied voltage to provide the low capacitance and dark current required for sensitive receiver operation. For the OEIC illustrated, the laser structure was grown on top of the FET channel with this material etched away except where required. This

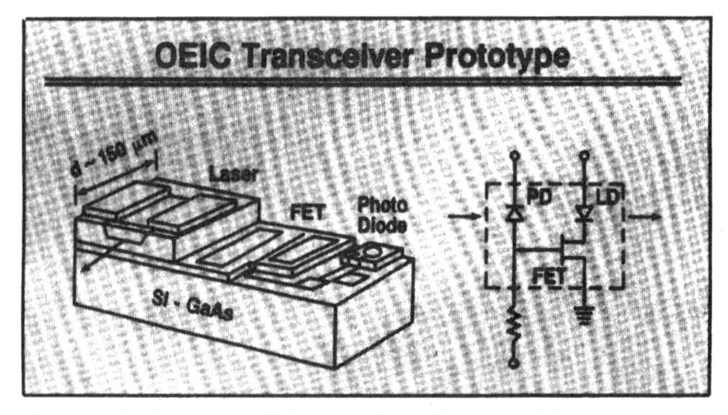

Fig. 1b Rudimentary OEIC transceiver chip reported in 1978 by Yust et al. [2] illustrating the integration of photodetector, preamp and laser.

Reprinted from *IEEE Circuits and Devices Magazine*, Vol. 5, No. 3, pp. 38-41, May 1989.

"vertical" approach accommodates the material incompatibility problem but results in a non-planar surface. Lack of planarity in turn creates difficulties for the sub-micro lithography required for high performance FETs.

These problems notwithstanding, the potential of OEICs for applications such as high speed optical interconnection of circuit chips in advanced high speed systems was widely recognized. In the U.S. this led the Defense Advanced Research Projects Agency (DARPA) to become an early sponsor of OEIC research, coupling this research with advancing high speed GaAs IC technology. DARPA programs at both Rockwell and Honeywell Corporate Research Centers [3,4] addressed most of the difficulties illustrated by the rudimentary device in Fig. 1b. Planarization of the OEIC chip was achieved by growing the laser structure in a "well" etched into the substrate. Incorporating at least one processed laser mirror removed restrictions on chip size, an approach that requires advanced materials processing capabilities and compromises laser threshold somewhat. Furthermore, the use of ion implanted electronics, and a novel metal-semiconductor-metal (M-S-M) photodetector compatible with FET processing (Fig. 2), greatly simplified overall processing. With this detector design, contact metallization is done at the same processing step as the FET gate metallization, facilitating the integration of detectors with electronics.

At Honeywell these efforts led to the demonstration of MSI levels of integration with more than 500 FET gates integrated with a photodetector and laser on the same 1 Gbit/sec transceiver chip (Fig. 3) [3]. Recently researchers at IBM [5] have demonstrated very high speed operation (5.2 Ghz BW) of a monolithic GaAs optoelectronic receiver circuit incorporating an MSM detector. In a separate IBM effort [6] even more complex GaAs OEIC receiver circuits, integrating over 2000 devices to perform complete optical receiver functions including detection and optical clock recovery at comparable bit rates, have been demonstrated.

In Japan, beginning in the early 1980s, the Ministry for International Trade and Industry (MITI) undertook sponsorship of an ambitious project to investigate a wide variety of optoelectronic technologies aimed at demonstrating applications for high speed LANs, including work on GaAs OEICs. This program involved a number of independent corporate research groups, as well as the establishment of a central Optoelectronic Joint Research Lab (OJRL) with the charter to investigate a range of more generic materials and materials processing problems all aimed at advancing technologies important for OEICs. This MITI program achieved

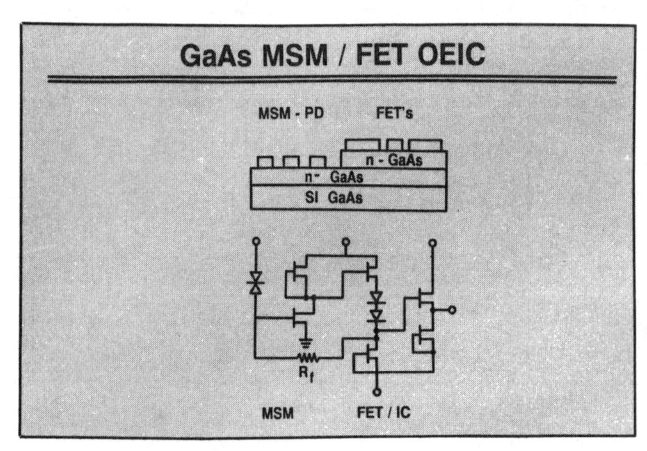

Fig. 2 Cross section of a GaAs metal-semiconductor-metal (MSM)/FET receiver OEIC with schematic of simple two-stage FET preamp.

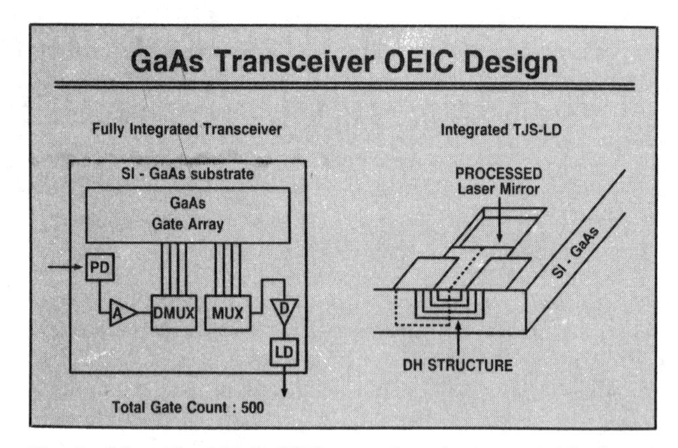

Fig. 3 Schematic of GaAs OEIC transceiver circuit reported by Ray et al. [3] of Honeywell illustrating the use of laser structure recessed in substrate to achieve planarization of wafer and layout of GaAs electronic functions. Typical of OEICs designed to perform complex electronic functions, the chip area is dominated by the electronics.

a number of impressive OEIC results that include the high performance multi-chip 4×4 optical receiver switch demonstrated by researchers at Fujitsu and illustrated in Fig. 4 [7]. Using three independent chips, the switching of four input optical bit streams into four output streams at rates up to a Gbit/sec is accomplished. For the receiver chip, Ito and co-workers [8] at Fujitsu were among the first to make use of the MSM photodetector which, as discussed above, has become the detector of choice for integration with GaAs MESFETs (Fig. 3). Today the initial five year MITI program has been completed and a new thrust, aimed at very high bit rate (10 Gbit/sec) and OEICs compatible with long wavelength telecommunication applications, has taken its place [9].

For telecommunications applications, where the operating wavelengths of fiber systems are at 1.3 and 1.5 μm, GaAs optical devices operating at 0.8 μm are not appropriate. Instead devices based on the quaternary alloy material InGaAsP, which can be grown lattice matched to InP, are the materials of choice for fabricating emitters and detectors. Research on OEICs based on these materials began in earnest sometime after the initial work on GaAs OEICs. Today the range of laboratories investigating long wavelength OEICs extends across the spectrum of telecommun-

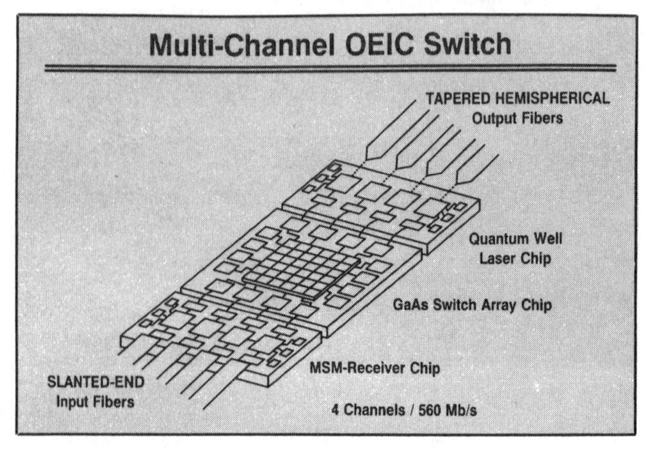

Fig. 4 Multichip OEIC 4×4 optical switch reported by Iwama et al. of Fujitsu [6] illustrating the use of a separate MSM-OEIC receiver array chip with a GaAs high-speed cross-point switch chip and a quantum-well laser array OEIC chip to achieve a chip set capable of high-speed switching of four input optical channels into any of four output optical channels.

ication research in the US, Japan and Europe. In Europe support for OEIC research is largely based on European Community-sponsored R&D programs such as ESPRIT and RACE which are supporting efforts teaming researchers in Germany, France and Great Britain [10].

While, as with the case of GaAs OEICs, both transmitter and receiver circuits are being researched as a means to achieve low cost, high performance components, for some time it has been recognized that in principle the electronic properties of InP-based materials offer advantages over GaAs that should translate into improved performance for OEIC receivers relative to hybrid pin/FET receivers (InGaAs pin detector and GaAs FET amplifier). However, realization of these advantages has been stymied by the fact that these materials do not lend themselves to a simple MESFET technology. One result is that there has been extensive research into alternative electronic device structures such as MISFETs, J-FETs [11,12] and heterojuction bipolar transistors for integration with optical devices. Recent reports on the excellent high frequency performance of discrete FET and HBT devices (f_t>200 Ghz for AlInAs/InGaAs MOD-FET [13] and correspondingly high transition frequencies for InP/InGaAs/InP HBTs [14, 15]) are dramatic confirmation of the potential for high performance electronic components using InP-based materials.

Considering that the realization of complex material structures represents the major technology hurdle to high performance OEICs, it is remarkable to note that impressive laser structures, fabricated with multiple material processing and regrowth cycles, have become routine, while in general the technology for InP OEICs has not advanced much beyond the small scale integration of the type illustrated in Fig. 1b. In large part this reflects the still relatively immature state of materials growth technologies for this system where until recently only liquid phase epitaxy (LPE) has been available to grow consistently high quality structures. As is well known, LPE produces some of the best material for lasers, but it in general lacks the control over layer thickness required for high performance electronic devices. Except for the isolated results on discrete electronic components recently reported, which represent the initial results of significant advances in materials preparation technologies, the general absence of high quality material structures has meant that InP and InGaAs electronic devices integrated with optical devices have not been able to match the commercially available high transconductance, low input capacitance of GaAs short channel MESFETs and HEMTs. One consequence is that receiver sensitivities achieved with InP-based OEICs have been disappointing when compared to results obtained with hybrid circuits.

Fig. 5 summarizes photoreceiver results obtained for OEICs and the best hybrid circuits. This figure illustrates the relationship between the number of detected photons required to achieve 10^{-9} bit-error-rate and bit-rate for an FET transimpedence photoreceiver incorporating high performance FETs (G_m = 50), with receiver sensitivity and net input capacitance as parameters [16]. As is readily evident, present OEICs have 10 dB poorer sensitivity than hybrid devices, while hybrid circuit performance is consistent with approximately 2 pf input capacitance. Most estimates are that an integrated pre-amp (OEIC) should be capable of achieving input capacitance a factor of four lower than this value. With such low input capacitance OEICs should outperform hybrid circuits by as much as 10 dB. Nevertheless,

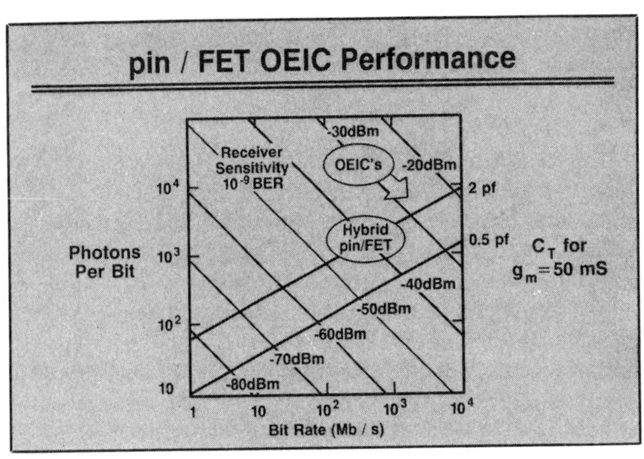

Fig. 5 The relationship between the number of photons per bit required to achieve 10^{9} bit error rate and system bit-rate, with receiver sensitivity and net input capacitance as parameters. The capacitance values are calculated assuming an FET transconductance of 50 mS. The two clouds indicate the range of sensitivities reported for hybrid pin/FETs and OEICs. The arrow indicates the improvement in OEIC performance recently reported for an OEIC using GaAs MESFET electronics fabricated using material grown non-lattice matched on an InP substrate [13].

at present the preparation of material structures compatible with such high performance electronic devices integrated with low capacitance, low leakage current detectors remains a research goal. One interesting advance has been the recent demonstration at Bellcore of an MSM detector technology for this material system [17] that should greatly simplify fabrication of integrated receivers.

On the transmitter side, for telecommunications applications, the performance advantages of OEICs are expected to be principally in the area of high bit rate systems. Integration of laser driver electronics, by minimizing lead inductance, offers advantages for very high speed modulation of lasers and, in the case of laser arrays, could greatly simplify driver electronics by minimizing cross talk. Yet it is interesting to note that at present only one commercially available OEIC exists, a Matsushita laser chip [18] incorporating a laser and three driver bipolar transistors. In the long term the integration of lasers and optoelectronic components is expected to make possible the integrated coherent receivers of the type illustrated in Fig. 1a and to provide some for complex frequency controlled lasers. Incorporation of on chip modulators and related electronics can also allow for the possibility of some novel bistable devices.

Particularly encouraging are recent advances in growth techniques for InP-based materials, including MOCVD and gas source MBE, each of which offers promise of greatly improved control over material structures for future OEICs. In addition, significant advances continue to be made in non-lattice matched epitaxy that is beginning to yield some excellent quality epitaxial material. While complete control over the deleterious effects of crystal defects has yet to be demonstrated, interest in these composite material structures remains very high because of the wide range of their potential applications. One such OEIC application is to circumvent the problems associated with InP electronics by separately optimizing the optical and electronic materials to achieve enhanced OEIC performance. Recent work of researchers at NEC illustrates this with their application of MBE to the heteroepitaxy of GaAs FET structures on InP substrates to demonstrate the first OEIC receiver with sensitivity approaching what has been achieved with hybrid circuits [19]. Similar material structures exhibiting excellent FET characteristics have been realized with MOCVD deposition at Bellcore [20].

This brief overview of work on OEICs has attempted to point up the promise of this emerging technology while highlighting the challenges that remain to be overcome. In anticipating future progress in this field it is helpful to be reminded that for the past two decades research in semiconductor optoelectronic device technologies has been marked by continued rapid advances in material growth techniques and device fabrication technologies. This momentum for innovation, combined with increasing pressure to extend fiber systems into the local loop, can be expected to maintain the pace of future achievements in OEICs.

References

[1] Yariv, A., *IEEE Trans. Electr. Devices*, EDL-31, 165, 1984.
[2] Yust, M. et al., *Appl. Phys. Lett.*, 35, 795, 1979.
[3] Ray, S. et al., *Proc. SPIE Conf.*, Vol 703, 100, 1986.
[4] Kilcoyne, M. K. et al., *Proc. SPIE Conf.*, Vol 703, 148, 1986.
[5] Harder, C. S. et al., *IEEE Electr. Device Lett.*, EDL-9, 171, 1988.
[6] Crow, J. D. et al., *IEEE Trans. Electr. Devices*, T-ED.
[7] Iwama, T. et al., *J. Lightwave Tech.*, LT-6, 772, 1988.
[8] Ito, M. et al., *IEEE Electr. Device Lett.*, EDL-5, 531, 1984.
[9] Hayashi, I., private communication.
[10] Shearman, C. *Physics Bull.*, 39, 152, 1988.
[11] Forrest, S. R. et al., *J. Lightwave Tech.* LT-3, 1248, 1985.
[12] Kim, S. J. et al., *IEEE Electr. Device Lett.*, EDL-9, 447, 1988.
[13] Mishra, U. K. et al., *IEEE Electr. Device Lett.*, EDL-9, 654, 1988.
[14] Nottenburg, R. N. et al., *IEEE Electr. Device Lett.*, EDL-10, 30, 1989.
[15] Mishra, U. K. et al., *Tech Digest, 1988 Electr. Device Meeting*, 873, 1988.
[16] Li, T., *IEEE J. on Selected Areas of Comm.*, SAC-1, 356, 1983.
[17] Shumacher, H. et al., *IEEE Electr. Device Lett.*, EDL-9, 609, 1988.
[18] Shibata, J. et al., *Appl. Phys. Lett.*, 45, 191, 1984.
[19] Suzuki, A. et al., *Electr Lett.*, 23, 954, 1987.
[20] Lo, Y. H. et al., *IEEE Electr. Device Lett.*, EDL-9, 383, 1988.

Photonic Time-Division Switching Systems

H. S. Hinton

Abstract

This paper will review some of the photonic time-division fabrics that have been either proposed or demonstrated.

Introduction

In telecommunications there are physical space channels used to connect users at point x to users at point y with an available channel bandwidth B_c. At each entrance point to these physical space channels are users requiring a bandwidth B_u between these two locations. When the available bandwidth of the channel equals the desired bandwidth of the user ($B_c = B_u$), one space channel should be assigned between each pair of users. On the other hand, when the available channel bandwidth is much greater than the user bandwidth ($B_c >> B_u$), it is desirable to share the available channel bandwidth between several users by allowing *multiple access* to the same space channel. This multiple access can be accomplished by multiplexing several users in either the time or spectral domain as illustrated in Fig. 1.

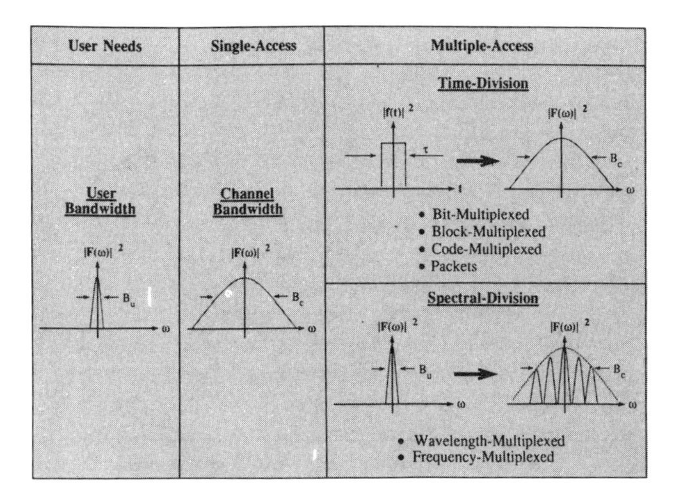

Fig. 1 Bandwidth utilization of a single space channel.

To multiplex several users onto the same channel in the time domain, the pulse widths of the information passing through the channel are shortened until they fill the available bandwidth. This multiplexing technique is referred to as time-division multiple access (TDMA). Several possible implementations of TDMA include bit or block-multiplexed data, code-division multiplexing (spread spectrum), and packet switching.

The second multiplexing method that can be used to fully utilize the available channel bandwidth is to operate in the spectral domain rather than the time domain. This can be accomplished through either wavelength-division multiple access (WDMA) or frequency-division multiple access (FDMA). WDMA refers to the case when the information from a single user is modulated onto an optical carrier. For the case of switching, each user is assigned a fixed transmitting (receiving) wavelength but has the capability to receive (transmit) the wavelengths of all the other users. As an example, for the case of a fixed transmitting wavelength per user, the information to be transported from one user to the other is modulated onto its assigned wavelength λ_1. The receiving user can then lock its tunable receiver onto the wavelength λ_1 and can receive the information.[1]

FDMA, on the other hand, electronically multiplexes several different frequencies together, and then uses this composite signal to modulate an optical carrier. This is also referred to as subcarrier multiplexing.[2]

This paper will focus on several of the proposed implementations of TDMA. The first section will begin by reviewing the two time-division multiplexing methods: bit-multiplexing and block multiplexing. This is followed by a description of code-division multiple access (CDMA). The next section will discuss several photonic switching systems that could be based on TDMA schemes. Finally, there will be a brief discussion on multi-dimensional switching systems that combine both TDMA and space switching.

TDMA Multiplexing Techniques

Prior to discussing photonic switching systems based on TDMA there needs to be a brief discussion on the different types of TDMA multiplexing techniques. This section will begin by discussing bit-multiplexed TDMA. This will be followed by a description of block-multiplexed TDMA. Finally, there will be a brief review of the spread spectrum approach to TDMA, commonly referred to as CDMA.

A bit-multiplexed data stream is created by interleaving the compressed or sampled bits from each of the users. This is illustrated in Fig. 2 where the bit information from four users are combined in an interleaved fashion into a

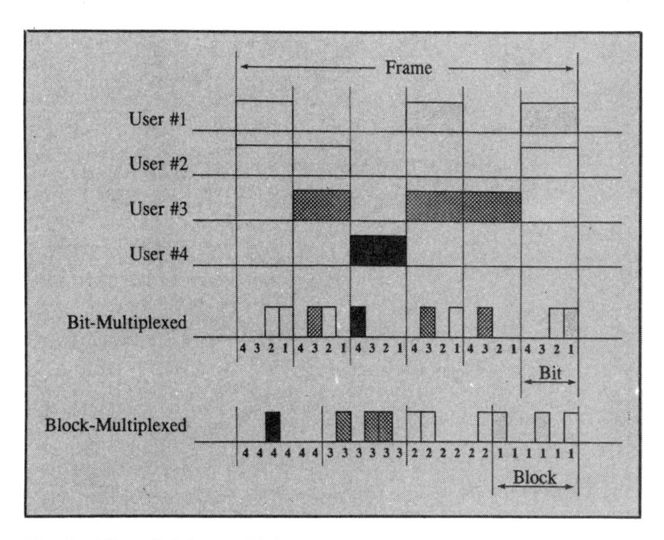

Fig. 2 Time-division multiple access for both bit and block multiplexing.

Reprinted from *IEEE Circuits and Devices Magazine*, Vol. 5, No. 4, pp. 39-43, July 1989.

time period equal to the duration of an uncompressed bit. Also note that this type of multiplexing requires that all of the users be bit synchronized. This type of multiple access is the multiplexing method of choice for most transmission systems since it only requires the storage of one bit of information for each user at any time. Unfortunately, most of the bit-multiplexed transmission systems are further complicated by adding pulse-stuffing and other special control bits to the data stream.

In the switching environment, this multiplexing scheme requires the capability to reconfigure the switching fabric in a time shorter than a single bit duration. For the case of nonreturn to zero (NRZ) formats, the reconfiguration will need to be significantly shorter than a single bit duration, while return to zero (RZ) formats allow one half of the bit time for reconfiguration.

Block-multiplexing, on the other hand, stores a frame's worth of information from each of the users and then orders the bits entering the channel such that each users data is contiguous. This also is shown in Fig. 2. When used in a switching environment, this multiple access method requires the switching fabric to reconfigure only at block boundaries. By allowing a small amount of dead time between the blocked-multiplexed information, the requirements on the reconfiguration time of the fabric can be relaxed. This can be attractive for switching systems such as lithium niobate systems that have slow reconfiguration times.[3]

The third method of multiplexing information onto a single channel in time is code-division multiplexing. This type of multiplexing is accomplished through the use of either orthogonal or pseudo-orthogonal codes to represent both the bits and the users.[4-6] In part (a) of Fig. 3, different code sequences, one associated with each user, are used to represent the bits. Each bit, then, is represented by the unique code of the user. When no bit is present, there will be no information present on the user's input channel.

In part (b) of this figure a conceptual implementation of multiplexing using CDMA is illustrated. Assuming bit-synchronized inputs, the CDMA scheme begins by the generation of a short pulse for every bit entering the system. In the figure this operation is labeled as the pulse generator. This pulse is then split among k fiber delay lines. The code sequence representing a given input channel is then composed of a unique collection of pulses (chips) of different delays. The code sequences from all the encoders are then combined and injected onto a single space channel. This CDMA channel is then the superposition of all the different code sequences generated by the encoders. The CDMA decoders, like the encoders, begin by splitting the optical energy among a group of fiber-delay lines. The decoder for user j has to undo the code sequence generated by its corresponding encoder. The fiber delay lines in the decoder are set at the appropriate lengths to combine all the individual pulses of a code sequence into a single pulse at the end of a normal bit duration. Since the bit-codes are either orthogonal or pseudo-orthogonal, a simple thresholding decision determines whether a bit is present or not. Finally, the output of the decoder, assuming a bit is present, has to be integrated or stretched to the appropriate bit duration to communicate with the outside world.

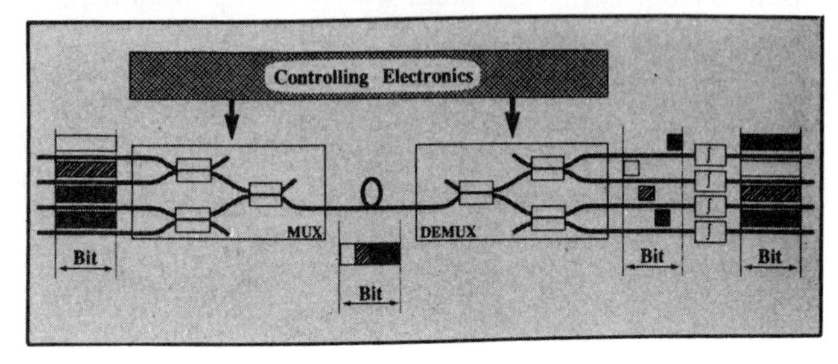

Fig. 4 A time-division bit switch.

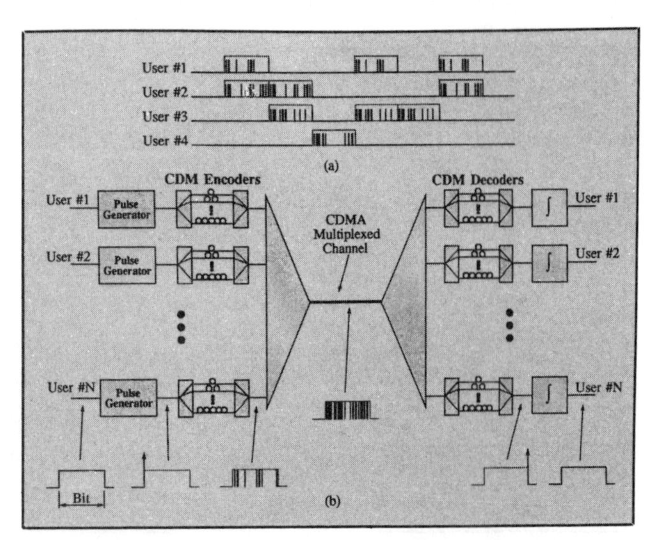

Fig. 3 Example of code-division multiple access multiplexing where a) illustrates the representation of users' bits and b) is a simple CDMA multiplexing implementation.

TDMA Switching Systems

The purpose of this section is to outline several photonic TDMA switching systems that have been either proposed or demonstrated. This section includes discussions on systems that perform bit-switching, block-switching, and CDMA switching. Finally, packet switching will be discussed.

Perhaps the simplest example of TDMA switching is bit-switching as shown in Fig. 4. [7-8] In this figure, bit-synchronized information is sampled by the multiplexor. In this case there are four inputs forcing the bit duration of the sampled data to be less than one-fourth of the original bit duration. Under the direction of the controlling electronics the demultiplexor then directs each of the sampled bits to the appropriate destination. These sampled outputs must

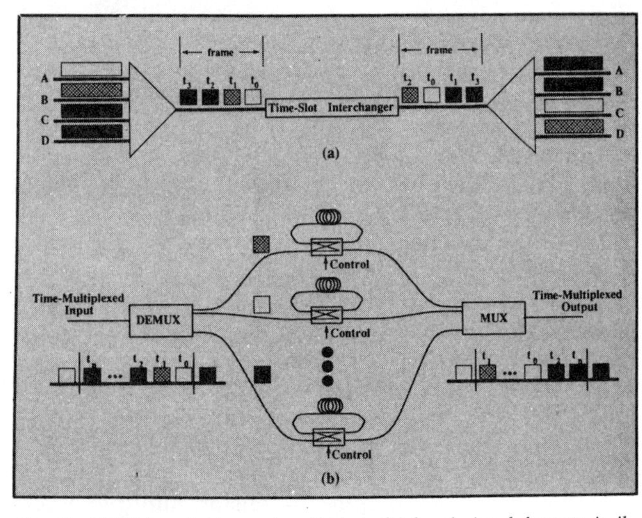

Fig. 5 TDMA switching using block-multiplexed signal formats is illustrated using a a) conceptual time-slot interchanger, and b) an example of a photonic implementation of a time-slot interchanger.

then go through a device that can stretch the sampled information to an entire bit duration. An example of this could be a bistable laser diode.[9]

The first example of TDMA block-switching is the sharing of a linear bus. An example of this occurs when all users have both read and write access to the same bus. Depending on the control scheme, a user can write information onto the bus. The receiving user, once knowing when to read the bus, can receive the information. The control for bus structures can be either centralized using pre-assigned time-slots, or distributed using a packet environment where all users continually monitor the bus, looking for information directed to them.

Another example of TDMA switching is the time-slot interchanger (TSI) illustrated in Fig. 5. In part (a) of this figure, the four input signals are time-multiplexed onto a single space channel. User A is put on the bus first, with user D being last. The TSI provides the function of interchanging these time-slots of information in time. For this example, user A's time-slot has been moved into the third time-slot. Since the TDMA demultiplexor will direct the first time-slot to user A, a connection has been made between users A and C. Also, notice the connections between users B and D, users C and B, and finally, users D and A. Part (b) of the figure illustrates a proposed photonic implementation of the TSI.[10] The input time-slots of the TSI are directed to fiber-delay lines where they can recirculate until needed at the output. The fiber-delay lines must create a time delay equal to the duration of a time-slot. As an example, the input time-slot t_0 will need to pass through the fiber-delay line $n+2$ times, while input time-slot t_n will pass through the fiber loop only once.

Regardless of the type of multiplexing used, whether bit or block, there will still be the need to synchronize the incoming data to bit, and in most cases, frame boundaries.[11] This is illustrated in Fig. 6. In part (a) of this figure, block multiplexed inputs are received through the input regenerators by the photonic switch. Since each of the inputs pass through a different amount of fiber, and fiber has a phase delay associated with temperature of 42 ps/km°C[12], each of the input blocks could arrive at a different point in time. Regardless of how fast the photonic switch can reconfigure itself, there will still be bit phase discontinuities in the switched output channels. As an example, if there is no bit-phase alignment of the channels entering the space switch, after switch reconfiguration there will be no phase relationship between adjacent bits on a given output channel. These phase discontinuities could force the regenerators downstream to begin their resynchronization process. This prevents any information from passing through the network for 100s of nanoseconds. This is unacceptable! To prevent this problem, elastic stores are required to line up both the bit and frame boundaries as illustrated in part (b) of the figure.

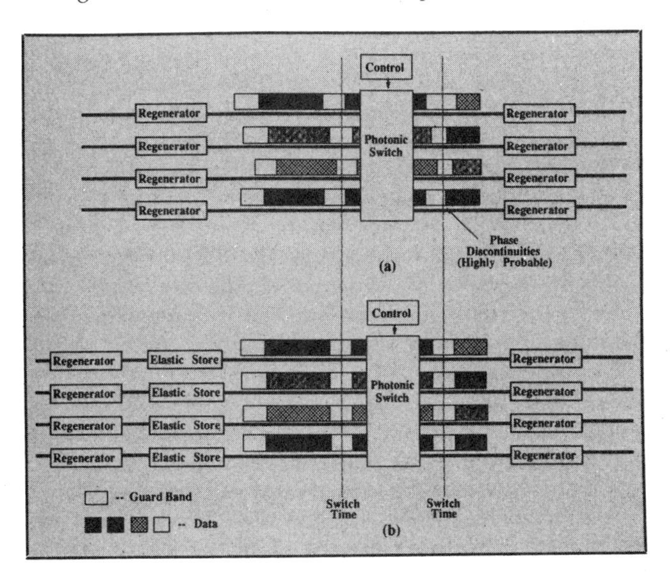

Fig. 6 (a) Unsynchronized photonic switching system, (b) synchronization through the use of an elastic store.

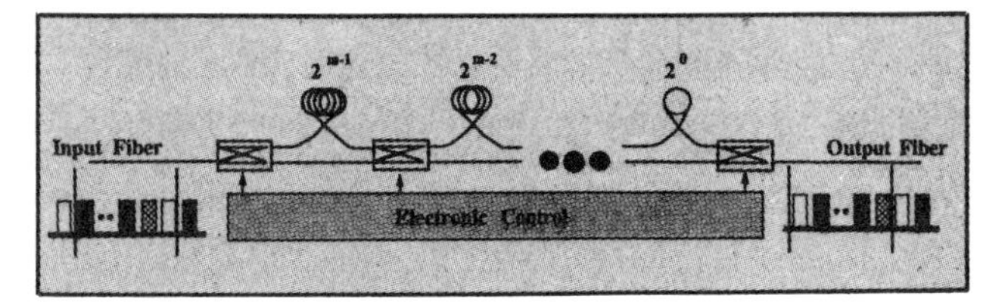

Fig. 7 An example of a photonic elastic store.[13]

A photonic elastic store can be implemented by connecting variable lengths of optical fiber with directional couplers as shown in Fig. 7.[14] For this sytem, when the directional coupler is put in the bar state the incoming light will not pass through the next fiber loop. On the other hand, if the coupler is put in the cross state the entering light will be forced to travel through the upcoming fiber loop. Each of these loops can be weighted to be integer multiples of minimum allowed bit error for the bit alignment section, or integer multiples of the bit duration for the frame alignment component of the elastic store. The delay required to line up both the bit and frame boundaries can be calculated by electronically monitoring the input data stream.

Another example of TDMA switching is ring networks.[15] For a synchronous ring structure, each user is assigned a unique piece of time (time-slot) that is used to read the information on the ring. Other users can send information to a user by entering information into the destination user's time-slot. Access to the time-slots is arbitrated by the centralized control. There are also many other schemes for using ring structures is switching applications both with centralized control and distributed asynchronous control schemes based on packet structures.[16]

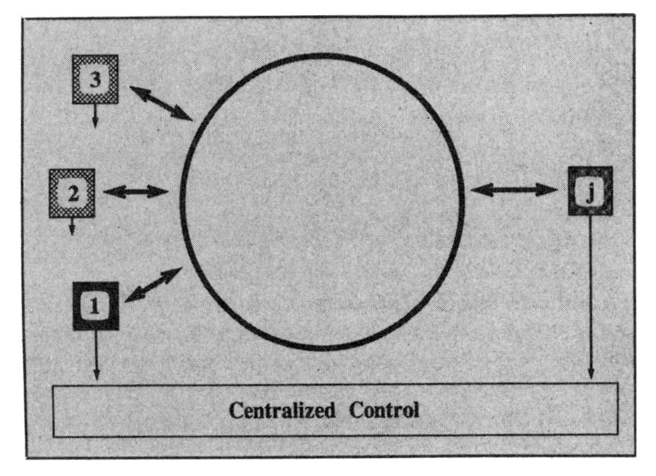

Fig. 8 Ring network.

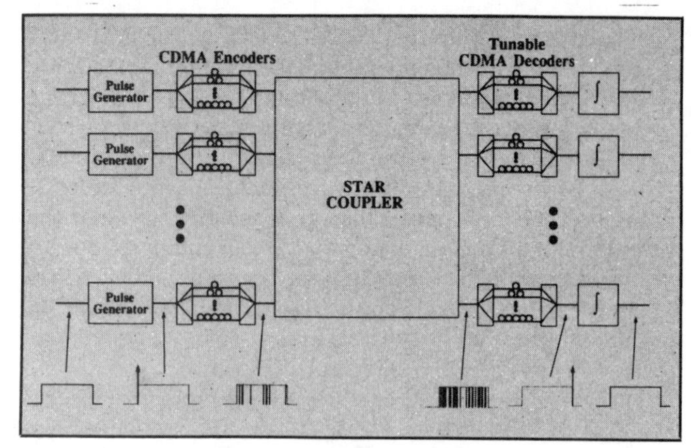

Fig. 9 CDMA switch.

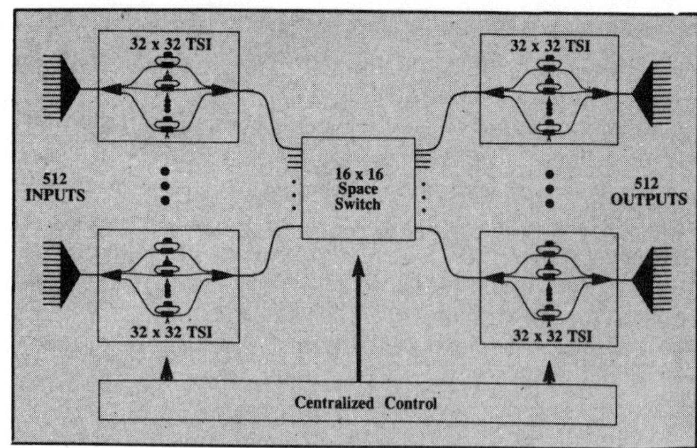

Fig. 10 512 × 512 time-space-time switch.

An example of a switching system based on CDMA is shown in Fig. 9. The difference between the CDMA multiplexor shown in Fig. 3 and the CDMA switch of this figure are the tunable decoders. For a switching system, each encoder is fixed for a particular code sequence while the decoders must be tunable to all of the input code sequences. This implies that each decoder must contain the inverse of all delay loops present in the encoders. Another implementation could have variable encoders and fixed decoders. The star coupler used in this figure could also be used in the CDMA multiplexors described previously since their function is to distribute the signals from each input channel to all the output channels.[17] This allows each output decoder to access any or all of the input channels, and to provide a strictly nonblocking switch.

The strength of this CDMA switching system is that the high speed portion of the system control is both distributed and photonic. This distributed control is the result of the code sequence being an effective address read by the designated decoder. The role of the controlling electronics is to determine which fiber delay lines are to be included for a given decoder. The weakness of these CDMA switching systems is that $k \ll N$, which limits them to smaller systems with low traffic environments, such as local area networks.

Packets are another method of allocating the available bandwidth of a channel. For this approach, the bit-rate of the information passing through the channel is set at its maximum possible value. The information from the users is collected and stored in small amounts that can be either fixed or variable lengths. When the channel is available, the data is injected into the channel with a special header directing the path to be followed by the data. The header normally provides the address information necessary for the transmitted data to get to its destination. One of the advantages of packet type systems is that their control structure is distributed rather than centralized.

Multi-Dimensional Switching

In the early days of telecommunications switching, the switching fabrics used in the switching systems were space division. With the advent of digitized voice it became apparent that electronic hardware in the fabric itself could be reduced by adding the dimension of time to the space-division fabric. As an example, if a 1024 × 1024 space-division switch were able to switch 128 time-slots per frame (1 frame = 125 μs), then a switching fabric with a dimensionality of approximately 128,000 × 128,000 could be made (4ESS™).

Another method of extending TDMA is by blending time and wavelength together. As in the case of time and space systems, time-wavelength division systems can be decomposed into wavelength-time (WT), time-wavelength (TW), time-wavelength-time (TWT), etc. Finally, switching systems can be made using time, space, and wavelength.

An example of a 512 × 512 TST switch is shown in Fig. 10. In this figure the input lines are partitioned into sections of 32 lines which are time-multiplexed onto a single space channel. Thus, each channel consists of 32 time-slots. If the bit-rate of the input signals is 150 Mb/s, then the time-multiplexed information stream will require a bit-rate > 4.8 Gb/s (≈200 ps/bit). This TDMA signal then enters the TSI where the 32 time-slots can be interchanged. From there the information enters the time-multiplexed space-division switch (the advantage of multi-dimensional switching is that the size of the space switch can be small). The output of the space switch is directed to the output TSI, which is then demultiplexed to the output space channels. The difficulty with TST configurations is the timing requirements imposed upon the centralized control. As an example, to avoid any phase discontinuities on the output channels from the space switch, there needs to be bit alignment of the TDMA information stream entering the 16 × 16 switch. Assuming a 5 Gb/s bit-rate implies that each bit has a pulse duration of 200 ps. Thus, to prevent these phase discontinuities on the output channels, all the input bits should be bit-aligned to within 10 ps of each other. This timing burden will be placed on the initial TDMA multiplexor or else an elastic store will have to be placed on the input to the space switch (This assumes that the controlling electronics can recognize variations of ≈10 ps). To illustrate the critical packaging problem, if the length of fiber from two TSIs differ by 1 cm (assuming an index of refraction of 1.5 in the fiber), there will be a 50 ps difference in the bit arrival times at the space switch. In addition to the bit and frame alignment required by the space switch, each TSI will require the alignment of bit and frame boundaries to prevent phase discontinuities on its output channel.

The strength of the multi-dimensional switching structures such as the TST switch previously shown is the minimal amount of hardware required to build them. An example of a pure space-division switch, a rearrangeably non-blocking interconnection network based on perfect shuffles, is

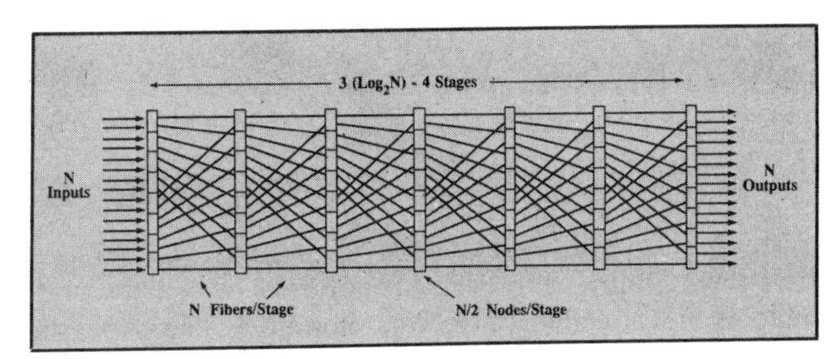

Fig. 11 512 × 512 rearrangeably non-blocking space-division interconnection network.

shown in Fig. 11. The basic building block is a 2×2 switch which could be a $LiNbO_3$ switch or an OEIC structure with fibers both entering and leaving the substrate. This space switch requires ≈13,800 fibers and ≈5000 2×2 switches. The cost of connectors for these fibers, at approximately $100/connector, is by itself enough to prevent it from ever becoming economically feasible. The TST switch of Fig. 10 requires ≈2000 fibers, 32 multiplexors (demultiplexors), one 16×16 space-division switch, and ≈1000 directional couplers for the TSIs. Even less hardware is required for a ring or linear bus network (512×512) which would require 512 couplers and 512 fibers. The disadvantage of the ring structure is the ≈80 Gb/s bit-rate on the single TDMA channel. Thus, the advantage of minimized hardware comes at the cost of increased timing complexity.

Conclusions

This paper has presented several different implementations of photonic switching systems that use time-division to more effectively exploit the available bandwidth of the optical domain. It presented TDMA structures that implemented bit, block, CDMA, and packet switching fabrics. There was also a discussion of the critical need of synchronization for most of these systems. Finally, it was pointed out that the advantage of hardware minimization comes at a cost of timing complexity.

References

[1] B. S. Glance, K. Pollack, C. A. Burrus, B. L. Kasper, G. Eisenstein, and L. W. Shultz, "WDM Coherent Optical Star Network," *Journal of Lightwave Technology*, vol. 6, no. 1, pp. 67–72, January 1988.

[2] T. E. Darcie, "Subcarrier Multiplexing for Multiple-Access Lightwave Networks," *Journal of Lightwave Technology*, vol. LT-5, no. 8, pp. 1103–1110, August 1987.

[3] K. Oshima, T. Kitayama, M. Yamaki, T. Matsui, and K. Ito, "Fiber-Optic Local Area Passive Network Using Burst TDMA Scheme," *Journal of Lightwave Technology*, vol. LT-3, no. 3, pp. 502–510, June 1985.

[4] P. R. Prucnal, M. A. Santoro, and T. R. Fan, "Spread Spectrum Fiber-Optic Local Area Network Using Optical Processing," *Journal of Lightwave Technology*, vol. LT-4, no. 5, pp. 547–554, May 1986.

[5] P. R. Prucnal, M. A. Santoro, and S. K. Sehgal, "Ultrafast All-Optical Synchronous Multiple Access Fiber Networks," *IEEE Journal of Selected Areas in Communications*, vol. SAC-4, no. 9, pp. 1484-1493, December 1986.

[6] G. J. Foschini and G. Vannucci, "Using Spread-Spectrum in a High-Capacity Fiber-Optic Local Network," *Journal of Lightwave Technology*, vol. 6, no. 3, pp. 370–379, March 1988.

[7] T. K. Gustafson and P. W. Smith, *Photonic Switching*, Springer-Verlag, New York, 1987. See article by T. Yasui and K. Kikuchi, entitled "Photonic Switching System/Network Architectural Possibilities," pp. 158–166.

[8] R. S. Tucker et al., "16-Gbit/s Optical Time-Division-Multiplexed Transmission System Experiment," *OFC '88 Technical Digest*, vol. 1 THB2, p. 149, OSA 1988.

[9] S. Suzuki, T. Terakado, K. Komatsu, K. Nagashima, A. Suzuki, and M. Kondo, "An Experiment on High-Speed Optical Time-Division Switching," *Journal of Lightwave Technology*, vol. LT-4, no. 7, pp. 894–899, July 1986.

[10] R. A. Thompson and P. P. Giordano, "An Experimental Photonic Time-Slot Interchanger Using Optical Fibers as Reentrant Delay-Line Memories," *Journal of Lightwave Technology*, vol. LT-5, no. 1, pp. 154–162, January 1987.

[11] T. K. Gustafson and P. W. Smith, *Photonic Switching*, Springer-Verlag, New York, 1987. See article by W. A. Payne and H. S. Hinton entitled "System Considerations for the Lithium Niobate Photonic Switching Technology," pp. 196–199.

[12] L. Cohen and J. Fleming, "Effect of Temperature on Transmission in Lightguides," *Bell System Technical Journal*, vol. 58, no. 4, April 1979.

[13] T. K. Gustafson and P. W. Smith, *Photonic Switching*, Springer-Verlag, New York, 1987. See article by R. A. Thompson, entitled "Optimizing Photonic Variable-Integer-Delay Circuits," pp. 158–166.

[14] R. Ian MacDonald, "Switched Optical Delay-Line Signal Processors," *Journal of Lightwave Technology*, vol. LT-5, no. 6, pp. 856–861, June 1987.

[16] A. Khurshid and D. M. Rouse, "Photonic Switching in Ring-Based Optic Networks," Submitted to *IEEE Journal of Selected Areas in Communications*.

[17] A. A. M. Saleh and H. Kogelnik, "Reflective Single-Mode Fiber-Optic Passive Star Couplers," *Journal of Lightwave Technology*, vol. 6, no. 3, pp. 392–398, March 1988.

Ultrashort Light Pulses

Peter W. Smith and Andrew M. Weiner

Abstract

In the past few years, there have been dramatic improvements in the ability to generate and control ultrashort light pulses. Researchers are now exploring ways in which such pulses may be used in future ultra-high capacity communications networks. This paper is a tutorial review of the field of ultrashort pulse generation and use. It is a revised version of an article that appeared in Bell Communications Research Exchange *magazine.*

Before the invention of the laser 25 years ago, the shortest pulses of light that could be generated were the strobe light flashes used for "stop-action" photography. Today, using lasers, researchers are able to generate light pulses 100,000 times shorter than these strobe light flashes and are using these ultrashort pulses to explore new frontiers in science and technology.

Light pulses under a picosecond in duration (see box: "What's a Picosecond?") were first observed 15 years ago. Since then, short pulse technology has advanced to the point where it is now possible to generate light pulses under 10 fsec long, Fig. 1. A 10-fsec pulse only lasts for about five periods of optical oscillation (cycles). Thus, we are rapidly approaching the fundamental limit of one optical oscillation cycle (about 2 fsec for visible light).

Researchers are working on the generation and control of ultrashort light pulses and are using these pulses to study new ultrafast optical switching elements. There are several motivations for this work.

First, ultrashort light pulses are a potential signal source in future high-bit-rate optical communication systems. The shorter the pulses, the more can be packed into a given time interval and the higher the data transmission rate. The use of ultrashort light pulses combined with all-optical switching and signal processing will allow future lightwave systems to take advantage of the tremendous bandwidth capacity of optical fiber transmission.

Second, ultrashort light pulses are opening up new areas in optoelectronics; for example, they may be used to generate, switch, and sample short electrical pulses. Thus, optical techniques can be used to measure the performance of very fast electronic structures and devices.

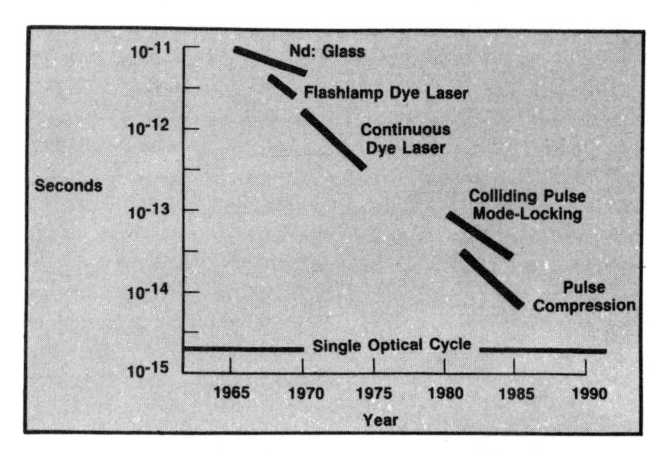

Fig. 1 Progress in short pulse generation.

Third, short optical pulses provide scientists with a tool for studying ultrafast processes in physics, chemistry, and biology. An analogy is strobe photography. Short bursts of light on the order of tens of microseconds in duration can "freeze" mechanical motions for stop-action photographs. By this method, we can photograph a bullet as it is shot through an apple or see a golf club hit and initially compress a golf ball. Strobe photography works because the light flashes occur so fast that no noticeable mechanical movement occurs during the flash. On a pico- or femtosecond time scale, the microscopic world is alive with motion. Atoms and molecules are vibrating; electrons are colliding with and scattering from the crystal lattice in metals and semiconductors; and rhodopsin molecules in the eye are undergoing complex photochemical changes in the process of converting incident light into a perceived image. Sophisticated measurement techniques using ultrashort light pulses make it possible to study this microscopic world.

To understand how short pulses are created, it helps to visualize how a laser operates. A typical laser consists of two essential elements: gain and feedback. The gain, or amplifying medium, is prepared by pumping its molecules up to a high energy level using an external power source such as an electrical discharge, a flashlamp, or another laser. A beam of light passing through such a material stimulates the molecules to release their extra energy in the form of additional light that adds to, or amplifies, the beam. Feedback is achieved by placing the amplifying medium within a resonator (a set of mirrors that reflects the beam back and forth, again and again, through the gain material). Each passage results in further amplification. As a result of this cumulative process, an intense, coherent beam of light (a laser beam) is produced.

Laser output occurs at a number of discrete wavelengths corresponding to different resonant frequencies (modes) of the resonator, Fig. 2a. If there is no fixed phase relationship between these modes, the various frequencies will interfere with each other, and the output will fluctuate over time, Fig. 2b. Mode-locking—fixing the relative phases of these modes—forces the laser to emit a train of narrow light pulses, Fig. 2c. The larger the band of frequencies over which the laser oscillates, the shorter the duration of the mode-locked pulses that can be produced. Mode-locking can be accomplished by placing a modulating element, either active or passive, within the laser resonator.

In active mode-locking, the modulator is driven by an external power source. The loss modulator that is commonly used may be pictured as a shutter that periodically opens and closes. When the modulation frequency is correctly adjusted, the shutter period is exactly synchronized to the resonator round trip time; a short pulse traveling back and forth within the laser may pass through the shutter, without loss, again and again. Pulses under 100 psec in duration have been obtained with this method.

Reprinted from *IEEE Circuits and Devices Magazine*, Vol. 4, No. 3, pp. 3-7, May 1988.

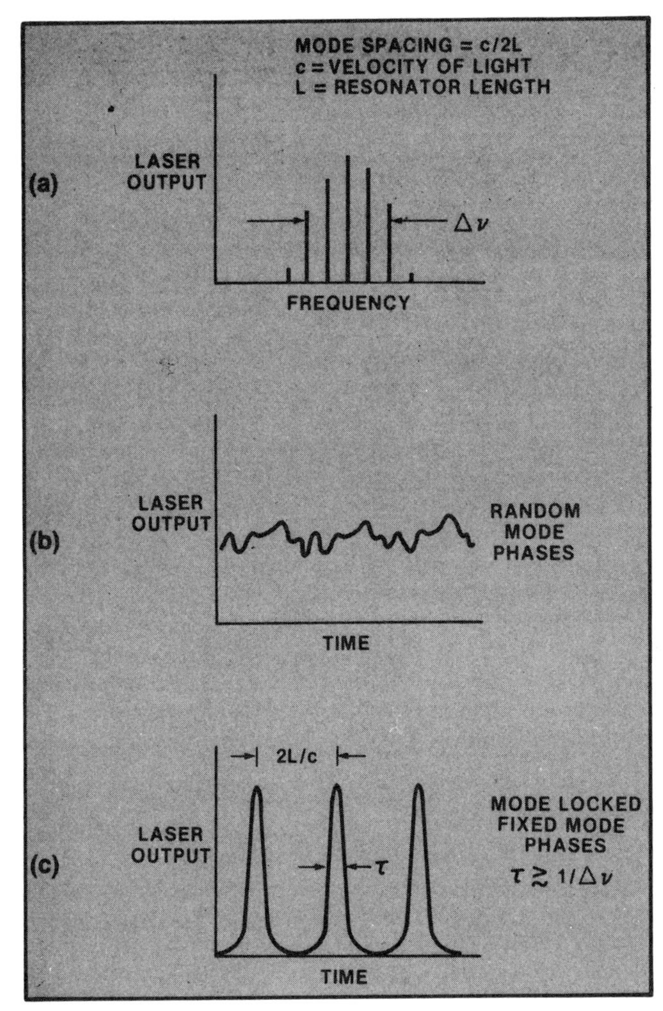

Fig. 2 Laser output—(a) Typical laser output as a function of frequency showing oscillating modes over frequency bandwidth $\Delta\nu$. (b) Laser output with randomly phased modes. (c) Mode-locked laser output with pulse duration $\tau \sim 1/\Delta\nu$.

Even shorter pulses have been produced using passive mode-locking, which requires no outside power source for its modulator. The modulator used in passive mode-locking is usually a saturable absorber dye. The absorption of such a dye decreases with increasing light intensity; i.e., the fraction of light transmitted by the dye increases with increasing intensity. When a pulse of light traveling back and forth within a laser passes through a saturable absorber, the absorber is bleached; hence, the transmission function of the absorber is modulated by the optical pulse. The saturable absorber in a laser that is mode-locked in this way produces a modulation that is automatically synchronized with the round trip time of the resonator.

This process of passive mode-locking is best exploited using a ring-shaped laser called a *colliding-pulse-mode-locked (CPM) ring dye laser*, Fig. 3. The ring geometry allows two pulses to exist in the laser at the same time; one travels around the ring in a clockwise direction, the other counterclockwise. The two pulses meet or "collide" in the saturable absorber. This collision leads to enhanced bleaching of the absorber, resulting in shorter pulses than those obtained with conventional mode-locking. With additional refinements that compensate for intracavity pulse broadening, such a laser has generated pulses as short as 27 fsec.

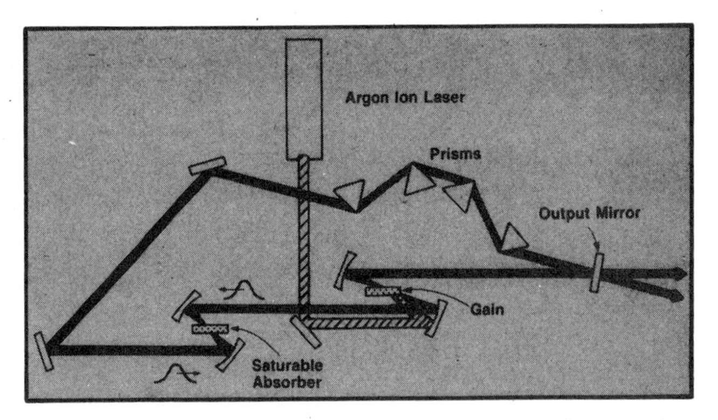

Fig. 3 Colliding-pulse-mode-locked (CPM) ring dye laser. Pulses "collide" in the saturable absorber but arrive singly in the gain medium, which uses an external argon ion laser as a power source. Femtosecond pulses are obtained through the partially transmitting output mirror.

Even shorter pulses can be produced when mode-locking is combined with a technique called *pulse compression*. Although this technique has roots in the radar technology of the 1950s, in the last several years, it has achieved new vitality in the optical field. The shortest optical pulses achieved to date—only 6 fsec long—have been obtained by applying pulse compression to the already very short pulses produced by a CPM ring dye laser. Relatively long pulses from actively mode-locked lasers may now also be compressed to durations previously accessible only with passively mode-locked lasers. For example, researchers have compressed neodymium:yttrium aluminum garnet (Nd:YAG) laser pulses by nearly two orders of magnitude: from 75 psec to only 0.8 psec (see box: "Pulse Compression").

Measuring the duration of ultrashort light pulses is no easy task. The shortest optical pulses today are more than three orders of magnitude shorter than the response of the fastest photodetectors and electronic sampling oscilloscopes. Thus, conventional electronic methods cannot be used to measure picosecond and femtosecond optical pulses. The pulses can be measured, however, using new optical techniques, which have been developed for this purpose. These techniques, known as autocorrelation techniques, use two synchronized, ultrafast pulses to sample each other (see box: "Measuring Ultrashort Light Pulses").

The ability to generate and control ultrashort light pulses has exciting implications for the design of future communications networks. We will mention briefly, here, two potential applications that appear to show great promise: high-speed signal processing and parallel processing.

In optical communications systems, the characteristics of large bandwidth, high speed, and the ability to process signals already in the form of light should be particularly useful. It should be possible, for example, to make an all-optical multiplexer, Fig. 4, that would take several optical data streams, each with the maximum data rate compatible with electronic devices, and multiplex them into a single data stream for transmission down an optical fiber at a much higher bit rate. At the other end, a similar all-optical device could demultiplex to get back to data rates that can be handled with electronic components.

Many optical devices are especially suited for parallel processing application, where many individual operations must be performed at the same time. One type of optical

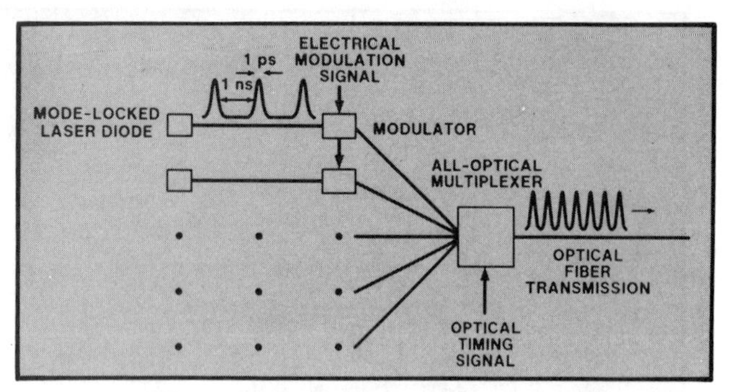

Fig. 4 A possible scheme for optically multiplexing signals from a number of mode-locked laser diodes using a fast all-optical multiplexer.

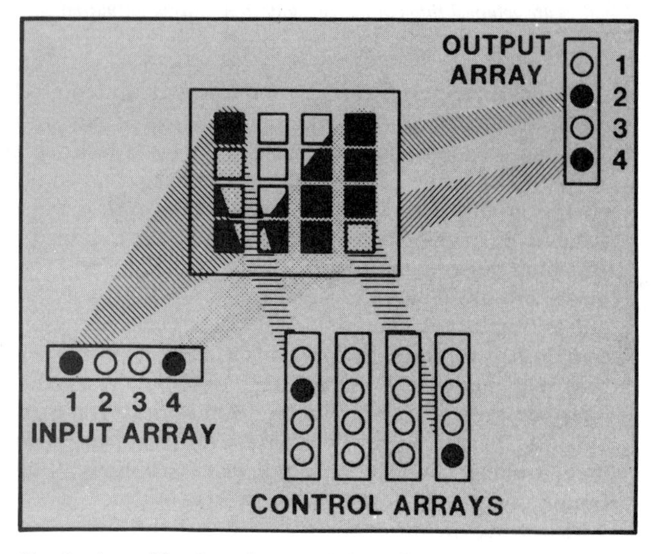

Fig. 5 A possible scheme for an optical crossbar switch array. The array is shown switched so that input 1 is connected to output 2, and input 4 is connected to output 4.

device that could be developed for this purpose is an optical ''crossbar,'' Fig. 5, in which each crosspoint is an all-optical switching element that can be switched from ''transmitting'' to ''not transmitting'' by a light beam from a control array. Each input beam illuminates one of the columns, and each output beam collects the light from one of the rows. By switching the crosspoints to the ''transmitting'' state, it is possible to connect any input to any output. This could, in principle, be done very rapidly and with very large bandwidth channels.

Before such applications can become a reality, however, further research is needed to develop ultrafast materials and devices that are compact and efficient. We mention below three examples of current research that will help lay the groundwork for telecommunications applications of ultrashort light pulse technology: mode-locking of semiconductor diode lasers, manipulation of picosecond and subpicosecond pulse shapes, and measurements of ultrafast semiconductor dynamics.

For ultrashort pulses to find widespread use outside of the laboratory, a simple, compact, and reliable pulse source is needed. Although the shortest pulses have been generated with dye lasers, such systems are large, costly, and inefficient. For this reason, researchers have been studying ways to mode-lock a semiconductor diode laser in order to produce a compact and efficient solid-state source of picosecond pulses. They have found that diode laser mode-locking can be achieved with the aid of a thin semiconductor wafer composed of many alternating layers of material, each of which is only a few atomic layers thick. This semiconductor structure, called *multiple quantum well (MQW) material*, has unique electrical and optical properties. The optical properties of MQW material make it potentially useful as a saturable absorber for passive mode-locking of diode lasers. By bombarding the MQW material with protons, researchers can tailor its absorption characteristics to produce optimum mode-locking.

Recently, Bellcore researchers reported a new record for the shortest pulses ever obtained in continuous train from a mode-locked diode laser: 0.8 psec.[1] This is an important step toward the practical application of picosecond light pulses in high-bit-rate communications systems.

Manipulating the shape of ultrashort light pulses is an area of research that grew out of pulse compression studies. Scientists have invented a way to produce arbitrarily shaped optical pulses, with picosecond or even subpicosecond features. This ability to synthesize arbitrary pulse shapes may well have important applications for optical communications systems, optical radar systems, and picosecond studies of optical interactions in nonlinear materials.

Pulses are shaped using a modified pulse compressor apparatus. Within this apparatus, the various frequency components (colors), which constitute the pulse, are separated in space. The amplitude and phase of the various spatially dispersed frequency components can be individually controlled. Since the temporal pulse shape is determined by the frequency spectrum, pulse shapes may be tailored by adjusting the spectrum. In this fashion, a wide variety of ultrashort optical pulse shapes have been generated.

Bellcore researchers have been successful in producing the world's first picosecond optical ''square'' pulse, Fig. 6. The rise and fall times of the square pulse are limited only by the available optical bandwidth. Such a shaped pulse should be useful in connection with digital optical signal-processing systems.

Other research has focused on studying the solid-state physics of semiconductor materials on a picosecond scale.

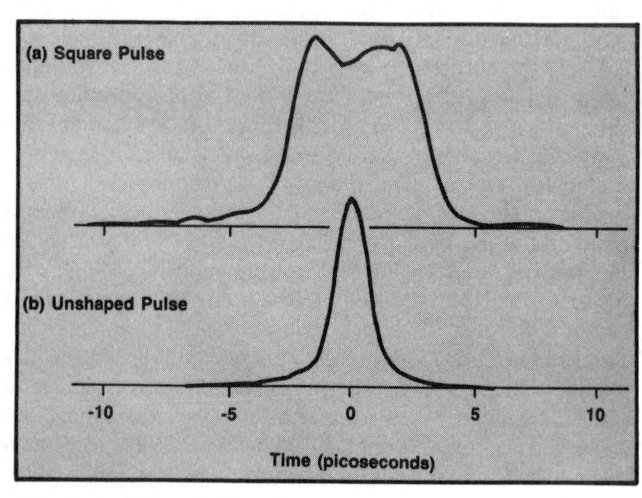

Fig. 6 A picosecond optical square pulse. An unshaped pulse is shown for comparison.

Specifically, researchers are investigating the dynamics of "hot" charge carriers in semiconductors. Carriers are said to be hot when they are driven out of thermal equilibrium with the surrounding crystal lattice to higher energy states.

In one type of experiment, picosecond light pulses from a mode-locked dye laser are used to produce the hot carriers within the semiconductor. The subsequent luminescence from the carriers provides information about their energy distribution. The electron temperature may be monitored by tracing out the time history of the luminescence spectrum. In other words, one can determine the way in which the carriers return to equilibrium. An understanding of carrier cooling mechanisms is important not only from a fundamental physical viewpoint but also because of its relevance to the design of high-speed optical and electronic switching devices using semiconductor materials.

Many challenges still face researchers before picosecond and femtosecond optics can be moved out of the laboratory and into telecommunications and signal-processing systems. Improved techniques for generating and controlling ultrashort light pulses are still being studied, and efficient solid-state materials for optical signal-processing devices are being developed. Parallel advances in other disciplines, such as microfabrication techniques, integrated optics, and network design, are rapidly taking place. Just as the research of a few years ago has led to today's technology, so will today's laboratory studies of ultrashort light pulses lead to tomorrow's ultrahigh capacity communications networks.

BOX 1 - What's a Picosecond?

In 1 sec, light travels 186,000 mi., so a 1-sec pulse would stretch three-quarters of the distance from the earth to the moon. Picoseconds and femtoseconds are miniscule fractions of 1 sec.

A picosecond is one trillionth of a second, or 10^{-12} sec. A light pulse of a picosecond duration has a length of about one-third of a millimeter—about the thickness of a business card.

A femtosecond is one quadrillionth of a second, or 10^{-15} sec. There are as many femtoseconds in 1 sec as there are seconds in 30 million years. One femtosecond is the time it takes for light to travel approximately 3000 Å, which is less than 1 percent of the thickness of a human hair.

BOX 2 - Pulse Compression

The technique of pulse compression exploits a nonlinear optical interaction that occurs naturally within single-mode glass fiber—the same sort of fiber now being used in optical communications systems. This interaction, known as self-phase modulation, occurs because the refractive index of the glass is modulated by the presence of an intense optical pulse. The change in refractive index in turn affects the optical phase. As a result of this process, the bandwidth of the optical pulse emerging from the fiber is increased dramatically. Since the minimum achievable pulse-width is limited by the bandwidth, the spectrally broadened pulses have the potential to be compressed by a substantial amount.

The additional bandwidth of the pulse emerging from the fiber is caused by a "chirp"—the instantaneous frequency increases with time. The leading edge of the pulse contains red-shifted frequency components, and the latter part of the pulse contains blue-shifted components. This chirped pulse is then directed through a pair of parallel diffraction gratings. The higher frequencies ("blue") travel through the grating pair faster than the lower frequencies ("red"). When the grating separation is properly adjusted, the "blue" catches up to the "red," and the pulse is compressed down to the bandwidth limit.

BOX 3 - Measuring Ultrashort Light Pulses

An autocorrelator is a device used to measure the duration of ultrashort pulses. In the autocorrelator, a single pulse is first split using a beam splitter (a partially transmitting, partially reflecting mirror) into two identical pulses. By causing each pulse to travel to different path lengths, the relative delay between these two pulses can be adjusted. (For example, a difference of 0.3 mm of path length in air generates a time delay of 1 psec.)

The two pulses are focused by a lens to a common spot in a nonlinear crystal, such as potassium dihydrogen phosphate, where second-harmonic generation takes place. This is a nonlinear optical process that produces output light at twice the frequency of the original light. Because the second-harmonic intensity is proportional to the square of the incident light intensity, more second-harmonic light is generated when both pulses are coincident than when the two pulses are separated in time. Thus, by measuring the amount of second-harmonic light as a function of the delay between the two pulses, researchers can infer the pulse durations.

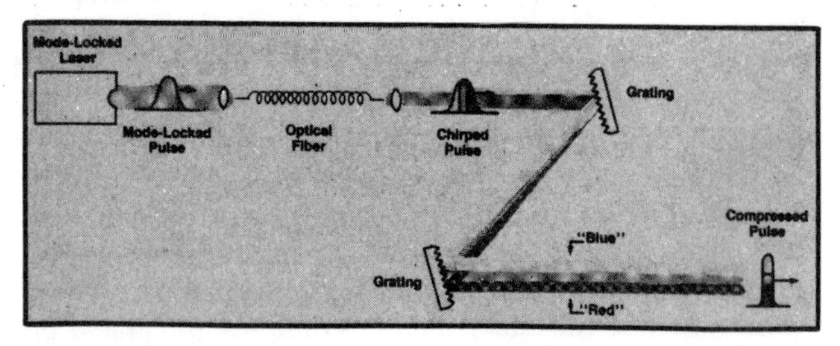

Sidebar 2 Pulse compression apparatus. Pulses from a mode-locked laser are spectrally broadened (chirped) in an optical fiber and then compressed with a pair of diffraction gratings.

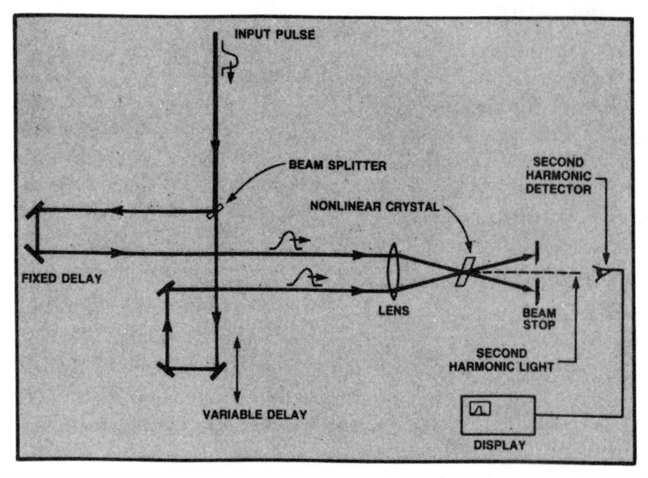

Sidebar 3 Autocorrelation apparatus used to measure the width of ultrashort light pulses.

For Further Reading

1. P. W. Smith, M. A. Duguay, and E. P. Ippen, "Mode-Locking of Lasers," in *Progress in Quantum Electronics,* vol. 3, J. H. Sanders and S. Stenholm, Eds. (Pergamon, New York, 1974). This book chapter gives an introduction to the basic mode-locking phenomena and techniques.
2. R. R. Alfano and S. L. Shapiro, "Ultrafast Phenomena in Liquids and Solids," *Scientific American,* June 1973, pp. 42–60. Discusses early measurements of picosecond phenomena in liquids and solids. Also an interesting discussion of short light pulses before the advent of lasers back through several hundred years.
3. S. L. Shapiro, Ed., *Ultrashort Light Pulses: Picosecond Techniques and Applications* (Springer Verlag, Berlin, 1977). This gives the background for the field and reviews the state of the art ca. 1977.
4. C. V. Shank and D. H. Auston, "Ultrafast Phenomena in Semiconductor Devices," *Science,* vol. 215, pp. 797–801, Feb. 12, 1982. Discusses application of ultrashort light pulses to the measurement of fast phenomena in semiconductor structures and devices.
5. C. V. Shank, "Measurements of Ultrafast Phenomena in the Femtosecond Time Domain," *Science,* vol. 219, pp. 1027–1031, March 4, 1983. This reviews the advances in pulse generation technology that occurred ca. 1981–1982 following the invention of the CPM laser.
6. G. R. Fleming and A. E. Siegman, Eds., *Ultrafast Phenomena V* (Springer Verlag, Berlin, 1986). This is a conference proceedings, which covers nearly every topic of current ultrashort light pulse research.
7. *Lasers and Applications,* pp. 79–82, Jan. 1985, and pp. 91–94, Feb. 1985. Covers the basic principles of active, passive, and synchronous mode-locking.

[1] A mode-locked diode laser with pulse widths below 0.6 psec was reported at the 1988 Conference on Lasers and Electro-optics by researchers from the University of California at Santa Barbara.

Optical Interconnects Speed Interprocessor Nets

When networking computers with wires is too slow or too error prone, optics may be the answer

by John D. Crow

The demand for more computer processing power continues to increase. This demand is being satisfied at a single location through increasingly powerful uniprocessors and through multiprocessor complexes. Within an entire enterprise, networks of mainframes, minicomputers, and workstations are being established to enhance computing power.

Improvements in the hardware technology and in the system organization (architecture) both contribute to increases in computing power. On the hardware side, increases in transistor speed, IC complexity, and packaging density have minimized the time needed to process instructions and data. But these advances have generated a secondary problem. It is becoming increasingly difficult to provide enough electrical-wiring speed and density to take full advantage of the high speed of logic circuits, and still keep noise and crosstalk on the lines to acceptable levels. On the architectural side, parallel processing and the sharing of memory and storage between processors are gaining popularity as ways to increase the aggregate computing power of a system. Again, high-speed wiring is needed.

An interprocessor network is a means of connecting the elements in such a multiprocessor environment. Unlike a data communications network, an interprocessor network must make connections quickly between the processors so it does not slow down the computing operations shared by the processors. Unlike inter-board and inter-chip networks, inter-processor networks may extend to hundreds of meters in length. The performance and cost objectives of inter-processor networks are difficult to meet with electrical wiring.

Thus, system designs that follow current trends in hardware or architecture have interconnection requirements that provide optical technology with an opportunity to play a critical role in computer-system performance. This is in marked contrast to the peripheral role that wiring has often taken in the past.

Network requirements

Optimizing the cost/performance of opto-electronic technology for inter-processor applications leads to a different chip and package technology than those being developed by the telephone and data communications industry [1]. In these inter-processor networks, the processors at the nodes may be mainframes or workstations with processing rates ranging from tens to hundreds of MIPS (million instructions per second). To exchange data quickly, e.g., less than 1K instruction cycles of delay, these processors must transfer data blocks in microseconds. These blocks can contain up to several thousand bytes, so data transfer rates can easily be in the range of gigabytes/s.

An interprocessor network may reside between boards in one equipment frame, between frames in a computer complex, or even between clusters of frames in a building. The desire for microsecond response times over the network will keep distances short (generally much less than 1 km) compared to telecommunications industry requirements. The electro-optic technology used in such a network should be compatible with the high packaging density, the power

Reprinted from *IEEE Circuits and Devices Magazine,* Vol. 7, No. 2, pp. 20-25, March 1991.

TABLE 1: Link Technology Objectives

Speed	Multigigabit/s
Optical Complexity	1-32 Devices
Electronic Complexity	50K Transistors
Dissipated Power	<5 W
Operating Temperature	60 C
Link Reliability	MAFR <0.01 percent/khr
Lifetime	Up to 10^5hrs
Power Supplies	5, 3.6 V
Cooling	Forced Air
Safety	OSHA Class I
Module Cost	<$100 per line

supply, and the cooling requirements of the processor and memory frames.

An interprocessor network will form the backbone of a computer complex, so the network's reliability is critical, in terms of very low component failure rate, network tolerance to faults without serious degradation in performance, and low error rates on the links. The functions that adapt the processor bus to the optical network (Fig. 1) contribute to this robustness, in addition to conditioning the data for transmission.

In designing the required network interface functions, there is a trade-off between the detrimental impact of the time taken to execute a function and the enhancement of network features. These features include reducing I/O count, improving reliability, and gaining flexibility. To obtain the lowest possible latency, for example, a designer would implement only limited serialization, but then each data link would require multiple optical lines. Typical requirements for an optimized optical link technology can be derived from these network interface requirements (Table I) [2,3].

Telephone Technology Won't Do

Optical link technology developed by the telecommunications industry for long-distance applications can meet the distance and data-rate requirements of interprocessor networks, but it falls short in almost all other technical requirements. Telecom technology is based on discrete-device O-E chips, low-integration-level ICs, and discrete components such as SAW filters. The devices are custom selected and tuned for optimum performance in long links. The packaging is optimized for the discrete devices used. EMI shielding, for example, is incorporated between analog and digital functions, and a laser diode must be aligned to an optical fiber with sub-micron precision while the laser is powered-on.

The telecomm industry has invested extensive R&D into this technology, which has lead to impressive demonstrations of links operating at high values of the often-quoted data-rate-times-distance figure of merit. In spite of this success, the technology is very expensive and is only produced in small volumes. The cost may be justified for long-distance applications, but it is far from cost-effective for data processing networks. For example, the Japanese Optoelectronics Industry Technology and Development Association (OITDA) estimated in 1989 that telecomm lasers based on InGaAsP and InP will cost approximately $280 in the mid-1990s when produced in 90K quantities. (Current commercial prices for such lasers in packages with single-mode-fiber-alignment connections are well over $1000.) Wiring for a computer complex that requires 100-1000 lines for interprocessor connections will not be cost-effective if it uses telecomm lasers for line drivers. (In comparison, GaAs consumer lasers are $3 today, and OITDA estimates that data-link GaAs lasers will cost $14 in the mid-90s.)

Currently, the computer industry is using fiber-optic links in LAN and I/O-channel network products, generally using LEDs and multimode fibers to achieve link robustness and lower cost. Merchant optical technology can meet the lower performance demands of these applications, but even this discrete-device technology produces link adapters whose cost is in the hundreds of dollars. This scheme limits the penetration of optical links into commercial network products. The recent movement toward adding a twisted-wire-pair option, or a GaAs OE-device-based option, to the ANSI FDDI standard reflects this growing concern about OE technology's cost. To compete with electrical wiring in interprocessor networks, as well as in the other short-distance data-communication networks, we need a more cost-effective optical technology. What are the elements of a cost/performance-optimized, optical-interconnect technology for interprocessor networks?

Opto-electronic Transducers

Quantum-well lasers are attractive OE components because they meet the requirements of gigabit/s modulation speed and high electrical-to-optical conversion efficiency. GaAs/GaAlAs lasers have been fabricated with drive-current requirements of only a few mA, energy conversion efficiencies of over 40 percent, and modulation frequencies in the multigigahertz range. These "optical line drivers" are less power consuming than their electrical counterparts.

Advanced facet-passivation techniques have greatly enhanced the reliability of GaAs lasers, and the yield of monolithic GaAs arrays is currently sufficient for commercial use. Thus, it is possible to secure link reliability from these components themselves, as well as through redundancy. For the future, planar processed facets are being developed for GaAs lasers. This could lead

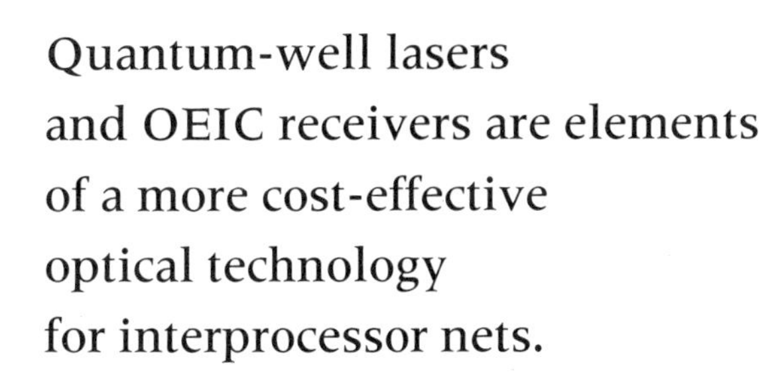

Quantum-well lasers
and OEIC receivers are elements
of a more cost-effective
optical technology
for interprocessor nets.

to device screening and characterization at the wafer level, which should reduce the cost of testing and packaging [4] (Fig. 2).

Another possible optical source is the InGaAsP LED array. NEC has recently demonstrated such arrays that have both low drive current (15 mA, p-p) and high data rate (150 Mb/sec). Although the LED is less efficient and slower than the laser, it has the advantages of a simpler structure, no need for device bias control, and longer demonstrated lifetime. The device's low power limits the maximum distance signals can be transmitted over a link (or the number of taps in a link), but the low optical power in the link is inherently safe for viewing by the human eye.

Currently at the research stage, the strained-layer, quantum-well, surface-emitting laser has demonstrated many attractive features for optical interconnections. Laboratory testing confirms the device's low drive currents, low divergence and easy coupling to lightguides, low susceptibility to noise from optical reflections, and an indication of enhanced reliability over non-strained lasers.

We need to integrate photodiodes on the same chip with the receiver electronics. This reduces the optical receiver's susceptibility to electrical noise pick-up from closely packaged digital electronic circuits without bulky and expensive shielding, and it reduces the parasitic capacitance at the input to the receiver's amplifier. With integration, the high-gain amplifier can be placed only 10-20 microns away from the photodiode, minimizing an electrical-noise-input coupling loop. Using OEIC receivers, IBM has demonstrated gigabit/s optical links with a bit-error rate of less than 10^{-15}, even with no shielding between the receiver and other on-chip digital switching circuits.

For fully integrated receivers, the photodiode must be process-compatible with the amplifier's on-chip electronic circuits. This requirement has made the planar, interdigitated, metal-semiconductor-metal (IMSM) photodiode a popular photodetector for integration. Other strengths are high bandwidth (exceeding 7 GHz), low total input capacitance (fF), and low noise (5 pA Hz). These specifications apply even when the IMSM photodiode is fabricated with a photosensitive area measuring 80 microns

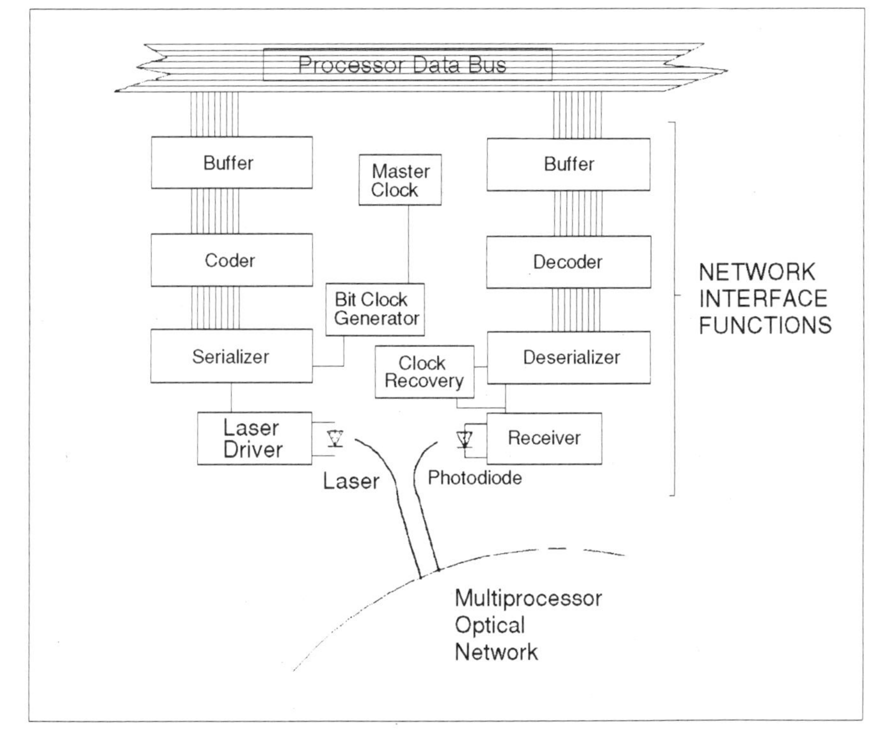

1. Several functions interface the data bus of a processor to an interprocessor network. Buffering provides speed matching and queueing. Serializing reduces the number of transmission lines from the tens or hundreds of electrical lines on the data bus to a few on the link. Link-speed clocks generate signals for data resynchronization or deserialization.

per side. This large area is a packaging advantage because it makes alignment to multimode or single mode fibers much easier. Note that packaging costs can dominate chip costs in OE modules.

These and related developments are extremely encouraging. Recent published research results show that optoelectronic integrated circuit (OEIC) receivers are as sensitive as their hybrid predecessors; receivers designed for manufacturing tolerances (in GaAs MESFET technology) and product environments should also have performance comparable to hybrid receivers. Furthermore, development-line results indicate that the yield of OEIC receiver chips is high enough, that the cost of packaged OEIC full receivers should be similar to the cost of today's packaged high-speed photodiode.

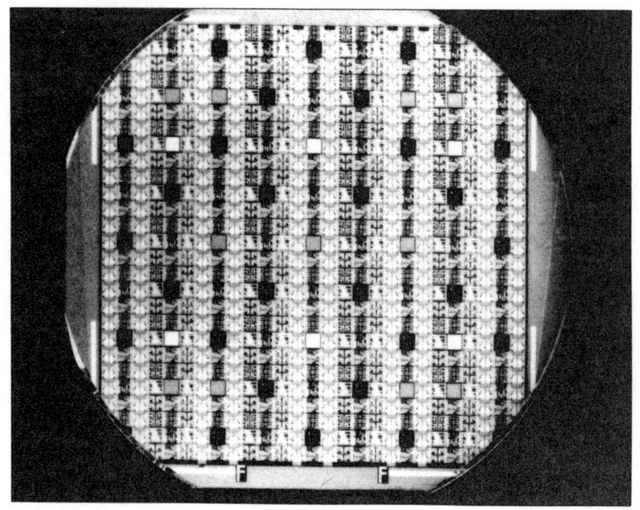

2. This 2-inch GaAs wafer contains 5000 etched-facet, quantum-well lasers on 1.5-mm-by-1.5-mm chip sites. The lasers are suitable for testing and screening using the automated probe-station techniques developed for IC technology. Photo credit: IBM Zurich Research Laboratories

Monolithic Integration of Functions

We would like to integrate all the functions of a link adapter onto a single chip with extendibility to arrays of link adapters on a single chip. For this, we need LSI IC technology.

An appropriate and commercially available IC technology uses the GaAs IC MESFET. Currently, the technology is producing devices with features of between 0.7 and 1 micron, and IC chips with between 5K and 10K circuits. GaAs-MESFET has the advantages of maturity and device simplicity over other III-V circuit candidates; and the advantages of OE integration and moderate power consumption over Si bipolar ICs. The MESFET technology uses refractory metals for device stability, self-aligned transistor processes for speed and uniformity, and multilayers of wiring. Its combination of enhancement- and depletion-mode devices keeps power consumption low, which allows chip modules to be air cooled. Both analog and digital circuits can be fabricated in the technology, and MESFET devices have good noise characteristics for fast receivers.

IBM has integrated all necessary link-adapter receiver functions onto a single GaAs MESFET chip using 1-micron gate FETs. The 3-by-4-mm chip (cover photo) incorporates four optical receivers, one clock recovery circuit for bit sync, and one deserializer circuit with byte sync and special character detection. Containing over 8K devices and dissipating 750 mW for a 1-gigabit/s transmission rate, this chip represents the state of the art in OEIC chips. NTT has demonstrated a 1300-device MESFET chip containing an optical receiver, switching circuits, and laser drivers that operates at 1.8 gigabits/s. It consumes about 2 W. With a circuit potential in this base IC technology of well over 5K devices, at least 4 parallel channels per chip with all link-adapter functions should be possible.

In the InGaAsP/InGaAs/InP material system, OEICs of about 80 devices have been demonstrated, providing functions such as a laser integrated with a driver circuit, or a photodiode integrated with a preamplifier. A four-channel photodiode-plus-preamp chip from STC (Fig. 3) measures 3.7-by-3.2 mm, and its receivers dissipate a total of about 200 mW when operating at 200 Mb/s. The IEEE/LEOS OEIC Workshop concluded that InP-based OEICs were about a decade behind their GaAs-based counterparts, but were nevertheless important for cost-sensitive, long-distance applications of fiber-optic links, such as metropolitan area networks and CATV distribution [7].

When OEIC chip development is started, meaningful reliability and reproducibility data will become available to users. Such an effort will also generate the data base required for better circuit models, which would lead to improved control over device parameters and a consequent improvement in circuit design margins.

A practical OEIC technology requires readily available design simulation and verification tools, as well as models of all the devices and circuits. These tools and models will have to incorporate the required optical and opto-electronic components, and account for the noise and distortion characteristics both of the analog circuits and the digital logic. Current CAD tools and models for GaAs MESFET IC technology have analog capability, models for OE devices, and some models for the coupling of OE devices to lightguides [8]. It still remains to merge these elements into an OEIC automated design system.

High Density Packaging

The density and cost advantages of chip integration will be lost if we do not develop manufacturable high-density packaging. Modules for the OEIC chip and its complement of passive optical and electrical components should take advantage of the base technologies developed by the IC industry. Multicomponent IC carriers demonstrate gigabit/s speed, and processed electrical wiring on the carriers, but optical wiring and packaging trail far behind their electrical counterparts.

The relative crudeness of optical wiring and packaging is demonstrated by the alignment of 4 optical fibers to 4 receivers on an OEIC (Fig. 4). The fibers are held in a common V-grooved piece of silicon, and the assembly is polished at the angle of total internal reflection for the glass-air interface. The fibers are kept in the plane of the chip to keep the profile of the package low and the light from the fibers is internally reflected from the polished end down onto the photodetectors. The fiber assembly is manipulated over the chip until alignment is observed and the assembly is epoxied in place. Although this is typical of current OE

The density and cost advantages of chip integration will be lost if we do not develop manufacturable high-density packaging

packaging techniques, the optics are bulky and the assembly procedure is costly. Two important needs for OEIC packaging are (1) planar-processed optical wiring for chips and chip carriers, and (2) self-aligned optical coupling between chips, chip carriers, and off-carrier optical transmission.

A number of candidate lightguide technologies are used for integrated optics, but few are compatible with the constraints of an IC chip carrier and the IC chip-attachment processes. Multi-chip carriers can be as large as 10 cm, so lightguide loss should be less than a few tenths of a dB/cm; branching and bends in the guide should have losses less than 1 dB. Once fabricated, the lightguide must maintain its low loss through all subsequent carrier processing, including the fabrication of electrical wiring and the chip-attachment soldering procedures. (Flip-chip solder-ball attachment of ICs, for example, can require temperatures over 300 C.) In addition, the materials used must be compatible with the carriers' temperature coefficient of expansion over the IC's operational temperature range.

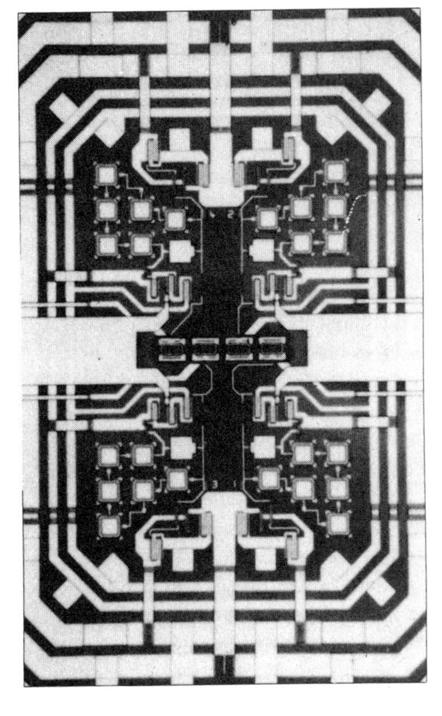

3. Indium phosphide photodiode receivers are far less highly integrated than gallium arsenide devices. This OEIC four-receiver array contains a receiver in each quadrant. The photodiodes are clustered at the center of the chip. Photo credit: STC Technology, Ltd. (Now part of Nortel Networks)

Both dielectric and polymer materials are candidates for lightguides. They can be directly deposited onto a chip-carrier substrate, and planar processed using photolithography and etching techniques to make lightguides, passive optical components, and features for mechanical alignment. An interesting class of polymer for these applications are the polyimides because of their popularity as a low-dielectric insulator for multilayer electronic wiring and their stability at high temperatures (greater than 350 C).

Honeywell has demonstrated polyimide lightguides etched with reactive ion beams that have 0.3 dB/cm loss and stability to 200 C. NTT has fabricated silica lightguides with low loss and high temperature stability. Lightguides and passive optical structures have been planar processed onto chip carriers, and an insertion loss of less than 0.5 dB has been measured between the lightguide and a fiber. Both polyimide and silica lightguide technologies are commercially available, but they have not yet been integrated into a multi-chip module with O-E components.

A potentially important development in OEIC chip attachment is the use of the solder-bump attachment used in high-performance flip-chip IC packaging, to self-align optical components to each other. Plessey has taken a $LiNbO_3$ lightguide that was initially misaligned from a fiber's optical axis by 30 microns, and aligned it to within a few microns when the solder bumps between the chip, carrier, and fiber were reflowed in a soldering operation. This alignment tolerance is sufficient for low chip-guide coupling losses between multimode lightguides, but would require improvement for low-loss coupling of single-mode lightguides — between a lightguide and laser, for example.

4. The state of the art in optoelectronic connections is crude compared to semiconductor IC practice. These 4 optical fibers are soldered into the grooves of a silicon alignment block on 250- micron centers and manually positioned over 4 photodetectors before being epoxied in place.

Future Directions

The computer industry is making increasing use of optical link technology and it is likely be a high-volume user of link adapter components. But the industry does not seem to be taking a leadership role in developing cost/performance-optimized link technology. Many system developers today rely on merchant technology for computer links. As a result, the overwhelming dominance of the telecommunication industry in directing OE component development today will thus necessarily lead to an OE technology being employed in computer systems that is poorly optimized for computer applications.

A more optimistic indication for the future comes from recent OE industry meetings between technology users and suppliers. Industry discussions of cost and performance trade-offs, the diversity of application requirements, and the need to be competitive with other interconnection technologies, could lead to more cost-effective options for OE components.

An optical interconnect technology based on GaAs OEIC and IC technologies, and on planar-processed optical and electrical wiring could potentially meet cost and performance objectives of the interprocessor network application. This is the only integratable optical technology ready for near-term development. Manufacturing tools and processes must still be developed, and it still remains to demonstrate the cost and long-term reliability of highly integrated link adapter modules.

The development of an integrated OE technology will likely require a large initial investment before the benefits of low part cost will be realized, just as occurred in IC development. It is encouraging that industrial consortia have formed to share the expense of this development. The early MITI-sponsored Opto-electronics Joint Research Lab and Optics Technology Lab in Japan; and the more recent EC-sponsored ESPRIT II/ OLIVES project in Europe are examples.

Another positive development is the emergence of standards activities for high-speed computer networks. (The ANSI X3T9.3 Fiber Channel Standard is an example.) Such standards could facilitate the development of an optimized technology for optical network adapters because standards would increase the volume of interface components used and amortize the cost of OEIC development.

The next few years should be interesting times for OEIC development. For data communications, the evolution from discrete to integrated components will likely be driven by the need for the higher performance levels of the B-ISDN networks. There seem to be political as well as technical reasons why this may be many years away. For interprocessor network applications, it could be a time for network technologists to take the lead in the development of a highly integrated optical technology. If a successful OEIC-based technology were to be developed for the computer-system marketplace, other optical components such as modulators and switches would probably be added in the longer term. We could then anticipate that the range of applications would broaden to encompass optical storage, printing, and display. In fact, such a low-cost link technology could well find widespread use in data communications itself.

Biography

John D. Crow [SM] is manager of the optical interconnect technology group at the IBM Thomas J. Watson Research Center in Yorktown Heights, New York. While at IBM, he has been responsible for research activities in computer optical I/O channel prototyping and OEIC chip development. ***CD***

References

1. R. F. Leheney, "Optoelectronic Integration: A Technology for Future Telecommunication Systems," IEEE Circuits and Devices Magazine, 5, 3, p. 38 (May 1989).

2. J. D. Crow, "Optical Interconnect Technology for Multiprocessor Networks," Critical Reviews of Optical Science and Technology, Vol. CR35: Optical Computing, SPIE Optical Engineering Press (1990).

3. T. A. Lane et al., "Digital System Applications of Optical Interconnections," Proceedings of SPIE O-E/Fiber Lase, Sept. 1988.

4. P. Vettiger et al., "Full Wafer Processing and Testing—The Future of Large Scale Laser Fabrication," Proc. of 12th IEEE Int'l Semiconductor Laser Conf., Sept. 1990., p. 144.

5. D. Rogers, "Integrated Optical Receivers Using MSM Detectors," Proc. of IEEE LEOS '90 Annual Meeting, Nov. 1990, paper OE10.1.

6. M. Dagenais et al., "Applications and Challenges of OEIC Technology: A report on the 1989 Hilton Head Workshop," IEEE Journal of Lightwave Technology, 8, 6, p. 846 (June 1990).

7. D. A. H. Spear, "Monolithic Integration of an InP/InGaAs 4 Channel Transimpedance Receiver Array," Proc. of IEEE/LEOS Topical Mtg. on Integrated Optoelectronics, Aug. 1990, p.69.

8. P. E. LaGasse, "Design and Modeling of OEICs," Proc. of IEEE/LEOS Topical Mtg. on Integrated Optoelectronics, Aug. 1990, p. 31.

AUTHOR INDEX

SUBJECT INDEX

G

H

I

L

M

O

P

Q

R

S

T

U

V

W

Y

Z

ABOUT THE EDITORS

Ronald W. Waynant received the B.E.S. degree from the Johns Hopkins University, Baltimore, MD, in 1962, and the M.S.E.E. and Ph.D. degrees from Catholic University, Washington, DC, in 1966 and 1971, respectively. In 1962, he joined the Westinghouse Electric Corporation where he worked on solid-state lasers and gas breakdown. In 1969, he joined the Naval Research Laboratory in Washington, D.C., where he worked on UV and VUV gas lasers, excimer laser kinetics, mid-infrared lasers, and novel laser pumping schemes resulting in RF pumped excimers and solid-state VUV lasers. In 1986, Dr. Waynant joined the Food and Drug Administration to work on laser interactions in surgery and medicine, fiber delivery, fiber optical diagnosis, and fiber dosimetry in the medical area. He is also an adjunct professor at Catholic University.

Dr. Waynant has published more than 60 journal papers, has edited three books, and has given more than 70 presentations. He has taught numerous short courses for both the IEEE and the Society of Photo-Optical Instrumentation Engineers (SPIE), and has organized and chaired conferences for the Engineering Foundation in both the laser medical and the short-wavelength laser areas. He holds seven patents.

Dr. Waynant is a Fellow of the IEEE, the Optical Society of America (OSA), the American Society of Laser Surgery and Medicine (ASLMS), the American Institute for Medical and Biological Engineering (AIMBE), and the Washington Academy of Sciences (WAS). He has served as editor-in-chief of the *IEEE Circuits and Devices Magazine* since 1985.

John K. Lowell received the Ph.D. in applied physics from the University of London. He has held technical and managerial assignments for United Technologies, Northern Telecom, Mostek, Texas Instruments, British Telecom/Dupont, AMD, and Applied Materials. He has also been a professor at Texas Tech University and the University of Texas, and has held consulting and visiting professorships at Arizona, Stanford, and Harvard University. Currently, Dr. Lowell is a visiting scholar at the National Science Foundation (NSF) Center for the Synthesis, Growth and Characterization of Electronic Materials at the University of Texas at Austin. His scientific interests include the surface chemistry of electronic materials, optical characterization of semiconductor devices, defects induced from plasma and thermal processing, and in-line electrical analysis of semiconductors.

Dr. Lowell has published over 100 papers, holds six patents, and has five patents pending. For the past 13 years, he has been the associate editor-in-chief of the IEEE Division I, *Circuits and Devices Magazine* and has been its guest editor twice.

Dr. Lowell is a senior member of the IEEE and is the current secretary of the Electron Devices Society (EDS) of the IEEE. He is also a distinguished lecturer of the EDS and has held AdCom-level positions previously within the Lasers & Electro-Optics (LEOS) and Circuits and Systems societies. At present, he is the vice-chairman of the American Physical Society (APS-Texas Section). Most recently, he was awarded a Technical Achievement award from the LEOS Central Texas chapter. Dr. Lowell is also a member of the MRS, and is an elected member of both the New York Academy of Sciences and the Texas Academy of Sciences.